D0982863

Genetics of Fitness and Physical Performance

Claude Bouchard, PhD
Université Laval

Robert M. Malina, PhD
Michigan State University

Louis Pérusse, PhD
Université Laval

Human Kinetics

Library of Congress Cataloging-in-Publication Data

Bouchard, Claude.
Genetics of fitness and physical performance / Claude Bouchard, Robert M. Malina, Louis Pérusse.
p. cm.
Includes bibliographical references and index.
ISBN 0-87322-951-7
1. Physical fitness--Genetic aspects. 2. Human genetics. 3. Medical genetics. 4. Phenotype. I. Malina, Robert M. II. Pérusse, Louis, 1957- . III. Title.
QP301.B76 1997
613.7--dc21 97-9083
CIP

ISBN: 0-87322-951-7

Developmental Editor: Kristine Enderle; **Assistant Editor:** Coree Schutter; **Editorial Assistants:** Laura Ward Majersky and Edward Sammons; **Copyeditors:** Brian Mustain and Karen Bojda; **Proofreader:** Kathy Bennett; **Indexer:** Diana Witt; **Graphic Designer:** Robert Reuther; **Graphic Artists:** Francine Hamerski, Tara Welsch, and Kim McFarland; **Cover Designer:** Jack Davis; **Printer:** Braun-Brumfield.

Printed in the United States of America 10 9 8 7 6 5 4 3 2 1

Human Kinetics
Web site: http://www.humankinetics.com/

United States: Human Kinetics, P.O. Box 5076, Champaign, IL 61825-5076
1-800-747-4457
e-mail: humank@hkusa.com

Canada: Human Kinetics, Box 24040, Windsor, ON N8Y 4Y9
1-800-465-7301 (in Canada only)
e-mail: humank@hkcanada.com

Europe: Human Kinetics, P.O. Box IW14, Leeds LS16 6TR, United Kingdom
(44) 1132 781708
e-mail: humank@hkeurope.com

Australia: Human Kinetics, 57A Price Avenue, Lower Mitcham, South Australia 5062
(08) 277 1555
e-mail: humank@hkaustralia.com

New Zealand: Human Kinetics, P.O. Box 105-231, Auckland 1
(09) 523 3462
e-mail: humank@hknewz.com

Contents

Chapter 4—Understanding the Methods 59

Chapter 5—Defining Performance, Fitness, and Health 89

Chapter 6—Genetics and Health Status 99

Chapter 7—Genetics of Metabolic Rate and Activity Level 119

Chapter 8—Genetics of Dietary Intake 131

Chapter 9—Genetics of Body Size and Physique Phenotypes 149

Chapter 10—Genetics of Body Fat and Fat Distribution 173

Chapter 11—Genetics and Skeletal Muscle Phenotypes 221

Chapter 12—Genetics of Cardiorespiratory Fitness Phenotypes 243

Chapter 13—Genetics of Metabolic Fitness Phenotypes and Other Risk Factors 267

Chapter 14—Genetics and Motor Development 299

Chapter 15—Genetics and Performance 311

Preface

We were delighted to be asked by Human Kinetics to write this book on a topic of growing interest in the sport science community. Interest in the genetic and molecular bases of physical fitness and sport performance is rapidly increasing as evidenced in almost weekly announcements concerning the genetics of various health phenotypes and worldwide interest in high-level sport performances. The genetic and molecular bases of fitness and performance will remain areas of research focus into the future. The time was ripe, therefore, to develop a compendium of presently available research results relating to genetics, fitness, and performance. The three of us have spent much time together as colleagues at many informal sessions discussing the scope and contents of the volume. In addition, we spent several weeks together over the past two years developing, writing, and critically evaluating each other's contributions to the final manuscript. As the volume progressed, we soon realized that the database dealing with genetics, fitness, and performance had many serious shortcomings. This book is thus a beginning; as some of the gaps in our knowledge will be filled, others will undoubtedly become more obvious as the body of research increases in the years to come.

We would like to express our thanks to several colleagues who provided feedback on various sections of the text. The contributions of Jean-Aime Simoneau, Université Laval, by his review of the chapter on skeletal muscle and of D.C. Rao and Treva Rice, Washington University, by their reviews of the chapter on methods must be highlighted.

Two individuals attached to the office of Claude Bouchard and the Laboratoire des sciences de l'activité physique at Université Laval contributed immensely to the development of the manuscript with editorial support and in production of the tables and figures. Our sincerest thanks and indeed a debt of gratitude are expressed to Diane Drolet, MSc, and Jean-Yves Dallaire for their assistance and fine contributions.

Finally, we would like to thank once again, as we have often done in the past, Dr. Rainer Martens for his continuous encouragement in this project. Thanks also to the staff at Human Kinetics who helped us to convert the manuscript into this volume.

Claude Bouchard
Robert M. Malina
Louis Pérusse

CHAPTER 1

Introduction

The field of human genetics is experiencing a complete revolution. The process is ongoing and will affect the lives of this generation and generations to come. The exercise sciences will be strongly influenced by these advances in human genetics as well as those in molecular biology. The body of knowledge is likely to grow at an extremely rapid rate over the next 5 to 10 years. Current knowledge about the role of genes and genetic individuality on health-related and performance-related fitness, and on the trainability of the human organism, will become largely obsolete. Indeed, there is a growing perception that the stage is now set for these changes to occur in the exercise sciences. We believe, therefore, that the time is right for a summary of the evidence, acquired largely over the last 25 years or so, regarding the genetics of fitness and performance. The next book on this topic, whether written by us or others, will likely be quite different from this one.

Current directions in the exercise sciences are remarkable and exciting. Those who entered the field 20 to 30 years ago could never have predicted that it would be possible one day to investigate the molecular basis of responses to exercise or exercise training and the genetic basis of individual differences in performance or fitness. However, these questions can now be asked and more importantly are being currently investigated. This is an exciting period for the beginning exercise scientist to embark on a career. Nevertheless, it does not mean that choosing an area of study and defining research topics will be any easier than before for either the student or the emerging scientist.

Statement of the Problem

Human variation is the central object of interest in human genetics. Variation is also the focus of those concerned with the genetic basis of phenotypes that are relevant to the exercise sciences. Indeed, DNA sequence variation in the human genome is astonishingly abundant. There are good reasons to conclude, as this book will demonstrate, that heterogeneity at the DNA level contributes immensely to observed interindividual differences in performance, fitness, and health.

In most areas of the exercise sciences, as in most fields of research, individual differences are often considered a necessary nuisance. Average values and mean treatment effects are important parameters. Studies in exercise physiology, exercise biochemistry, and other domains of the exercise sciences typically emphasize the central tendency and generally give little attention to the variability of individual scores. Variability in data is too often viewed as "noise," which has been described as the tyranny of the "golden mean" (1). To a large extent, such dismissal of individual differences has become a self-perpetuating phenomenon; it is the norm in nongenetic literature, because of the passive nurturing of external referees and editors. As a result of this trend in the exercise sciences, many effects and relationships have gone unnoticed and have been ignored even by the scientists who generated the data. A long-overlooked effect, for example, is the response of $\dot{V}O_2$max to regular exercise. Even though mean responses of sedentary children, adolescents, and adults of both sexes have been reported in scholarly journals for a long time, it was not until 1983 that attention was drawn to the heterogeneity of response in $\dot{V}O_2$max to a standardized training regimen (2). The tacit notion that variability of scores necessarily reflects an abnormal response or technical problems must be abandoned.

Human variability represents the substrate of human genetics, which aims to define relationships with DNA sequence variation, and gene interactions and coactions with lifestyle factors and environmental variables. In other words, individual differences are the focus of attention. The interest of human genetics for human variation is of particular relevance for the exercise sciences. Understanding of genetic variability can shed light on commonly encountered situations and can contribute to a better understanding of the relationship between humans and exercise, performance, and physical activity. For instance, why is there a wide range of motor ability? Are individuals gifted in motor performance genetically different from others? Why are all fit and physically active persons not protected against hypertension, noninsulin dependent diabetes mellitus, or coronary heart disease? Why can some people reach a $\dot{V}O_2$max of 75 ml O_2 per kg per min or more with training, while others are not even able to attain 50 ml? Why are some people adversely affected by a diet rich in cholesterol and saturated fat, while others are not? Such questions and a variety of others are of interest to the exercise scientist, but they cannot be addressed thoroughly without the contributions of molecular biology and human genetic paradigms.

Nature Versus Nurture

The debate over nature (genetics) versus nurture (environment) is as old as the species. It is no surprise that discussion of the causes of similarities and differences among individuals has been lively across the centuries. However, the debate over whether inherited factors are more important than environment in molding phenotypes is largely sterile. One cannot exist without the other. Both are intrinsically linked in the sense that it is impossible to understand the genetic basis of a trait without making assumptions about nurture, and vice versa.

Over the last 100 years or so, the debate has taken on new dimensions for a variety of reasons. First, research on differences within and among populations in several anthropometric, physiological, fitness, and skill phenotypes has been regularly reported. Among the probable causes of human variation, both nature and nurture arguments have been proposed. Second, the influence of strong scientific personalities, such as Francis Galton, William Bateson, and Karl Pearson, was instrumental early in launching the debate and keeping it alive. Galton in particular, with his publications *Hereditary Genius* (9) and *Natural Inheritance* (10) and his views on eugenics, put the debate in the forefront of the scientific agenda of the day. Third, genetics became progressively established as a science with initially two major but competing and often adversarial schools, the biometric and the Mendelian schools. Fortunately, rapid progress in genetics in the 20th century brought the two groups together and established genetics as a science with several disciplines. Fourth, proponents of either school of thought in the nature versus nurture controversy often had philosophical and political agendas. Thus, it became increasingly difficult to separate scientific arguments from personal convictions and political correctness.

The debate assumed unexpected prominence in the late 1960s and 1970s, particularly after the publication of a paper by Arthur Jensen in the Harvard Educational Review in 1969 (14). Jensen raised the issue of differences in IQ between blacks and whites and discussed whether the apparent racial difference in IQ observed at that time could be eliminated by broad educational measures. Jensen concluded that it could not. The paper was vigorously criticized by scholars, the lay press, and the political establishment. Moderate views were often held by fellow scientists who became involved in the controversy. The episode has certainly allowed discussion of various arguments on both sides of the debate, which has surfaced again with the publication of *The Bell Curve* (12). In reality, it is not an issue. For every human phenotype and every human situation, the nature and nurture elements are interwoven. The nature-nurture dyad varies from phenotype to phenotype, from situation to situation, and from population to population. The controversy can be put to rest, and will eventually be put to rest, only with further advances in understanding the real genetic and nongenetic bases of human variation.

The Growing Power of the Biological Sciences

Over the last 10 years, spectacular progress in molecular biology has given new impetus to the study of the human genome. The rapid transfer of new technologies for the genotyping of human beings has extended the range of questions that could be contemplated in genetic investigations. This has now reached the point where one can define problems in terms of molecular epidemiology (20), molecular nutrition (21), and molecular medicine (3). There is no indication that the pace of progress is slowing. Major advances are being reported at an increasing frequency.

Combining the strengths of molecular biology with the paradigms of human genetics has generated a dynamic and vibrant new field with powerful and continuously expanding technologies. Single-gene diseases are being elucidated at the DNA and biochemical levels. For example, the genes causing a disease or associated with marked predisposition for a disease have been isolated or localized to a region of a given chromosome for conditions such as muscular dystrophies, myotonic dystrophy, cystic fibrosis, Huntington's chorea, amyotrophic lateral sclerosis, breast cancer, colon cancer, Alzheimer's disease, and others as well.

Complex entities such as atherosclerosis, noninsulin dependent diabetes mellitus, obesity, and hypertension are being scrutinized in depth for single-gene effects and susceptibility genes. This is done through a variety of animal and human models. Complex segregation analysis applied to multifactorial phenotypes, such as those of interest in the exercise sciences, can provide statistical evidence for the effect of a single gene and can be helpful in defining its mode of inheritance. The availability of literally thousands of polymorphic markers (4,16,19,26) distributed across the 22 pairs of autosomes and the sex chromosomes makes it possible to undertake a broad genome search for association and linkage with a greater probability of success than in the past. Further, a large number of candidate genes for a particular phenotype or the physiopathology of a disease can now be tested.

Development of the quantitative trait locus approach for the study of complex multifactorial traits has enabled geneticists to begin to define the genetic basis of a trait without a priori knowledge of the genes involved. The method can also be applied to human phenotypes. Moreover, its use in rodent crossbreeding experiments is generating extremely valuable information about putative genetic loci on human chromosome areas equivalent to the rodent chromosome areas of interest (23). More will be presented on these powerful methods in chapter 4. This approach has been aided because the mouse genetic map has been greatly advanced (7), the definition of homologous regions on chromosomes of mice and humans has been expanded (17,18,27), and appropriate statistical tools have been developed for interval mapping (15).

It is increasingly feasible to begin genetic studies with a phenotype for which nothing or little is known about the genes involved. Using a variety of techniques such as those just listed and others such as positional cloning and breeding of transgenic animals for a transgene of interest, it is possible to identify a gene or isolate a specific chromosomal region encoding a gene associated or linked with a phenotype. Such approaches are analogous to reverse genetics that was initially described in experimental organisms. The approach has been greatly facilitated by advances in DNA technologies, particularly those that probe sequence variation, such as Southern blotting, polymerase chain reaction amplification, and techniques for identifying tandem repeats and typing microsatellites. DNA banking methods and the relative ease of establishing lymphoblastoid cell lines for long-term studies are also contributing to an environment in which it is becoming easier to probe human genetic differences at the molecular level.

Necessary and Susceptibility Genes and Other Effects

Fitness and performance phenotypes cannot be easily reduced to simple Mendelian traits. They are complex multifactorial traits that have evolved under the interactive influences of dozens of affectors from the social, behavioral, physiological, metabolic, cellular, and molecular domains. Segregation of the genes is not easily detected in familial or pedigree studies; whatever the influence of the genotype on etiology, it is generally attenuated or exacerbated by nongenetic factors.

Efforts to understand the genetic causes of such phenotypes in human beings will succeed only if they are based on an appropriate conceptual framework, adequate measurements of phenotype and intermediate phenotype, appropriate samples of unrelated persons and nuclear families or of extended pedigrees, and extensive typing of candidate genes and other molecular markers. In this context, the distinction between "necessary" genes and "susceptibility" genes recently proposed by Greenberg (11) is particularly relevant. Although there are several examples of necessary loci resulting in impaired physical performance (e.g., inherited disorders of glycogen storage), they occur in only a small fraction of the population. In such conditions, carriers of one or two copies of the deficient alleles have the disease (depending on the mode of inheritance).

The susceptibility gene concept (11) is especially important for the human fitness and performance phenotypes. A susceptibility gene is one that increases susceptibility or predisposition for high or low values for a given phenotype or risk level for a disease, but one that is not necessary for the expression of the phenotype or disease. In the case of a disease phenotype, an allele at a susceptibility gene locus may make it more likely that the carrier will become affected, but the presence of the allele is not sufficient by itself to explain the occurrence of the disease. It merely lowers the threshold for a person to develop the disease (11). In the case of fitness and performance phenotypes, the concept implies that there are genes that have an effect on individual differences observed for one or several such phenotypes. In general, susceptibility or predisposition genes are defined as each having a small effect on the total variance of a phenotype. As proposed by Greenberg (11), if susceptibility genes are neither necessary nor sufficient for disease expression, linkage analysis is likely to yield little more information than is obtained from association studies. This is particularly true when the deficient allele at the locus under consideration carries less than about 10 times the risk for a disease phenotype in comparison with the normal allele. The same rationale applies to non-disease-related fitness and performance phenotypes.

Figure 1.1 illustrates these concepts in a schematic manner. The figure allows for the contribution of necessary and susceptibility genes to human variation in health- and performance-related phenotypes. Note also that the figure incorporates two types of interaction effects, namely gene-environment interactions and gene-gene interactions. The gene-environment interaction effect (e.g., exercise, nutrition, drugs, climate, bacteria, viruses, etc.) results from the fact that the response to environmental conditions, changes in lifestyle, or exposure to new elements is influenced by genes. Gene-gene interaction effects also need to be considered. When a large number of genes are involved in a phenotype or its response to environmental exposures, gene-gene interactions may be ubiquitous and complex. There is some evidence that gene-exercise interaction effects are quantitatively large for several fitness and perfor-

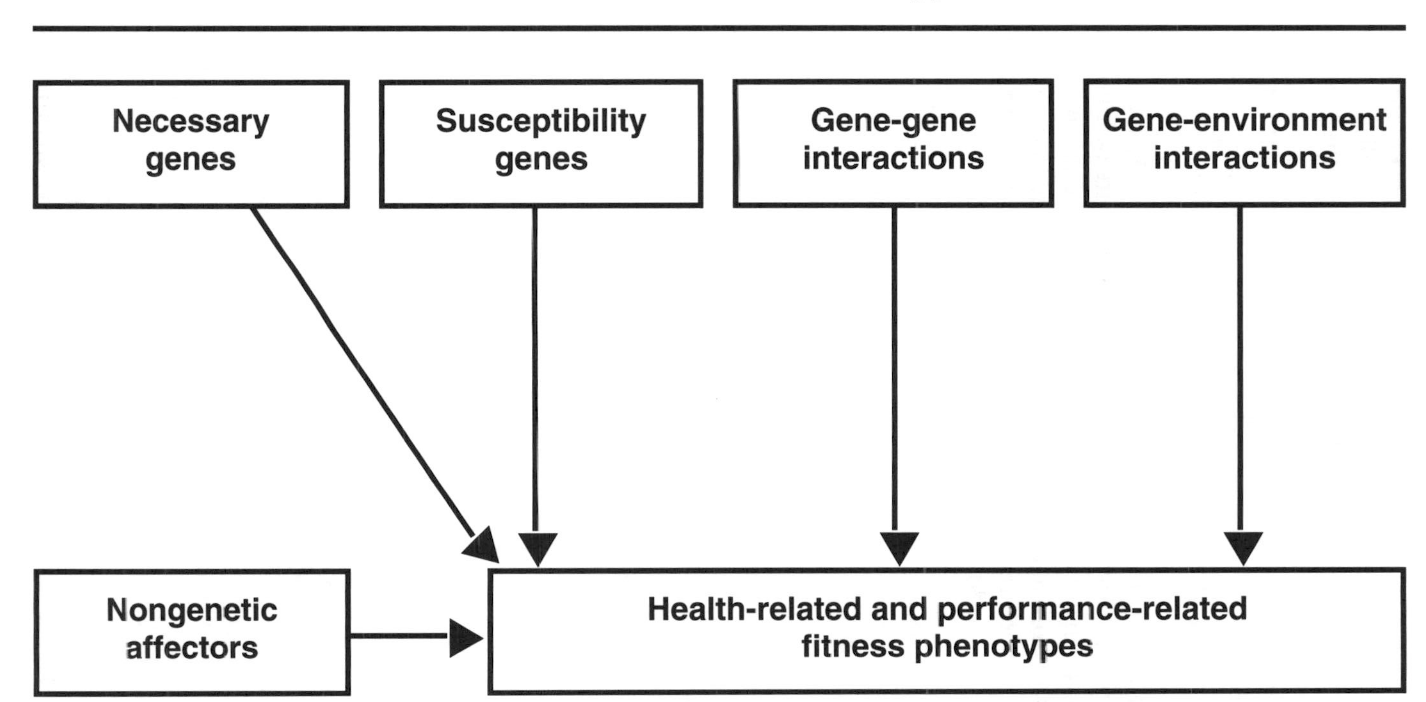

Figure 1.1 A schematic overview of the various genetic effects that need to be considered in the study of complex multifactorial phenotypes.

mance traits. This topic is addressed in subsequent chapters. At present, there are no data that allow assessment of the importance of gene-gene interactions for health- and performance-related phenotypes.

The International Effort to Study the Human Genome

In one of the most amazing stories in the history of science, James Watson and Francis Crick described the structure of DNA in 1953 and concluded that it met all of the requirements to be the molecular basis of heredity (24). This insightful contribution earned them the Nobel Prize in 1962, which they shared with Maurice Wilkins. Even more remarkably, their short report in *Nature* (24) unleashed a true revolution in the biological sciences that culminated in the human genome initiatives that were launched in several countries in the early 1990s. Less than 40 years after it was recognized that DNA was the prime molecule of genetic information, efforts are currently underway to sequence the DNA of all chromosomes of the human genome. These spectacular advances will affect the lives of future generations even more than our own.

Scientists from several countries, including the United States, France, Great Britain, Japan, Germany, Russia, Canada, and others, are involved in a gigantic effort to map a series of physical and genetic markers along the chromosomes of the human genome and to sequence all of the 3 billion DNA base pairs. The worldwide enterprise has been spearheaded by the Human Genome Project of the United States under a joint research plan of the National Institutes of Health (NIH) and the Department of Energy (DOE). The Human Genome Project, as it is called, was officially launched in 1991 after several years of discussion and debate. Several papers have described the genesis of the project (6,25). The two United States agencies (NIH and DOE) involved in the Human Genome Project are spending about 300 million dollars on this initiative at this time.

The Human Genome Project is an ambitious venture. It has been described as the equivalent in biological research of the project Man on the Moon in space research. Major advances have occurred in the first three years of the Human Genome Project, and some of the initial goals have already been achieved. A new five-year plan has been developed for the United States Human Genome Project for the period 1994-98 (5). Table 1.1 provides a listing of the five-year goals as summarized in the November 1993 issue of *Human Genome News.* The goals pertain to genetic mapping, physical mapping, and DNA sequencing. The genetic map is based on a set of polymorphic markers distributed over the whole genome; the initial aim is to have a set of markers within two to five centimorgans or recombination units at meiosis. A centimorgan (cM) is an indicator of the frequency with which two traits or markers are inherited. Two markers that are 10 cM apart are separated by recombination at meiosis 10% of the time. A distance of 1 cM is equivalent, on average, to about 1 million DNA base pairs. A genetic map with a series of markers within 2 to 5 cM would require approximately 1,500 to 2,000 polymorphic markers.

With the 1992 and 1993 reports from the Paris-based Centre d'étude pour le polymorphisme humain (CEPH) and Généthon, this goal was essentially achieved (4,26). However, in September 1994, an international team composed mainly of scientists from France and the United States reported the construction of a genetic map with 5,840 loci spaced at a mean interval of 0.7 cM (16). Thus the original goal was achieved within three years, and the map is already at least five times more detailed than anticipated. The current genetic map includes over 3,600 short tandem repeat markers (di-, tri-, and tetranucleotide repeats), over 400 genes, and about 1,800 other markers, such as restriction fragment length polymorphisms and anonymous DNA fragments.

It must be appreciated that the 5,840 markers are defined in terms of linkage with respect to one another on the relevant chromosomes. The number of genes and DNA fragments identified is, at present, much higher than the number appearing on the latest genetic map. For instance, it is estimated that more than 3,500 genes and 100,000 anonymous segments of DNA are recorded in the human genetic database (8).

The plan calls also for a physical map of the various chromosomes with an average resolution of about 100 kilobases (kb). The physical map is analogous to a road map with metric posts at a standard distance along the road (e.g., every kilometer). In December 1995, it was announced that a physical map of the human genome with 15,086 sequence-based landmarks had become accessible (13). The average distance between markers was about 200 kb. It is anticipated that the 100-kb target will be met in about one more year of work and will require approximately 30,000 sequence-based markers.

The plan states that DNA sequencing should continue at an unabated rate over the next few years. Considerable resources will be devoted to the development of new and more efficient techniques to improve the accuracy and speed of sequencing. The ultimate goal is to determine the DNA sequence of the entire human genome, a task that may require another 10 years to be attained.

Other specific goals of the revised five-year plan include the development of methods to recognize gene sequences in long strings of sequenced DNA; continuing efforts to advance the genetic and

Table 1.1
Revised Five-Year Research Goals of the U.S. Human Genome Project (1994-98)

Mapping and sequencing the human genome

Genetic mapping

- Complete the 2- to 5-cM map by 1995*.
- Develop technology for rapid genotyping.
- Develop markers that are easier to use.
- Develop new mapping technologies.

Physical mapping

- Complete a sequence tagged site (STS) map of the human genome at a resolution of 100 kb.*

DNA sequencing

- Develop efficient approaches to sequencing one- to several-megabase regions of DNA of high biological interest.
- Develop technology for high-throughput sequencing, focusing on systems integration of all steps from template preparation to data analysis.
- Build up a sequencing capacity to allow sequencing at a collective rate of 50 Mb per year by the end of the period. This rate should result in an aggregate of 80 Mb of DNA sequence completed by the end of FY 1998.

Gene identification

- Develop efficient methods for identifying genes and for placement of known genes on physical maps or sequenced DNA.

Technology development

- Substantially expand support of innovative technological developments as well as improvements in current technology for DNA sequencing and for meeting the needs of the Human Genome Project as a whole.

Model organisms

- Finish an STS map of the mouse genome at a 300-kb resolution.
- Finish the sequence of the *Escherichia coli* and *Saccharomyces cerevisiae* genomes by 1998 or earlier.

(continued)

- Continue sequencing *Caenorhabditis elegans* and *Drosophila melanogaster* genomes with the aim of bringing *C. elegans* to near completion by 1998.
- Sequence selected segments of mouse DNA side by side with corresponding human DNA in areas of high biological interest.

Informatics

- Continue to create, develop, and operate databases and database tools for easy access to data, including effective tools and standards for data exchange and links among databases.
- Consolidate, distribute, and continue to develop effective software for large-scale genome projects.
- Continue to develop tools for comparing and interpreting genome information.

Ethical, legal, and social implications (ELSI)

- Continue to identify and define issues and develop policy options to address them.
- Develop and disseminate policy options regarding genetic testing services with potential widespread use.
- Foster greater acceptance of human genetic variation.
- Enhance and expand public and professional education that is sensitive to sociocultural and psychological issues.

Training

- Continue to encourage training of scientists in interdisciplinary sciences related to genome research.

Technology transfer

- Encourage and enhance technology transfer both into and out of centers of genome research.

Outreach

- Cooperate with those who would establish distribution centers for genome materials.
- Share all information and materials within six months of their development. This should be accomplished by submission of information to public databases or repositories, or both, where appropriate.

*Goals for map resolution remain unchanged.

Adapted from *Human Genome News*, November 1993.

physical maps of selected model organisms in parallel with the human genome; progress in informatics, technology development, and transfer; training of scientists; fostering of the study of ethical, legal, and social implications; and promotion of outreach measures. It is generally estimated that the Human Genome Project will be able to meet its objectives in about 10 to 15 years.

Needless to say, the Human Genome Project will have an enormous impact on biological and behavioral research. The exercise sciences will also be affected in ways that are hard to predict. It is reasonable to assume that the information and technologies generated by the various human genome initiatives will profoundly influence the health status and presumably fitness levels of future generations. The field of human athletic performance, particularly at the elite level, will also be affected by the emerging body of knowledge about the human genetic makeup. These topics are addressed in more detail later in the volume.

Content of the Book

The main objective of this book is to summarize the evidence currently available regarding the genetic basis of health-related and performance-related fitness phenotypes. A second aim is to present the biological basis of heredity and the concepts and methods of genetic epidemiology and molecular biology that are necessary for understanding the results of the various studies. A third aim is to identify several complex and sensitive ethical issues that may challenge the exercise scientist, the fitness specialist, and the coach in the near future.

The book is divided in three parts and includes 18 chapters. The first part discusses concepts and genetic methods that are relevant to the aims of the book and specifically to the exercise sciences. The second part summarizes the evidence for genotypes' role in human phenotypes of interest to the exercise science community. Finally, the third part explores some of the scientific, practical, and ethical issues that are likely to confront exercise scientists as progress is made in molecular biology, human genetics, and the biology and technology of reproduction.

References

1. Bennett, A.F. Interindividual variability: An underutilized resource. In: Feder, M.E.; Bennett, A.F.; Burggren, W.W.; Huey, R.B., eds. New directions in ecological physiology. New York: Cambridge University Press; 1987:147-69.

2. Bouchard, C. Human adaptability may have a genetic basis. In: Landry, F., ed. Health risk estimation, risk reduction and health promotion. Proceedings of the 18th Annual Meeting of the Society of Prospective Medicine. Ottawa: Canadian Public Health Association; 1983:463-76.
3. Caskey, C.T. Molecular medicine. A spin-off from the helix. JAMA. 269:1986-92; 1993.
4. Cohen, D.; Chumakov, I.; Weissenbach, J. A first-generation physical map of the human genome. Nature. 366:698-701; 1993.
5. Collins, F.; Galas, D. A new five-year plan for the U.S. Human Genome Project. Science. 262:43-46; 1993.
6. Cook-Deegan, R.M. The genesis of the Human Genome Project. Molec. Genet. Med. 1:1-75; 1991.
7. Copeland, N.G.; Jenkins, N.A.; Gilbert, D.J.; Eppig, J.T.; Maltais, L.J.; Miller, J.C.; Dietrich, W.F.; Weaver, A.; Lincoln, S.E.; Steen, R.G.; Stein, L.D.; Nadeau, J.H.; Lander, E.S. A genetic linkage map of the mouse: Current applications and future prospects. Science. 262:57-66; 1993.
8. Craig, I.W. Organization of the human genome. J. Inher. Metab. Dis. 17:391-402; 1994.
9. Galton, F. Hereditary genius. London: Macmillan; 1869.
10. Galton, F. Natural inheritance. London: Macmillan; 1889.
11. Greenberg, D.A. Linkage analysis of "necessary" disease loci versus "susceptibility" loci. Am. J. Hum. Genet. 52:135-43; 1993.
12. Herrnstein, R.J.; Murray, C.A. The bell curve: Intelligence and class structure in American life. New York: Free Press; 1994.
13. Hudson, T.J.; Stein, L.D.; Gerety, S.S.; et al. An STS-based map of the human genome. Science. 270:1945-54; 1995.
14. Jensen, A. How much can we boost I.Q. and scholastic achievement? Harvard Educ. Rev. 39:1-123; 1969.
15. Lander, E.S.; Botstein, D. Mapping Mendelian factors underlying quantitative traits using RFLP linkage maps. Genetics. 121:185-99; 1989.
16. Murray, J.C.; Buetow, K.H.; Weber, J.L.; et al. A comprehensive human linkage map with centimorgan density. Science. 265:2049-54; 1994.
17. Nadeau, J.H. Maps of linkage and synteny homologies between mouse and man. Trends Genet. 5:82-86; 1989.
18. Nadeau, J.H.; Davisson, M.T.; Doolittle, D.P.; Grant, P.; Hillyard, A.L.; Kosowsky, M.; Roderick, T.H. Comparative map for mice and humans. Mamm. Genome. 1:S461-515; 1991.
19. NIH/CEPH Collaborative Mapping Group. A comprehensive genetic linkage map of the human genome. Science. 258:67-86; 1992.
20. Schulte, P.A.; Perera, F.P. Molecular epidemiology. San Diego: Academic Press; 1993.
21. Scriver, C.R. Changing heritability of nutritional disease: Another explanation for clustering. World Rev. Nutr. Dietetics. 63:60-71; 1989.

22. U.S. Human Genome Project updates goals. Hum. Gen. News. 5:1-3; 1993.
23. Warden, C.H.; Daluiski, A.; Lusis, A.J. Identification of new genes contributing to atherosclerosis: The mapping of genes contributing to complex disorders in animal models. In: Lusis, A.J.; Rotter, J.I.; Sparkes, R.S., eds. Molecular genetics of coronary artery disease. Monographs in human genetics. Basel: Karger; 1992:419-41.
24. Watson, J.D.; Crick, F.H.C. Genetic implications of the structure of deoxyribonucleic acid. Nature. 171:964; 1953.
25. Watson, J.D.; Cook-Deegan, R.M. Origins of the human genome project. FASEB J. 5:8-11; 1991.
26. Weissenbach, J.; Gyapay, G.; Dib, C.; Vignal, A.; Morissette, J.; Millasseau, P.; Vaysseix, G.; Lathrop, M. A second-generation linkage map of the human genome. Nature. 359:794-801; 1992.
27. Yamada, J.; Kuramoto, T.; Serikawa, T. A rat genetic linkage map and comparative maps for mouse or human homologous rat genes. Mamm. Genome. 5:63-83; 1994.

CHAPTER 2

The Human Genome

Understanding the fundamentals of the human genome and the concepts of human genetics is essential in order to elucidate the role of genetic factors in physical fitness and performance. The biology of the gene and characteristics of the human genome are complex and cannot be adequately covered in a few pages. Nevertheless, a brief outline of key concepts is presented. The interested reader is referred to contemporary textbooks of genetics for more comprehensive discussions.

The Human Genome

The complement of genes in the nucleus of each cell is the genome. The genetic material of the human genome consists of tightly coiled threads of deoxyribonucleic acid (DNA) and nucleoproteins organized into microscopic units called chromosomes. Genes are arrayed linearly on the chromosomes. All normal human cells contain 23 pairs of chromosomes, one member of each pair inherited from each parent at the moment of conception. Members of chromosome pairs are known as homologs. There are 22 pairs of autosomes and one pair of sex chromosomes (XX for females and XY for males). The normal human diploid number (2n) of chromosomes is thus 46.

Human chromosomes can be isolated from blood lymphocytes or other cells during cell division. Each chromosome contains a constricted region called the centromere, which establishes the general appearance of each chromosome. When stained with specific dyes, chromosomes reveal reproducible patterns of light and dark bands. Differences in size, location of the centromere, and banding patterns allow the chromosomes to be distinguished from each other and in turn represented as a karyotype (figure 2.1, *a* and *b*). In a karyotype the chromosomes are arranged by size and position of the centromere, autosomes are numbered from 1 to 22, and the sex chromosomes are noted as X and Y. The short arm of a chromosome is denoted as p and the long arm as q. Each arm is subdivided into regions numbered consecutively from the centromere to the telomere (tip of the chromosome), and each band within a given region is identified by a number. With this nomenclature, it is

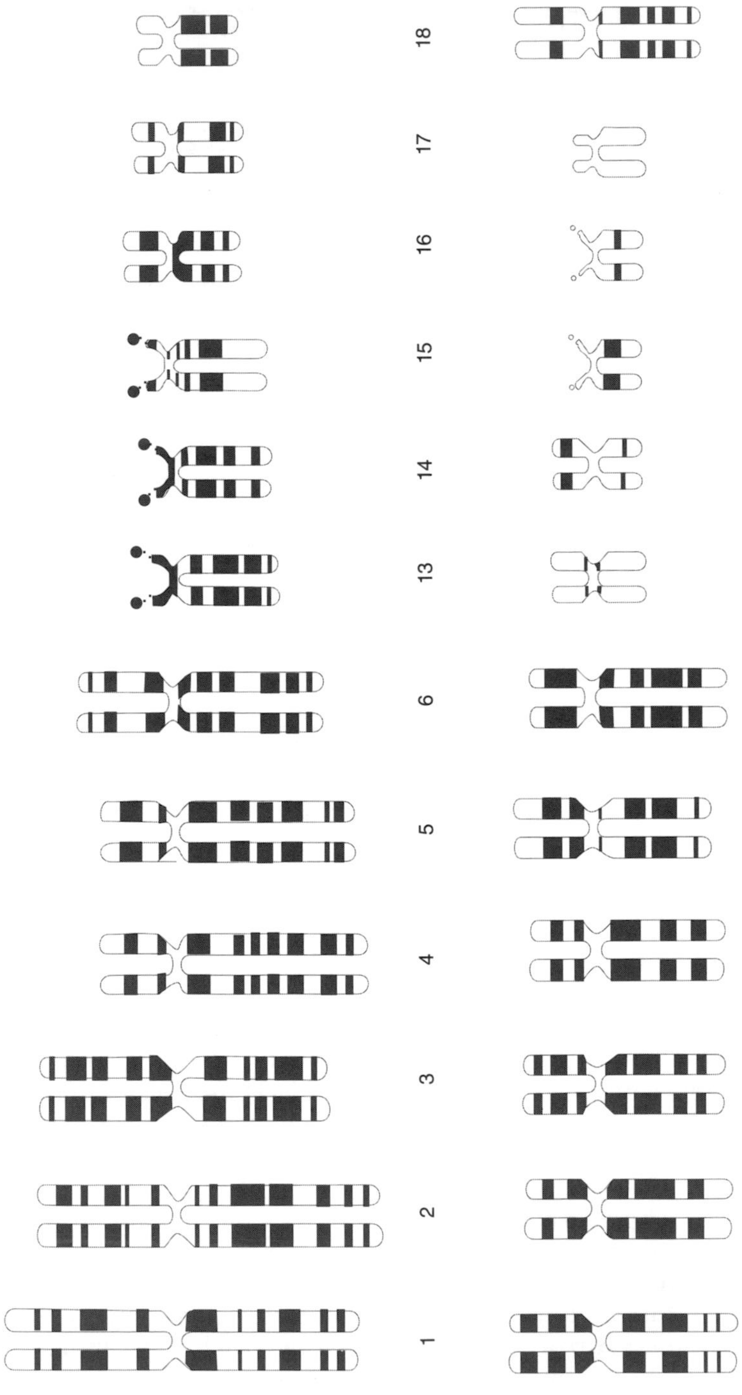

Figure 2.1*(a)* Human karyotype, showing banding patterns of chromosomes. Redrawn from Weiss (8). Adapted, by permission, from J.S. Thompson and M.W. Thompson, 1986, *Thompson & Thompson's Genetics in Medicine* (Philadelphia: W.B. Saunders).

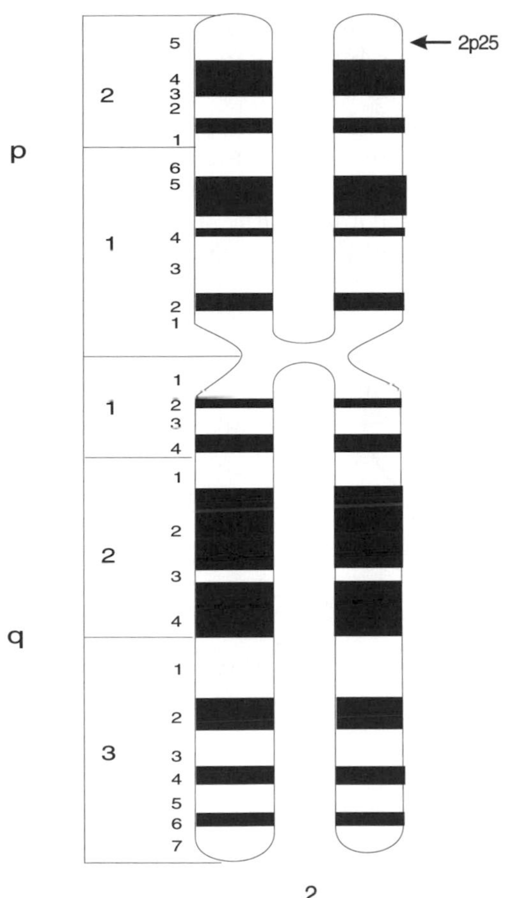

Figure 2.1 ***(b)*** Numbering system based on chemical staining of a chromosome preparation shown for chromosome 2. Adapted from Cummings (2).

possible to specify any chromosomal region by an "address" (e.g., 2p25 refers to chromosome 2, p arm, region 2, band 5).

Chromosomes and mitosis

Chromosomes are not visible in the nuclei of nondividing cells and appear as strands of a material known as chromatin. When somatic cells begin to divide during mitosis, the chromatin condenses into the units recognizable as chromosomes. Mitosis is the process of cell division. It produces two genetically identical daughter cells. Two events take place during mitosis: first, duplication of chromosomes and distribution of a complete set of diploid chromosomes to each daughter cell; second, division of the cytoplasm by splitting of the cell volume into two parts, a process known as cytokinesis.

Mitosis can be divided into four phases (figure 2.2): prophase, metaphase, anaphase, and telophase. The interphase represents the interval between each mitotic division. During prophase the duplicated chromosomes become visible and appear as two structures, chromatids, joined together at the centromere. The two chromatids

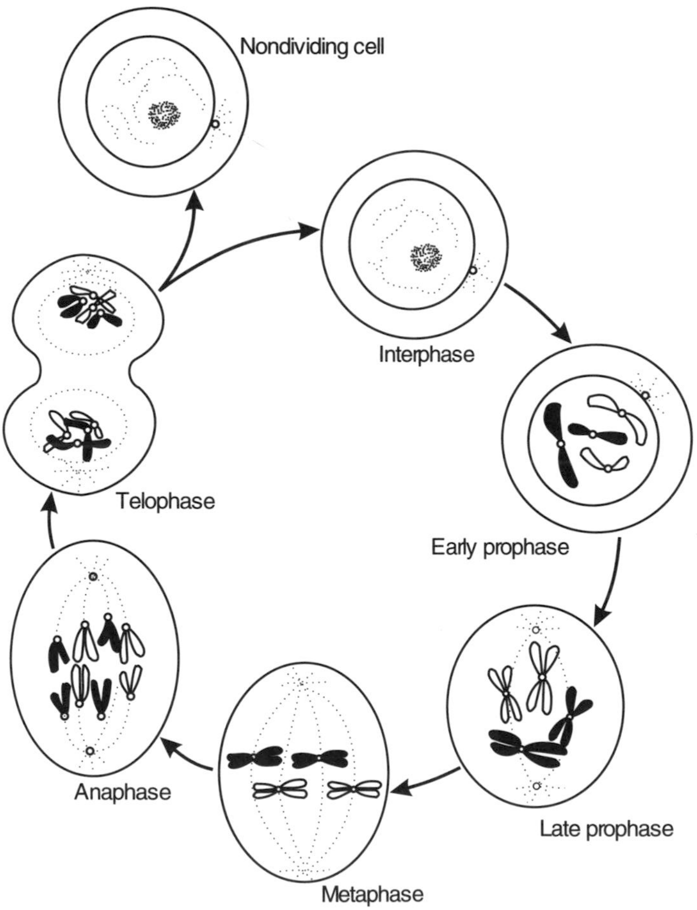

Figure 2.2 Phases of mitosis illustrated with two pairs of homologous chromosomes. Paternally and maternally derived chromosomes are shown in black and white, respectively. Reprinted, by permission, from M.R. Cummings, 1991, *Human Heredity: Principles and Issues* (St. Paul: West Publishing Co.) © Wadsworth Publishing Co.

joined by a common centromere are known as sister chromatids. Near the end of the prophase, the nuclear membrane starts to disappear and the spindle fibers become visible, forming a framework along which chromosomes are pulled toward each pole of the cell. During metaphase the sister chromatids align themselves at the midline of the cell, forming what is known as the metaphase plate. Anaphase begins with the longitudinal division of each centromere, allowing the sister chromatids to separate from each other and the chromosomes to move to opposite sides of the cell. At the end of anaphase, a full complement of chromosomes is present at each pole of the cell. During telophase, the final phase of mitosis, cytokinesis begins and a nuclear membrane appears around each set of chromosomes to reform the nucleus. The net result of mitosis is the formation of two diploid daughter cells, each with 46 chromosomes derived from the parent cell. Mitosis is an essential biological process that ensures somatic cell replacement throughout the life span.

Chromosomes and meiosis

During the active reproductive years, sex cells are produced. They are called oocytes in females and spermatocytes in males. Gametogenesis, the production of gametes from primordial sex cells, occurs through the process of meiosis, during which the duplication of the diploid (2n) set of chromosomes is followed by two successive cell divisions to produce four haploid (n) cells, each containing one member of every chromosome pair. The two divisions are referred to as meiosis I and meiosis II, and each can be divided into prophase, metaphase, anaphase, and telophase (figure 2.3).

During prophase of meiosis I, which can be subdivided into five stages (leptonema, zygonema, pachynema, diplonema, and diakinesis), the duplicated chromosomes become visible and homologous chromosomes pair with each other to form a tetrad, the two sister chromatids arranged side by side. Prophase I is also characterized by exchange of chromosomal material between nonsister chromatids within the tetrad, a process known as crossing-over. In metaphase I, the tetrads move to the center of the cell and line up at the metaphase plate. The orientation of each tetrad at the metaphase plate is random, which means that chance alone determines which of the paternal or maternal chromosomes migrates to a given pole and, in turn, is included in a gamete. This process is known as the independent assortment of homologous chromosomes. During anaphase I, homologous chromosomes separate and migrate to opposite poles. Meiosis I ends with telophase I, during which cytoplasm divides to produce two cells, each containing the duplicated

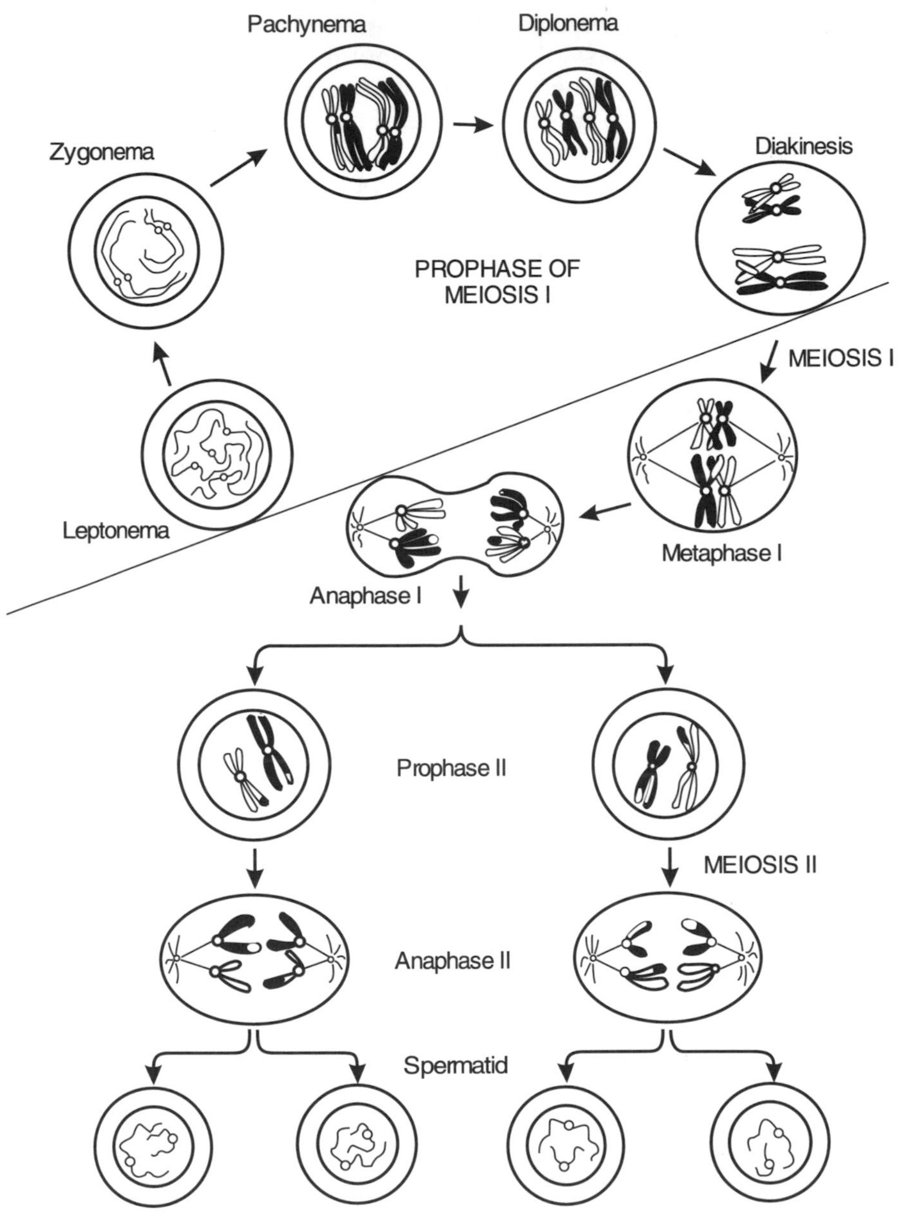

Figure 2.3 Schematic representation of meiosis. Only two pairs of chromosomes are shown. Reprinted, by permission, from M.R. Cummings, 1991, *Human Heredity: Principles and Issues* (St. Paul: West Publishing Co.) © Wadsworth Publishing Co.

set of haploid chromosomes. Meiosis I is also referred as the reductional division, as it results in cells that contain only one-half the number of chromosomes, that is, one member of each pair of chromosomes (see figure 2.3, upper section).

Meiosis II can be viewed as a mitotic division of the haploid set of chromosomes of the two resulting cells of meiosis I (figure 2.3, lower section). The 23 sister chromatids line up at the metaphase plate (metaphase II) and move to opposite poles (anaphase II) to form two haploid cells. Meiosis II is also known as the equational division.

The production of gametes through meiosis in males and females is referred to as spermatogenesis and oogenesis, respectively (see figure 2.4). Spermatogenesis differs from oogenesis in the timing and the number of gametes produced. Spermatogenesis begins at puberty and continues throughout life; a cycle of spermatogenesis lasts approximately 64 days. In females, meiosis starts during embryogenesis and stops at prophase I. At birth, a female has already produced a full complement of her oocytes. Most will degenerate, while others will remain in this stage until puberty. With the attainment of menarche (first menses) or shortly thereafter, one oocyte will complete meiosis approximately every month and will be ovulated. The process is repeated on a more or less monthly basis until menopause. In females, therefore, the process of meiosis becomes activated at about age 12 and persists to about age 50.

The number of mature gametes resulting from meiosis is also different between males and females. In males, all four meiotic products, known as spermatids, develop into mature sperm. During this period of maturation, the nuclear material becomes concentrated in the head of the sperm and the cytoplasm is lost. In females, only one of the four cells resulting from meiosis becomes a mature oocyte. The others are nonfunctional small cells known as polar bodies. As a result of the unequal division of the cytoplasm during oogenesis, the cell that develops into a mature oocyte receives a large amount of cytoplasm. Cytoplasmic constituents such as mitochondria and their DNA are inherited only from the female.

At fertilization, the nuclear content of both gametes fuses, and the diploid number of chromosomes (23 pairs) is restored: 22 pairs of autosomes, with either a pair of X chromosomes (XX, a female zygote) or a pair comprising an X and a Y chromosome (XY, a male zygote).

Two important events of meiosis contribute to the extraordinary amount of genetic diversity characteristic of humans and other sexually reproducing species. The first event is the independent assortment of

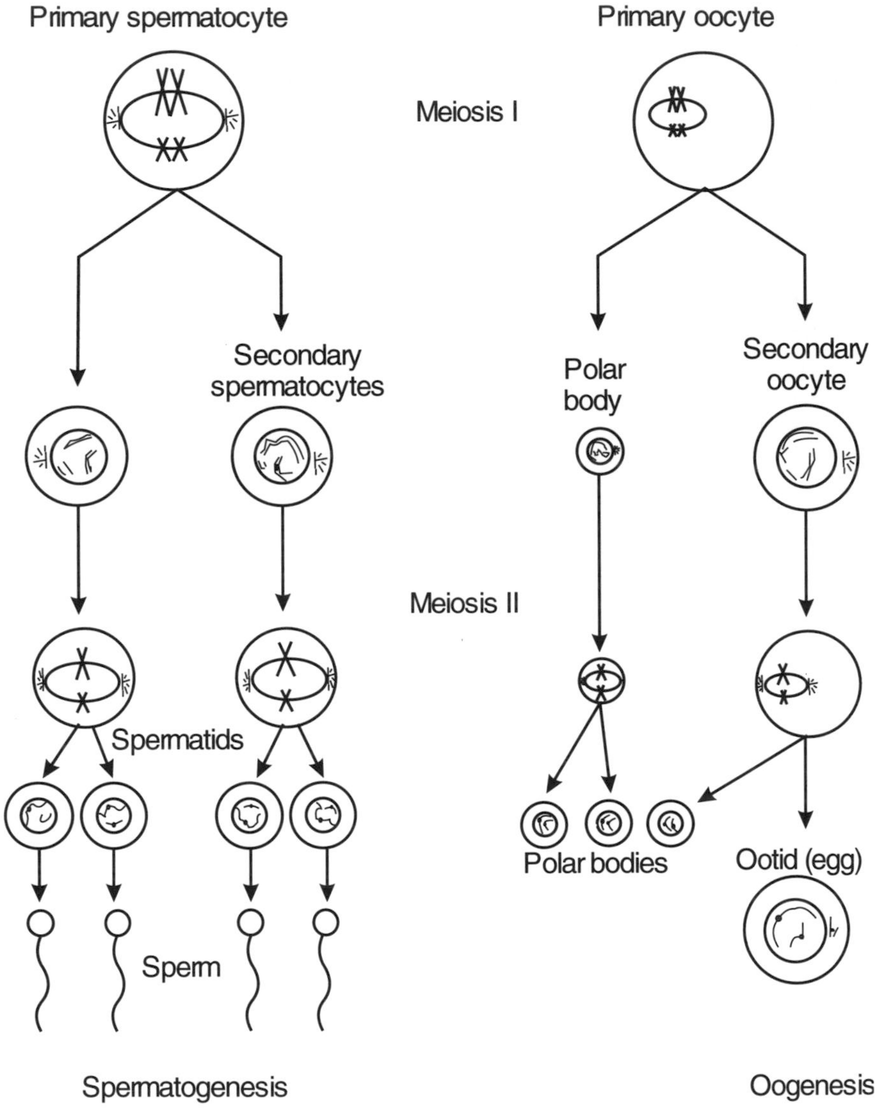

Primary spermatocyte

Primary oocyte

Meiosis I

Secondary spermatocytes

Polar body

Secondary oocyte

Meiosis II

Spermatids

Polar bodies

Ootid (egg)

Sperm

Spermatogenesis

Oogenesis

Figure 2.4 Meiosis in the genesis of sperm (spermatogenesis) and egg (oogenesis) cells. Reprinted, by permission, from R.M. Malina and C. Bouchard, 1991, *Growth, Maturation and Physical Activity* (Champaign, IL: Human Kinetics).

chromosome pairs in metaphase I. With 23 pairs of chromosomes in humans, there are 2^{23} or 8,388,608 different combinations of paternal and maternal chromosomes that can occur in a gamete.

The second event that enhances genetic variation is recombination. During prophase I of meiosis, when homologous chromosomes are paired, crossing-over occurs. This exchange of chromosomal segments results in the recombination of alleles (alternative forms of genes) between the homologous chromosomes of maternal and paternal origins. This is illustrated in figure 2.5.

If a pair of chromosomes carries three genes, each existing in two different forms in the population (e.g., A and a, B and b, C and c), a crossing-over between loci B and C will result in two recombinant chromosomes with new gene combinations and two nonrecombinant chromosomes carrying the parental gene combination. It is estimated that about two to three recombination events take place between each pair of homologous chromosomes (i.e., between pairs of chromosomes of maternal and paternal descent) during meiosis. The existing genetic variability in a population is thus amplified by independent assortment and recombination during meiosis to yield an almost infinite number of different gametes, ensuring the genetic uniqueness of each individual (except monozygotic twins).

The Structure of DNA

The structure of DNA is shown in figure 2.6. The DNA molecule consists of two strands that wrap around each other forming a kind of twisted ladder. The sides of the ladder are made of sugar and phosphate molecules (sugar-phosphate backbone). The two strands are held together by relatively weak hydrogen bonds between pairs of molecules called nucleotides. Nucleotides consist of a sugar (deoxyribose), one phosphate, and one of the following four nitrogenous bases: adenine (A), guanine (G), thymine (T), and cytosine (C). The nucleotides are linked by covalent bonds between the phosphate group of one nucleotide and the sugar of another, forming polynucleotides. Each polynucleotide strand carries a phosphate group at one end (referred to as the 5′ end) and a sugar group at the other end (called the 3′ end).

The entire human genome contains about 3 billion base pairs. Only two different base pairings occur in DNA: adenine always pairs with thymine (A-T pair) and cytosine always pairs with guanine (C-G pair). Since the sequence of nucleotides in one strand determines the sequence in the other, the two strands are complementary to each other. DNA thus consists of two polynucleotide strands oriented in opposite directions and held together by hydrogen bonds between complementary base pairs. The two polynucleotide strands are coiled around a central axis to form a helicoidal structure. The order of the bases

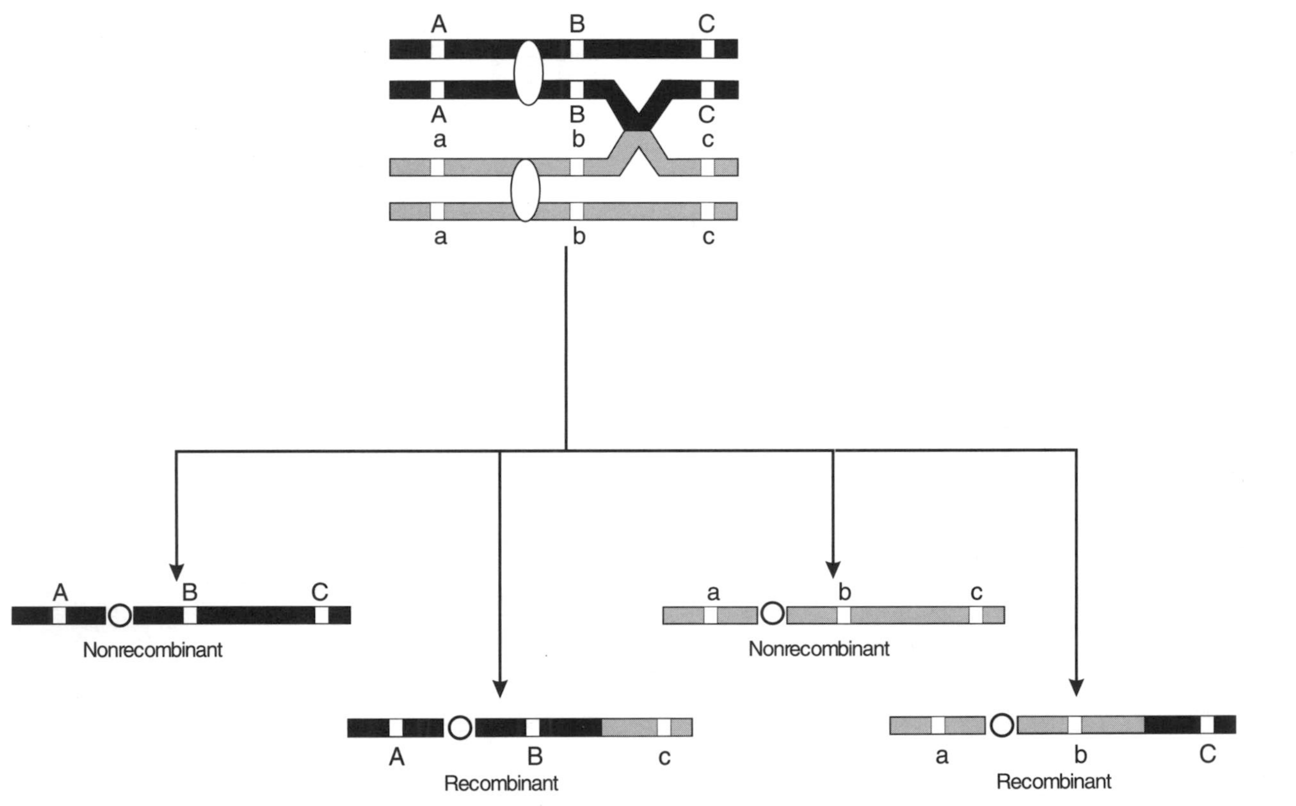

Figure 2.5 Consequences of crossing-over between two homologous chromosomes during meiosis. Two nonrecombinant and two recombinant chromosomes are produced.

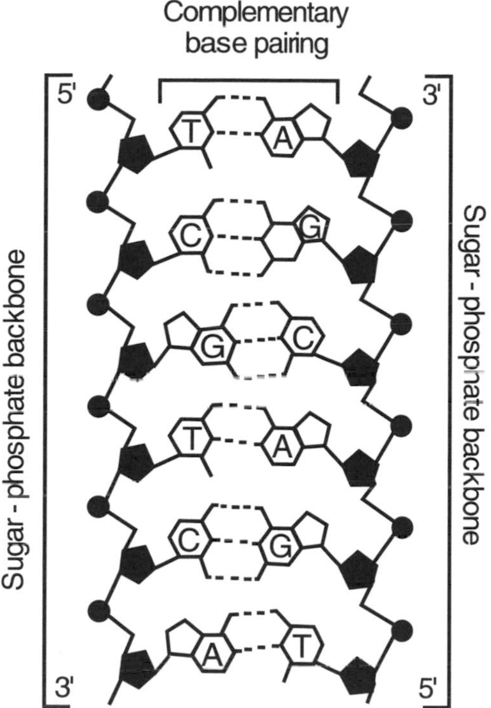

Figure 2.6 Schematic representation of DNA structure. Two polynucleotidic chains running in opposite directions with sugar-phosphate backbones on the outside and the nitrogenous bases inside paired to each other by hydrogen bonds. Adenine (A) pairs with thymine (T), while cytosine (C) pairs with guanine (G). Reprinted, by permission, from R. Roberts, J. Towbin, T. Parker et al., 1992, *A Primer of Molecular Biology* (New York: Chapman & Hall).

arranged along the sugar-phosphate backbone determines the sequence of nucleotides and, consequently, the genetic information carried in all cells of an individual.

The sequence of nucleotides in the DNA molecule is duplicated as cells reproduce themselves. The double-helical structure of the DNA molecule immediately suggests how the genome is duplicated to permit the transmission of genetic material from one generation of cell to the next. During cell division, the weak hydrogen bonds between base pairs break, allowing the two strands of DNA to separate. Each strand serves as a template to direct the synthesis of a complementary new strand, with free nucleotides matching up with their complementary bases on the template strand. Two identical DNA molecules containing the same genetic information are thereby generated. DNA replication is said to be semiconservative, since one old strand of DNA is conserved in each newly synthesized DNA molecule (figure 2.7). DNA replication is a complex process that is accomplished under the action of an enzymatic complex known as DNA polymerase.

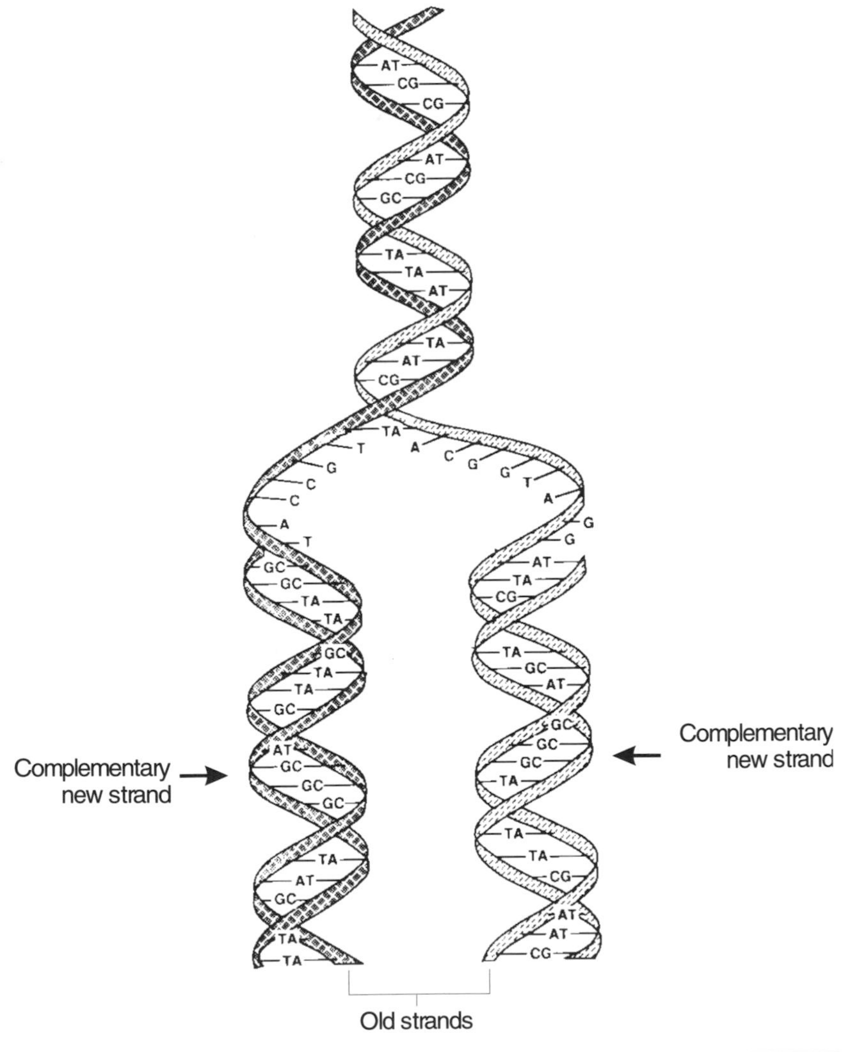

Figure 2.7 DNA replication. During DNA replication the DNA molecule unwinds and each single strand becomes a template for the synthesis of a new complementary strand. Redrawn from *Primer on Molecular Genetics* (5).

From DNA to RNA to Proteins

The genetic information stored in DNA is not directly used by the cells to synthesize proteins. Rather, it must first undergo transcription, a process by which messenger RNA (mRNA) is produced. Messenger RNA can be considered a working copy of the gene. This mRNA then undergoes translation into the sequence of amino acids to be incorporated in the protein. The flow of genetic information from DNA to RNA to protein is the central dogma of molecular genetics. The processes of transcription and translation are thus essential to the expression of genetic information (figure 2.8).

Production of proteins based on the genetic information stored in DNA requires the interaction of several cellular components, including a special type of nucleic acid called ribonucleic acid or RNA. Several characteristics distinguish RNA from DNA. First, ribose is the sugar in RNA compared to deoxyribose in DNA. Second, the base uracil is found in RNA instead of the thymine in DNA. Third, RNA is a single-stranded molecule that is transcribed from DNA.

Three classes of RNA are involved in protein synthesis: messenger RNA (mRNA), transfer RNA (tRNA), and ribosomal RNA (rRNA). Messenger RNA is the complementary copy of the base sequence in DNA that constitutes a gene.

Transcription is the process by which genetic information contained in a specific base sequence of DNA is copied into the mRNA that serves as a template to direct protein synthesis in the cytoplasm. In this process, a complementary RNA is inserted for each DNA base; thus, RNA C is inserted when G is specified in the DNA and vice versa. However, in the case of DNA A, uracil (U) is inserted in the mRNA. Transcription is accomplished by the enzyme RNA polymerase. The primary product of transcription, which is also called pre-mRNA or heterogeneous RNA (hnRNA), is spliced out of intervening and some flanking sequences and then further processed into mature mRNA.

After transcription, mRNA migrates to the cytoplasm, where it binds to ribosomes to direct the synthesis of the protein, a process known as translation. During translation, the base sequence in mRNA is converted into the specific linear sequence of amino acids that are linked together by peptide bonds to form a polypeptide. The process requires transfer RNAs (tRNAs) and ribosomal RNAs (rRNAs). Transfer RNAs are small RNA molecules containing both a binding site for a specific amino acid and a three-base segment, the anticodon, that recognizes a complementary three-base sequence in the mRNA, the codon. Transfer

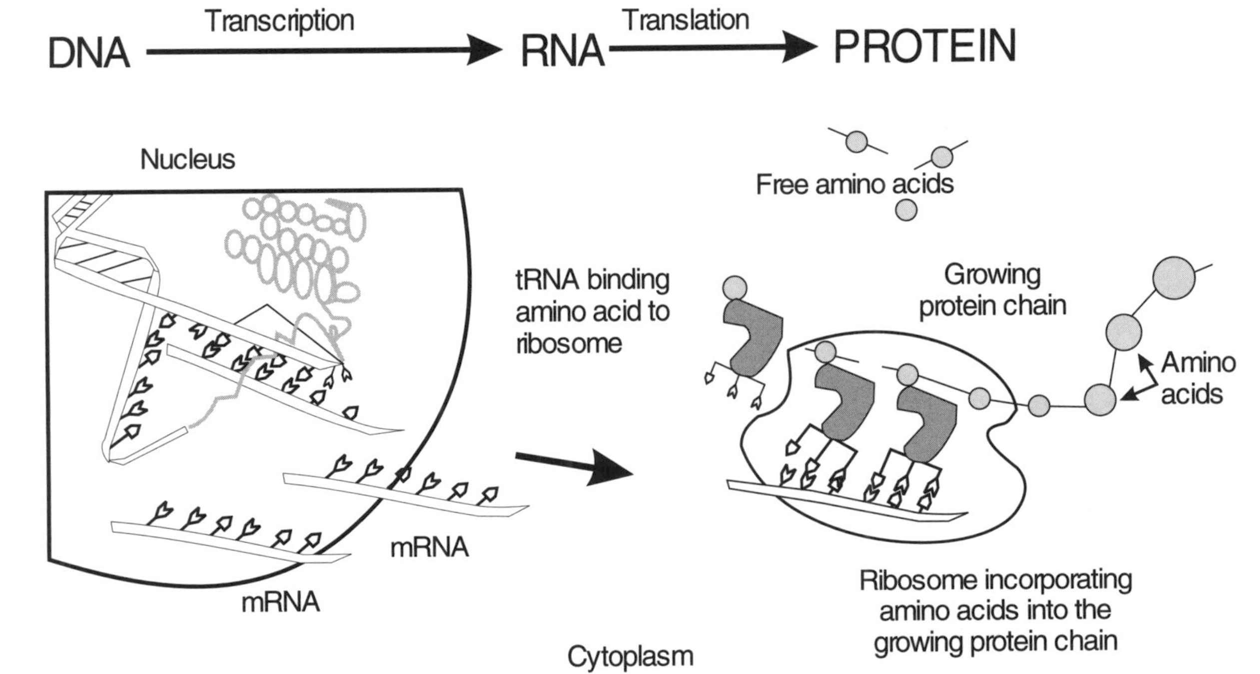

Figure 2.8 Expression of genetic information. When genes are expressed, the DNA found in the nucleus is first copied into messenger RNA (mRNA) by a process known as transcription. The mRNA migrates to the cytoplasm and is incorporated into ribosomes, where the synthesis of proteins by a process known as translation will take place. The genetic information contained in the mRNA is read by triplets of bases known as codons and requires transfer RNAs (tRNAs) that ensure transport of amino acids to the ribosomes for incorporation into the growing protein chain. Adapted from *Primer on Molecular Genetics* (5).

RNAs are responsible for the insertion of the proper amino acid in the growing polypeptide. Ribosomal RNA is a special type of RNA, associated with ribosomes, the structures in the cell where translation takes place.

The relationship between the sequence of bases in DNA and the sequence of amino acids in the protein is specified by the genetic code (table 2.1). The three bases in the mRNA specify the amino acid to be incorporated in the protein. The bases in DNA complementary to the codon are the code words. Since genetic information is determined by the sequence of the four bases (A, T, C, and G) in DNA and proteins are made up of 20 different types of amino acids, the code words should be composed of at least three letters to generate a sufficient number of words ($4^3 = 64$) to represent each amino acid. The genetic information is thus encoded in 64 triplets of DNA bases, each forming a code word that corresponds to a given amino acid. For example, the triplet TAC specifies the incorporation of the amino acid methionine, which also signals the initiation of a genetically significant sequence of DNA. The triplets ATT, ATC, and ACT, on the other hand, signify the end of a genetic message or a gene. The other 60 triplets encode for the remaining 19 amino acids that are incorporated into polypeptide chains and proteins.

The Human Gene

The gene is the basic physical and functional unit of heredity. The site on the chromosome where the gene is located is the locus. Since a gene is often thought of as a discrete unit located at a specific site on a given chromosome, the terms gene and locus are often used interchangeably. A gene is a specific sequence of nucleotides that codes for either an enzyme or a structural protein. Enzymes are the catalysts of all biochemical reactions; structural proteins are important building blocks of cells or tissues. The human genome is estimated to include perhaps as many as 100,000 genes. This estimate has been recently supported by the findings that there may be as many types of RNA transcripts in the various tissues and organs of the body. This is schematically depicted in figure 2.9 for some cell types, tissues, and organs of the body based on the results of a major effort directed by the Institute for Genomic Research.

The typical structure of a human gene is shown in figure 2.10. Genes in higher organisms consist of protein-coding sequences of DNA called exons, interrupted by intervening sequences called introns that are not translated into proteins. Within a given gene there can be

Table 2.1
DNA, mRNA Triplets and Corresponding Amino Acids in the Genetic Code

			Second base										
	DNA	**mRNA**	**A / U**		**G / C**		**T / A**		**C / G**		**mRNA**	**DNA**	
	A	U	UUU	Phenylalanine	UCU	Serine	UAU	Tyrosine	UGU	Cysteine	U	A	
			UUC		UCC		UAC		UGC		C	G	
			UUA	Leucine	UCA		UAA	Stop codon	UGA	Stop codon	A	T	
			UUG		UCG		UAG		UGG	Tryptophan	G	C	
First base	G	C	CUU	Leucine	CCU	Proline	CAU	Histidine	CGU	Arginine	U	A	Third base
			CUC		CCC		CAC		CGC		C	G	
			CUA		CCA		CAA	Glutamine	CGA		A	T	
			CUG		CCG		CAG		CGG		G	C	
	T	A	AUU	Isoleucine	ACU	Thereonine	AAU	Asparagine	AGU	Serine	U	A	
			AUC		ACC		AAC		AGG		C	G	
			AUA		ACA		AAA	Lysine	AGA	Arginine	A	T	
			AUG	Initiation Methionine	ACG		AAG		AGG		G	C	
	C	G	GUU	Valine	GCU	Alanine	GAU	Aspartic acid	GGU	Glycine	U	A	
			GUC		GCC		GAC		GGC		C	G	
			GUA		GCA		GAA	Glutamic acid	GGA		A	T	
			GUG		GCG		GAG		GGG		G	C	

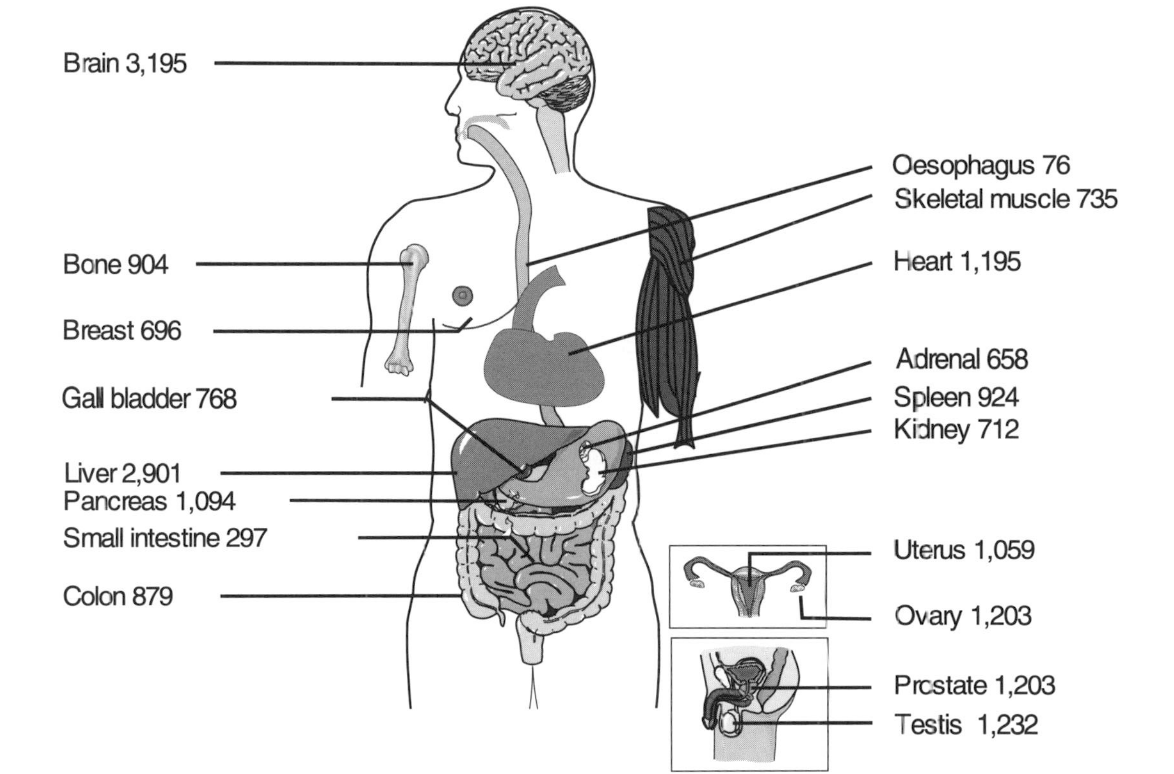

Figure 2.9 Number of RNA transcripts in selected tissues and organs of the body. Adapted from Adams et al. (1).

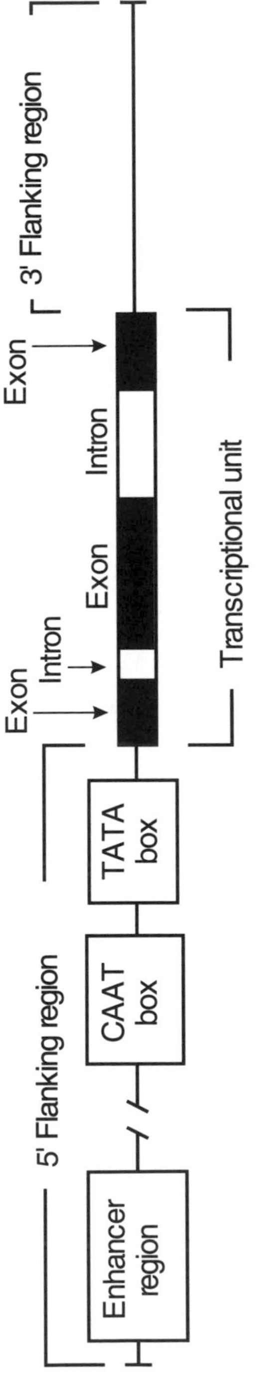

Figure 2.10 Structure of a typical human gene. Reprinted, by permission, from M.R. Cummings, 1991, *Human Heredity: Principles and Issues* (St. Paul: West Publishing Co.) © Wadsworth Publishing Co.

many exons interspaced by introns. On both sides of the gene, flanking regions containing specific DNA sequences are important for the control of gene expression. In the 5' flanking region adjacent to the transcriptional unit, such controlling DNA sequences include the TATA box and the CAAT box (named after the sequence of bases they contain). The TATA box and the CAAT box constitute the essential elements of the promoter that serves as the binding site for the RNA polymerase. In addition, farther from the promoter, there are enhancers, which contain regulatory DNA sequences that stimulate transcription; but there are also silencers, DNA sequences that may inhibit transcription. Flanking regions of genes also contain sequences known as responsive elements, which are capable of receiving signals from signaling molecules (such as hormones) to initiate, enhance, or repress transcription of the gene. The transcriptional unit consists of the DNA sequence, including coding and noncoding regions, that will be transcribed.

In humans, DNA is transcribed into a precursor mRNA (pre-mRNA) that must be modified to remove introns. During the normal processing of pre-mRNA, known as RNA splicing, nucleotides are added to both the 5' and 3' ends of the transcript, and introns and some of the flanking segments are spliced out so that mature mRNA contains only the exons or coding sequences. The splicing process is schematically illustrated in figure 2.11. The first step in processing the pre-mRNA, the modification step, consists of the addition of nucleotides to both sides. In the 5' end, a series of about 30 nucleotides known as the cap is added, while a series of 100 to 200 nucleotides containing adenine (poly-A tail) is added to the 3' end. After this modification step, introns are removed and exons are joined together to form the mature mRNA that will migrate to the cytoplasm for translation into protein.

Genes and Proteins

The role of genes is to specify the nature of proteins through the process of transcription and translation previously described. Proteins are thus central in the relationship between genes and phenotypes. Proteins are ubiquitous in the chemical reactions of the body and constitute over 50% of the dry weight of a typical cell. Proteins are best understood in the context of their functions, which are summarized in table 2.2. Nine types of proteins are indicated in the classification. Several subdivisions could be added, but this classification is adequate to demonstrate the central role of proteins in

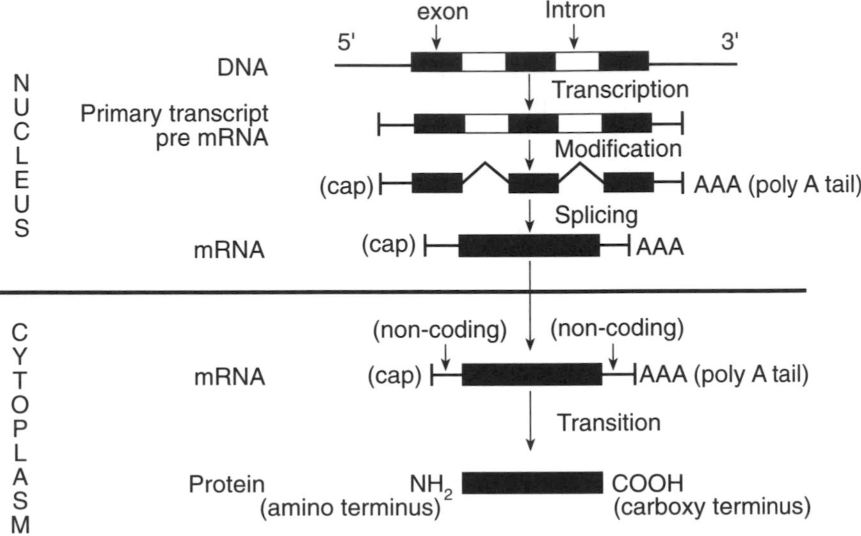

Figure 2.11 Steps in the formation of mRNA. The primary RNA transcript (pre-mRNA) of each gene undergoes processing where the RNA is capped at the 5' end and polyadenylated at the 3' end (poly A tail). The noncoding introns are spliced out and discarded to form a mature messenger RNA (mRNA). The mRNA is used to synthesize the protein in the cytoplasm. Adapted, by permission, from R. Roberts, J. Towbin, T. Parker et al., 1992, *A Primer of Molecular Biology* (New York: Chapman & Hall).

mediating the chain of events between genetic specifications and the expression of phenotypic characteristics. For example, enzyme proteins are a very diverse class that have in common the capacity to increase the rate of biochemical reactions in cells. An enzyme typically has the property of accelerating a specific chemical reaction, although there are several exceptions. More than 1,000 enzymes have been identified to date, and there may be as many as 10,000 human enzymes specified by different genes.

The relationship between genes and proteins is well illustrated in human hemoglobin, a transport protein. The most abundant protein in red blood cells, hemoglobin transports oxygen from the lungs to the tissues. Adult hemoglobin is a protein composed of four polypeptide chains: two alpha chains of 142 amino acids each, and two beta chains of 146 amino acids each. Hemoglobin also contains four heme groups that are capable of binding oxygen. The alpha chain is coded by two identical genes (duplicated) on chromosome 16. An individual therefore has four genes for the alpha chain, two each of maternal and paternal origin. The beta chain is coded by the beta

Table 2.2
A Classification of Proteins by Function

Proteins	Example
Structural	Collagen, the human body's most abundant protein, is found in various kinds of connective tissues.
Storage	Ovalbumin is a major source of materials and energy during embryonic development.
Transport	Hemoglobin transports oxygen from areas of high concentration in the lungs to areas of lower concentration in the tissues.
Receptor	Insulin receptors are proteins embedded in the cell membrane and exposed on the surface of the cell. When insulin and the receptor combine, glucose molecules enter the cell.
Hormone	Growth hormone, released by the pituitary gland, stimulates growth of most body tissues and has widespread metabolic effects.
Protective	Antibodies are produced in response to the presence of foreign substances, organisms, or tissues in the body.
Contractile	Actin and myosin arranged in orderly arrays in muscle fibers produce shortening by sliding past each other in a controlled manner.
Regulatory	Regulatory proteins influence which genes are expressed. Steroid hormone receptor proteins in target cells are an example.
Enzymes	Enzymes are the largest and most diverse class of proteins. Creatine kinase, which allows the phosphorylation of ADP into ATP using creatine phosphate as the substrate, is an example.

Adapted from Singer (7).

globin gene on chromosome 11. The beta globin gene has three exons and three introns. The primary transcript thus requires extensive processing before the mature mRNA stage is achieved. Finally, after the mRNA is translated into polypeptides in the red blood cell, two alpha chains and two beta chains fold together, binding the heme groups in a characteristic manner to produce mature hemoglobin.

The final shape of hemoglobin is dictated by the physical and chemical properties of its polypeptide chains. The chains of amino acids fold and assume a spatial configuration that is reproduced every time such polypeptides are generated. This is true for all gene products, including enzymes. The three-dimensional structure of an enzyme is dictated by the physical and chemical properties of the sequence of amino acids of the particular enzyme and is thought to occur in the presence of molecules known as chaperons. Under such conditions, two individuals with the same gene for the beta globin chain will have exactly the same gene product. They may show differences in the amount of protein produced, but if the structural genes are identical, they will have the same gene product.

Regulation of Gene Expression

The mechanisms regulating gene expression, that is, control of timing and rate of both transcription and translation, are complex. They represent an active area of research that is important for understanding the processes of normal growth and aging, as well as abnormal cellular processes such as occur in cancers.

All normal cells of the body have the same 23 pairs of chromosomes with the same array of genes. After the zygote stage, some cells differentiate into bone cells, muscle cells, nerve cells, and so on. Differentiation into distinct lines of cells occurs during the embryonic period, about the second through eighth weeks after fertilization. Scientists believe that differentiation occurs as some genes are repressed and others activated, leading to the development of specific cell lines and tissues. Some genes (e.g., those coding for the enzymes of glycolysis) are expressed in almost all tissues, because they are essential for the normal metabolic maintenance of cells.

An illustration of variability in gene expression is provided by the myosin molecule: Cell lines that evolve into bone tissue do not need to express the genes for heavy-chain myosin at any stage of development. Precursor cells of myogenesis and future muscle fibers, however, express the genes for myosin heavy chains. There are several forms of myosin, including embryonic, neonatal, and slow or fast muscle type. Thus, in a well-regulated manner, myogenic cell lines express genes that are not expressed in bone and other tissues.

Gene expression during embryonic differentiation is distinct from the subsequent gene regulation that is necessary to meet the needs of growth, environmental demands, and so on. During cellular differentiation, genes are selectively and apparently permanently repressed in some cell lines. Later in life, specific genes may be in an active (on) or inactive (off) state in some tissues, depending on the needs of the cells. Activation and repression of genes are dynamic states that are modulated by neural and hormonal events. It appears that membrane and nuclear receptors, which are unique to a given cell, determine the cellular responses to particular neural or hormonal stimuli and trigger the activation of a gene or a coordinated set of genes. Gene regulation to meet the specific demands of cells or tissues is largely a function of the regulation of the transcription of the genes involved, even though some regulation may occur at the level of translation.

Mitochondrial DNA

Not all of the genetic information of human cells is encoded by nuclear DNA. Mitochondria of all cells contain several copies of a double-stranded circular molecule of 16,569 base pairs known as mitochondrial DNA (mtDNA). The structure of human mtDNA is shown in figure 2.12. Each mtDNA molecule codes for 13 protein subunits of four biochemical complexes associated with the activity and function of the mitochondrion, as well as for two rRNAs (12S and 16S) and 22 tRNAs.

The oxidative phosphorylation pathway located within the mitochondrial inner membrane comprises five multiple enzyme complexes whose genes are dispersed between the mtDNA and nuclear DNA. The mtDNA encodes for the following 13 enzyme subunits: 7 (ND-1, ND-2,

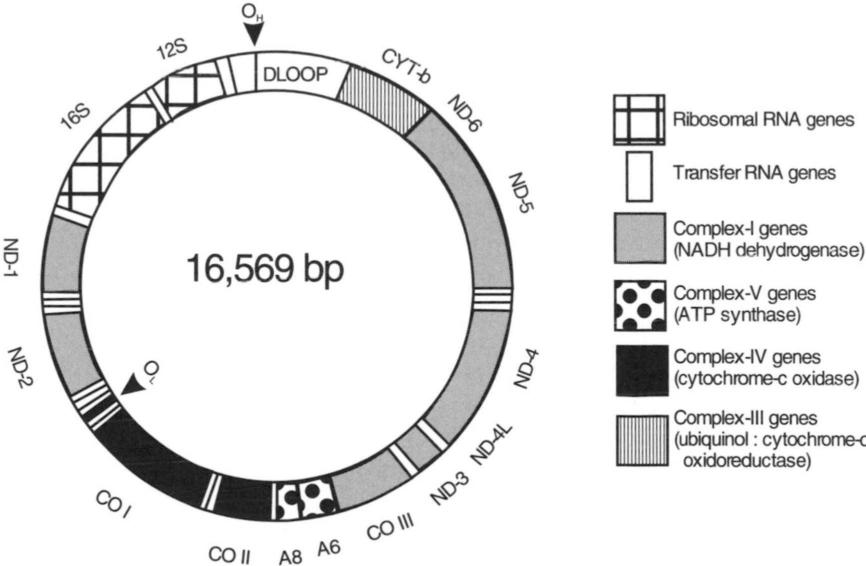

Figure 2.12 Structure of mitochondrial DNA. O_H denotes the origin of the heavy-stranded DNA replication located within the noncoding D (displacement) loop; O_L denotes the origin of the light-stranded DNA replication; CYT-b denotes the cytochrome-b subunit; ND-1, ND-2, ND-3, ND-4, ND-4L, ND-5, and ND-6 are NADH dehydrogenase subunits; CO I, CO II, and CO III are cytochrome-c oxidase subunits; 12S and 16S are ribosomal RNA subunits; and A6 and A8 are ATPase subunits. Adapted from Johns (3).

ND-3, ND-4, ND-4L, ND-5, and ND-6) of the 42 polypeptides of complex-I NADH dehydrogenase (ND); 1 (cytochrome-b) of the 11 polypeptides of complex-III (ubiquinol: cytochrome-c oxidoreductase); 3 (CO I, CO II, and CO III) of the 13 polypeptides of complex IV (cytochrome-c oxidase); and 2 (A6 and A8) of the 16 polypeptides of complex V (ATP synthase). The mtDNA regions between the polypeptide-coding genes usually encode the tRNAs.

Mitochondrial DNA is replicated independently of nuclear DNA, and it has its own system of transcription and translation. Human mtDNA is inherited maternally through the cytoplasm of the ovum. It also has a high mutation rate. It is estimated that the mutation rate of mtDNA is about 10 times greater than that of nuclear DNA. In contrast to nuclear DNA, mitochondrial DNA has no introns, so that mutations generally affect coding DNA. In addition, mtDNA has neither protective histones nor an effective repair system and is exposed to oxygen free radicals generated by oxidative phosphorylation. All of these features explain the frequently observed association between the small mtDNA and diseases.

Summary

The genetic material found in the nucleus of each cell constitutes the human genome. This genetic material is carried in chromosomes that occur in pairs known as homologs (22 pairs of autosomes and one pair of sex chromosomes), one member of each pair coming from the mother and one from the father. The replication of each chromosome and distribution of a complete set of diploid chromosomes to each daughter cell occur during mitosis. Meiosis is another form of cell division in which the duplication of the diploid (2n) number of chromosomes is followed by two successive cell divisions to produce four haploid (n) sex cells (or gametes) that contain only the paternal or the maternal copy of each chromosome pair. The diploid number of chromosomes is restored at fertilization by the fusion of the female (oocyte) and male (spermatocyte) gametes. Two important events during meiosis are responsible for the genetic uniqueness of each individual. The first, known as independent assortment, results from the random distribution of paternal and maternal chromosomes into gametes; the second, known as recombination, results from the exchange of chromosomal segments between homologous chromosomes of maternal and paternal origins.

Deoxyribonucleic acid (DNA), made of phosphorylated sugar molecules connected by pairs of nitrogenous bases, constitutes the genetic material. The entire human genome contains about 3 billion base pairs. The order of these bases along the sugar-phosphate backbone of DNA determines the genetic information found in the nucleus of all cells. The expression of this genetic information into proteins occurs through the processes of transcription and translation that take place in the nucleus and the cytoplasm, respectively. The genetic code describes the relationship between the sequence of bases in DNA and the sequence of amino acids, which are the building blocks of proteins. These proteins are coded by about 100,000 different genes, which is the estimated number of genes in the human genome. In addition to this nuclear genetic information, mitochondria have their own DNA that codes for proteins associated with the function of the mitochondria.

References

1. Adams, M.D.; Kerlavage, A.R.; Fleischmann, R.D.; et al. Initial assessment of human gene diversity and expression patterns based upon 83 million nucleotides of cDNA sequence. Nat. Genet. 377:3-20; 1995.
2. Cummings, M.R. Human heredity: Principles and issues. St. Paul: West Publishing Company; 1991.
3. Johns, D.R. Mitochondrial DNA and disease. N. Engl. J. Med. 333:638-44; 1995.
4. Malina, R.M.; Bouchard, C. Growth, maturation and physical activity. Champaign, IL: Human Kinetics; 1991.
5. Primer on molecular genetics. Washington, DC: U.S. Department of Energy, Office of Energy Research, Office of Health and Environmental Research; 1992.
6. Roberts, R.; Towbin, J.; Parker, T.; Bies, R.D. A primer of molecular biology. New York: Elsevier; 1992.
7. Singer, S. Human genetics: An introduction to the principles of heredity. 2nd ed. New York: W.H. Freeman; 1985.
8. Weiss, K.W. Genetic variation and human disease: Principles and evolutionary approaches. Cambridge: Cambridge University Press; 1993.

CHAPTER 3

Detection and Extent of Human Genetic Variation

Mutations are at the origin of human genetic variation. A mutation is a modification of genetic material at the DNA level. Variation in the genetic material can also result from alterations at the chromosomal level, called chromosomal anomalies. This chapter reviews the different classes of mutations as well as the various methods to detect genetic variation in humans. The extent of human genetic variation is also discussed.

Chromosomal Anomalies

Chromosomal anomalies may affect the number and structure of chromosomes. Numerical aberrations can involve the entire complement of haploid chromosomes (polyploidy) as when, for example, there are three haploid sets of chromosomes (triploidy), or they can involve an alteration in chromosome number that is not a multiple of the entire diploid set of chromosomes (aneuploidy). In humans, aneuploidies most often involve one chromosome. They can result from the presence of an extra chromosome, which is referred to as trisomy ($2n + 1$), or from the deletion of one chromosome, which is referred to as monosomy ($2n - 1$). Nondisjunction, which is the failure of chromosomes to separate during meiosis, is one cause of aneuploidy. Aberrations in the number of chromosomes are quite common in humans. It is estimated that one in every two conceptions may be aneuploid, that aneuploidy is responsible for about 70% of spontaneous abortions, and that about one in every 175 live births is at least partly aneuploid (4). Individuals born with aneuploidies present various physical and behavioral defects, but severity depends on the specific chromosome involved.

Variations in chromosome structure (structural aberrations) result from breakage and rearrangement of chromosomal segments. Such chromosomal mutations are produced either spontaneously through errors in duplication of DNA or by various environmental agents such as radiation or ultraviolet rays. Alterations in the structure of chromosomes include deletion or loss of a chromosomal segment; duplication, in which a segment of a chromosome is present in more than one copy on the chromosome; translocation, in which a segment of a chromosome is exchanged with a segment of a nonhomologous chromosome; and inversion, which involves a 180° position reversal of a chromosomal segment.

Mutation and Genetic Variation

Mutations occur in both somatic and germinal cells (sex cells), but only mutations in the latter can be transmitted from parents to offspring. Many mutations are genetically lethal, while less deleterious mutations that do not affect the individual's ability to survive and reproduce may be transmitted across generations and occur at quantifiable frequencies in a population. These are the mutations that provide the genetic diversity upon which natural selection operates to produce the substantial genetic variation characteristic of human populations.

Gene mutations affect the sequence of bases within a DNA molecule. Gene mutations involving substitutions, deletions, or insertions of one or more bases in the DNA are point mutations. Mutations resulting from the insertion or deletion of a base within a coding region of the gene are called frameshift mutations: They alter the reading frame of the genetic code, such that the sequence of all subsequent triplets is modified. Frameshift mutations may greatly alter the amino acid sequence of the resulting polypeptide and protein product.

Three categories of gene mutations are defined according to their effects on the gene product: (1) in silent mutations, the base substitution does not result in change of the amino acid sequence of the gene product; (2) in missense mutations, the base substitution results in replacement of one amino acid by another; (3) in nonsense mutations, there is an extended or shortened gene product due to elimination or production, respectively, of a stop codon.

In addition to point mutations, gene mutations can involve deletions of large segments of a gene, an entire gene, and even a set of contiguous genes. These deletions can interrupt or remove some of the flanking and/or coding regions and considerably alter the function of the gene. Examples of such mutations have been described in the gene coding for the LDL-receptor, the receptor involved in the removal of LDL cholesterol from the blood. These mutations are responsible for familial hypercholesterolemia (FH), a genetic disease that leads to premature coronary heart disease. Among the mutations responsible for FH among the French Canadian population of Quebec, two deletions have been identified. One involves a deletion of 10 kilobases (kb) that removes the promoter and the first exon of the gene and is responsible for about 60% of cases (10). The other involves a 5-kb deletion that removes exons 2 and 3 of the gene and accounts for about 3% of FH cases (14,15).

Genetic Polymorphism

The study of human genetic variation is based on the chromosomes, proteins, and DNA of somatic cells. Genetic variation at the chromosome level is generally studied for prenatal diagnosis or in a pathological context. For instance, aberrations in the structures of some chromosomes have been identified in certain types of cancers. Detection of this type of genetic variation is usually based on cytogenetic studies. Such variations are observed in particular pathological conditions, which are beyond the scope of this volume.

The concept of polymorphism is used to define genetic variation in structural genes or at the DNA level. A polymorphism for a given gene exists when an allele is present in a population at a frequency of at least 1%. Genetic polymorphisms have been extensively studied in the recent past for human gene products associated with red blood cell antigens, tissue antigens, serum proteins, and red blood cell enzymes.

Two measures are generally used to assess the degree of polymorphism at a given locus. One is the index of heterozygosity, defined as the probability that an individual is a heterozygote, that is, has two different versions of a structural gene at a given locus. The other measure is the polymorphism information content (PIC), defined as the probability that the genotype of an offspring at a given locus will permit detection of which of the alleles was received from a given parent. These two measures of polymorphism are a function of allele frequencies and have generally similar values. High PIC values and high levels of heterozygosity are associated with a high degree of polymorphism.

Since the flow of genetic information is from DNA to RNA to proteins, proteins can be used to detect genetic variation. Protein polymorphisms can be detected by a technique known as electrophoresis, in which protein charge variants can be distinguished by migration in an electric field. If a gene mutation results in the replacement of an amino acid with a different charge property, the net charge of the protein will be affected and will influence the pattern of migration of the protein in the electric field. Proteins can also be separated according to their mass, by a technique known as two-dimensional electrophoresis (16). Two-dimensional electrophoresis separates proteins on the basis of both charge and mass. Advances in molecular biology over the past 20 years, however, permit detection of genetic variation not just with proteins but at the DNA level.

Recombinant DNA Technology

New methods permit more detailed examination of genetic variation at the DNA level so that exons, introns, and the flanking regions of genes can now be systematically studied in search of new genetic variants. These methods have revolutionized the analysis of human genetic variation in modern genetics and are briefly reviewed in this section. Detailed description of the methods can be found in molecular biology textbooks. The methods are collectively referred to as recombinant DNA technology (2,20).

Isolation of DNA

Although human DNA can be isolated from any cell in the body, white blood cells or leukocytes represent the major source of DNA for the study of genetic variation in humans. Leukocytes are isolated from whole blood by centrifugation, treated with a detergent (sodium dodecyl sulfate) to liberate DNA from the nucleus, and then treated with proteinase to remove the attached proteins from the DNA. DNA is extracted with phenol and chloroform, then precipitated with ethanol. Lymphocytes can also be transformed into lymphoblasts with the Epstein-Barr virus, so that they can be grown in culture (19). These transformed lymphoblast cell lines can be frozen and stored to provide a renewable supply of DNA for an individual.

Restriction endonucleases

The key step in the development of recombinant DNA technology was the discovery in the mid-1970s of restriction enzymes. These enzymes, found in bacteria, are endonucleases and act as "molecular scissors." They are characterized by their ability to recognize and cleave DNA at specific sites of base sequences. Each of these enzymes recognizes a unique DNA sequence, called a restriction site, and cuts DNA whenever this sequence is encountered along the DNA. These enzymes are named after the bacteria from which they were isolated. For example, EcoR1, which recognizes and cuts DNA at the six base sequence GAATTC, is the first restriction enzyme that has been isolated from *Escherichia coli,* the bacteria found in the intestine. More than 300 different restriction enzymes have been identified thus far.

When cleaving DNA, many restriction enzymes leave fragments with single-stranded complementary tails at both ends. Cleavage of DNA from two different sources with this type of restriction enzyme generates the same complementary ends, which allows a DNA fragment from one source to join the complementary fragment from the other, creating a recombinant DNA molecule. The basic steps in recombinant DNA technology are illustrated in figure 3.1. These include (1) generation of DNA fragments by use of restriction enzymes; (2) insertion of DNA fragments into other DNA molecules (known as vectors, usually DNA from bacteria, viruses, or yeast) to create a recombinant DNA molecule; (3) insertion of the recombinant molecule into a host cell (generally bacteria) for replication, which will produce several copies, or clones, of the inserted DNA fragment; and (4) isolation of the cloned DNA fragments from the host cell for further study.

Southern blotting

One widely used procedure to locate a specific gene or DNA fragment is Southern blot analysis (figure 3.2). When DNA is extracted from human cells (generally white blood cells) and digested with a particular restriction enzyme, the DNA is cut into several thousand small fragments. The DNA fragments are placed on an agarose gel and subjected to electrophoresis. Since DNA molecules are negatively charged, DNA fragments migrate through the gel from the negative to the positive pole according to their length: Smaller fragments migrate faster, since they move more easily than larger ones in the gel. DNA fragments are therefore separated according to their size: Large fragments are found at the top of the gel and small fragments at the bottom.

The DNA fragments in the gel are then denatured into single-stranded fragments and transferred in a buffer from the gel to a suitable filter (nitrocellulose or nylon) by capillary action. This process is known as blotting. It was first discovered by E.M. Southern (21), hence it was named Southern blotting. The nitrocellulose or nylon filter provides a solid support for the DNA fragments so they can be more easily manipulated. The DNA fragments are fixed permanently to the filter by baking (for nitrocellulose filter) or irradiation with ultraviolet light (for nylon filter). Once on the filter, the fragments are exposed to a specific DNA fragment containing the sequence of interest. This specific single-stranded DNA fragment, the probe, has been made radioactive using ^{32}P. The probe hybridizes to the complementary DNA sequence on the

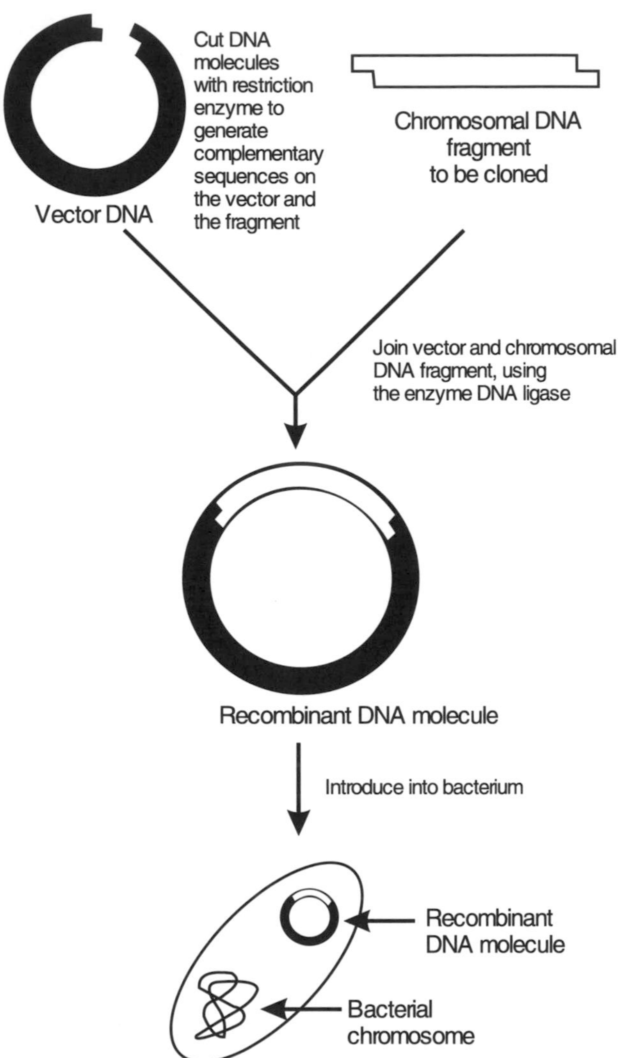

Figure 3.1 Basic steps in recombinant DNA technology. Redrawn from *Primer on Molecular Genetics* (18).

filter. The nonhybridized probe is washed away from the filter, while the remaining hybridized fragments are detected by exposure of the filter to an X-ray film, a technique known as autoradiography.

Conventional Southern blot analysis has several limitations. It is most useful to analyze DNA fragments no longer than 20 kb, since larger fragments cannot migrate in agarose gel. Several methods such as pulse-

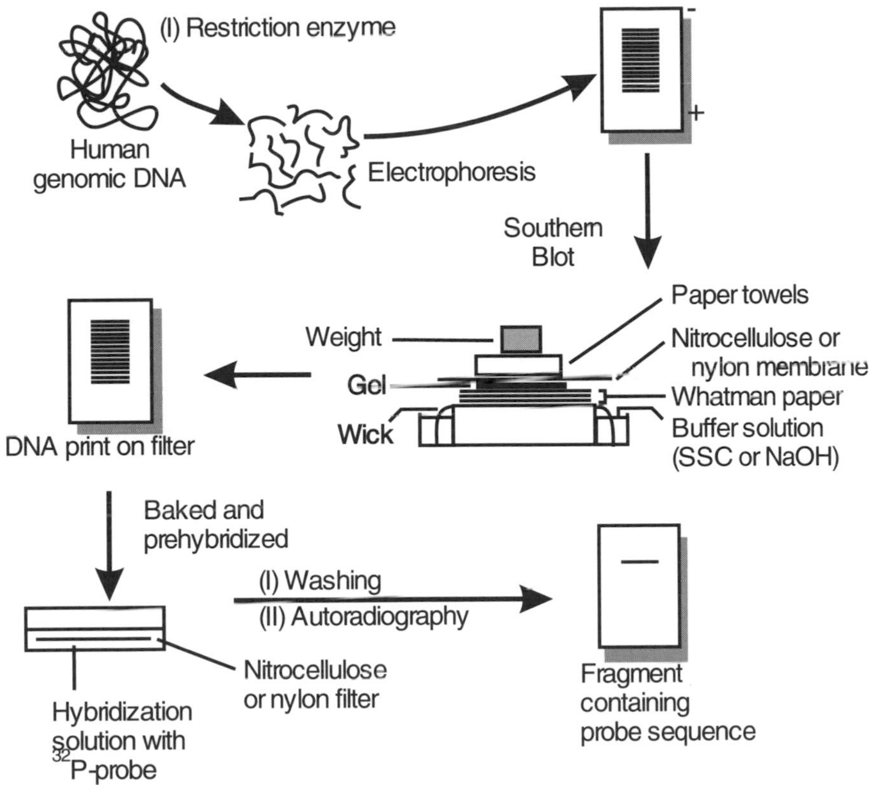

Figure 3.2 Schematic representation of Southern blot analysis. Reprinted, by permission, from R. Roberts, J. Towbin, T. Parker et al., 1992, *A Primer of Molecular Biology* (New York: Chapman & Hall).

field gel electrophoresis, field inversion gels, and contour-clamp homogeneous electric field electrophoresis are now available to study larger DNA fragments up to 1,000 kb. The Southern blot technique is also expensive (several reagents are needed) and not amenable to automation. Finally, its major limitation is the length of time necessary for the different steps; it may take several days before DNA fragments can be visualized.

Polymerase chain reaction

A powerful alternative approach for DNA amplification is the polymerase chain reaction (PCR). This method has completely transformed the way genetic analyses are done today, because of its ability to rapidly amplify very small amounts of DNA (5). It is illustrated in figure 3.3. PCR is based on the annealing and extension of two oligonucleotide

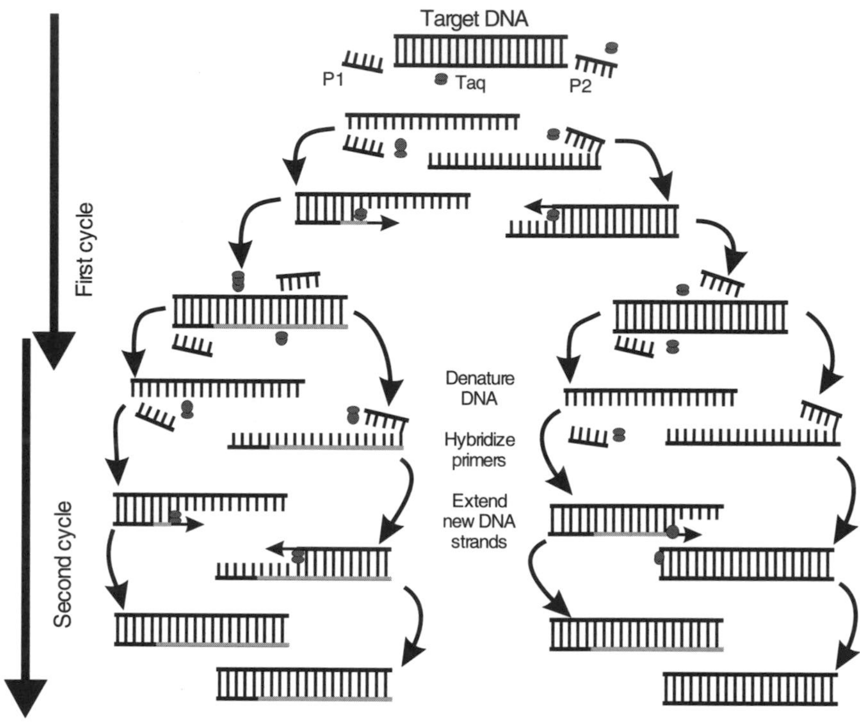

Figure 3.3 DNA amplification using polymerase chain reaction (PCR). P1 = primer 1; P2 = primer 2; Taq = thermus aquaticus enzyme. Redrawn from *Primer on Molecular Genetics* (18).

primers (single-stranded DNA of 15 to 30 bases in length) that flank a specific target DNA segment that is to be amplified. The first step of the PCR reaction involves denaturation of DNA in order to separate the two strands. This is accomplished by heating the DNA at a temperature around 95°C for about 20 seconds to 1 minute. In the second step the temperature is lowered to approximately 55°C, which allows the primers to anneal to the target DNA. The third step of the reaction is the synthesis of DNA complementary to the strand template by extension of the annealed primers, which is done at a temperature of about 72°C. DNA synthesis is ensured by a heat-stable DNA polymerase isolated from the bacterium *Thermus aquaticus* (Taq polymerase). The use of this thermostable polymerase has allowed the PCR reaction to be automated by avoiding the use of fresh polymerase at each denaturation step. The three steps (denaturation, annealing, and DNA amplification or extension) represent one PCR cycle. Each cycle results in a doubling of the number of strands of DNA found at the previous cycle. If this process is repeated n times, the DNA sequence between the primers will be amplified 2^n. For example, after about one hour and 20 PCR cycles, the original DNA segment will have been amplified a millionfold.

PCR presents several advantages for the study of human genetic variation (5). Its rapidity is obviously a major advantage. Another advantage is that very small amounts of DNA (in the order of picograms, i.e., 10^{-12} g) can be studied. Many genetic loci can also be amplified at the same time in one PCR reaction by using multiple primer pairs. The PCR can be fully automated, allowing a large number of samples to be processed with little intervention. With all these advantages, PCR has surpassed bacterial cloning and is now a standard procedure to amplify DNA fragments. Finally, it is important to remember that the sequence of the DNA fragment to be amplified must be known in order to synthesize specific primers flanking the DNA fragment of interest.

Detection of Genetic Variation

The developments brought about by DNA recombinant technology and PCR have provided geneticists with new tools to detect genetic variation directly at the DNA level. With these tools, different types of DNA polymorphisms in both coding and noncoding regions of the human genome were discovered. These polymorphisms and the methods used to detect them are discussed below.

Restriction fragment length polymorphisms (RFLPs)

The number and length of DNA fragments resulting from digestion of genomic DNA with a restriction enzyme vary according to the number of restriction sites along the DNA molecule and the distance between the restriction sites. Variation in a base sequence can create or destroy a restriction site, giving rise to a detectable variation in the length of the DNA fragment that is inherited in a Mendelian codominant fashion (figure 3.4). Such genetic variation is known as a restriction fragment length polymorphism (RFLP) and represents the first type of DNA polymorphism that was identified with the Southern blot method.

The A and B alleles in the figure represent a DNA segment of 10 kb from homologous chromosomes. The thick region represents a segment that can be detected using a probe. The arrows indicate the pattern of restriction enzyme cutting sites that define the A and B alleles. The A allele has three restriction sites generating fragments of 6 kb and 4 kb. The absence of a restriction site in the B allele generates only one fragment 10 kb in length. Because these alleles are codominant, three genotypes (AA, AB, and BB) are thus generated. The fragment patterns

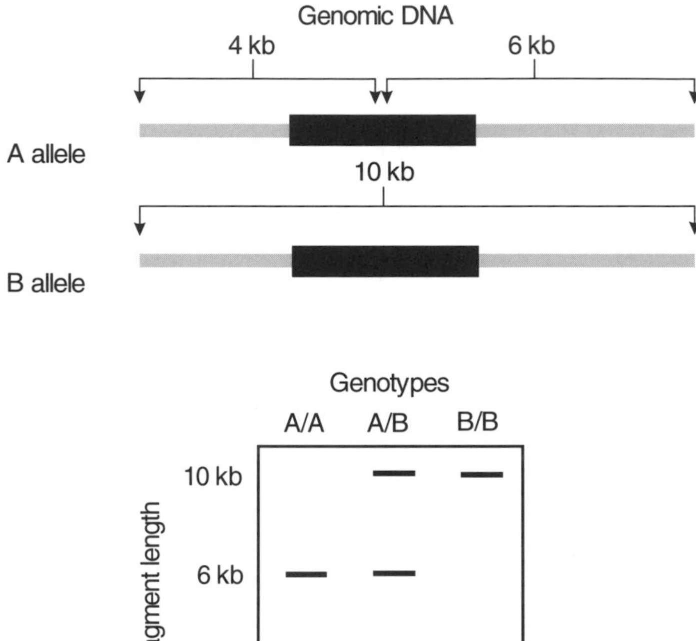

Figure 3.4 Detection of a restriction fragment length polymorphism (RFLP). *Essentials of Genetics, 2/E* by Klug/Cummings, ©1996. Reprinted by permission of Prentice-Hall, Inc., Upper Saddle River, NJ.

corresponding to each genotype can be visualized on a Southern blot. The DNA fragment can also be amplified by PCR if the sequence surrounding the restriction site is known. In such case, the PCR product is digested with the same restriction enzyme and analyzed by electrophoresis. The major limitation of RFLP is that these polymorphisms are limited only to changes in DNA affecting restriction sites.

Tandem-repeat polymorphisms

A second category of polymorphisms found in the human genome results from the repetition of a short DNA sequence a variable number of times (11). Such sequences are known as tandem-repeat sequences. When tandemly repeated sequences are replicated during cell division, the number of repeats can change. Variations in the number of tandem-repeats will result in variable-length DNA fragments that can be detected by Southern blotting or PCR, depending of the length of the tandem-repeat. If the repetitive sequence involves more than 5 base pairs (generally around 10 to 15 base pairs), these polymorphisms are called variable number of tandem-repeats (VNTR) and are visualized by Southern blotting. On the other hand, if the repetitive sequence is shorter, these polymorphisms are called short tandem-repeats or microsatellites and are best visualized by PCR. The most frequent type of microsatellites are the dinucleotide repeats involving cytosine and adenosine (CA repeats), whose lengths vary approximately between 24 base pairs (12 repeats) and 80 base pairs (40 repeats). This type of polymorphism is illustrated in figure 3.5: The number of copies of tandem-repeats is indicated by the number of boxes and determines the size of the DNA fragment measured between the sites of the two primers. It is visualized by separating the PCR products by electrophoresis (11). These short tandem-repeat polymorphisms are abundant, evenly distributed within the genome, and highly polymorphic (PIC ≥ 0.7), making them the most widely used markers to map genetic diseases and to study the molecular basis of multifactorial phenotypes.

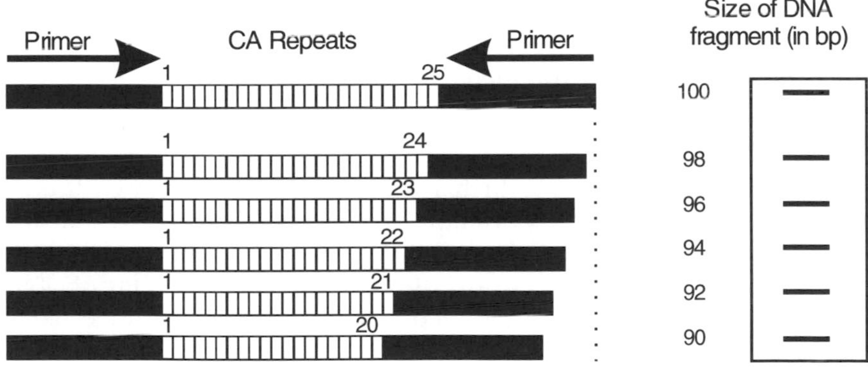

Figure 3.5 Short tandem-repeat polymorphism. Adapted from Kaplan and Delpech (12).

PCR-based analysis of genetic variation

Most of the approaches used to detect DNA polymorphism are based on the analysis of PCR products. Detection of DNA polymorphism with the PCR method involves amplification of a segment of DNA surrounding the site of the mutation, which requires a priori knowledge of the DNA sequence in order to synthesize specific primers. The method used depends on whether identification of a known pathogenic mutation or an unknown mutation is desired. In the first situation, methods like RFLP analysis (if the mutation perturbs a restriction site), allele-specific oligonucleotide (ASO) hybridization, or allele-specific enzymatic amplification can be used. In the second situation, where no specific mutation is investigated, four different methods can be used: denaturing gradient gel electrophoresis, single-strand-conformation polymorphism, chemical cleavage of base-pair mismatches, and DNA sequencing. This is the case for the detection of DNA polymorphism.

Denaturing gradient gel electrophoresis

Denaturing gradient gel electrophoresis (DGGE), which allows the separation of DNA fragments differing by as little as a single base pair, is based on the melting properties of DNA molecules in solution (19). DNA molecules melt in discrete segments known as melting domains that vary in size from 25 bp to hundreds of base pairs. Each melting domain has a distinct melting temperature (T_m) that is highly dependent on its nucleotide sequence but that can be altered by a change in the sequence. Mutations in DNA will alter the T_m of the PCR-amplified DNA fragments. Electrophoresis of DNA fragments in a polyacrylamide gel containing a linear gradient of increasing DNA denaturant will cause fragments with decreased thermal stability to denature first and alter the mobility of the fragments in the gel.

Single-strand-conformation polymorphism (SSCP)

The detection of mutations with this technique, developed by Orita et al. (17), is based on the mobility of single-stranded DNA in nondenaturing polyacrylamide gel electrophoresis (9). Any mutation modifying the sequence, even a single base substitution, will affect the folding of DNA and consequently the conformation of the molecule. When subjected to nondenaturing polyacrylamide gel electrophoresis, the mutant DNA will have an altered electrophoretic mobility and will be visualized as a band at a different position on the autoradiograph. The sensitivity of SSCP to detect mutations varies according to the length of the DNA fragment and tends to decrease with increasing fragment length (9).

Chemical cleavage of base-pair mismatches

This method, described by Cotton et al. (3), takes advantage of the formation of heteroduplexes. These are double-stranded DNA fragments resulting from pairing with a mismatch of two single-stranded DNA fragments, one with a mutation and the other without it. The heteroduplex DNA sequences are incubated either with osmium tetroxide for the mismatches involving thymine (T) or cytosine (C), or with hydroxylamine for the C mismatches. This is followed by piperidine incubation that cleaves the DNA at the modified mismatched base (19).

Products of the cleavage are then analyzed by electrophoresis on a denaturing polyacrylamide gel. This technique can detect all types of mutations, including deletions, insertions, and substitutions.

DNA sequencing

The techniques previously described permit the determination of sequence variation in PCR products. In the presence of positive results, DNA sequencing (determination of the nucleotide sequence of the amplified DNA segment) is undertaken to precisely identify the mutation. Two basic methods, differing in the way the nested DNA fragments are produced, are used for DNA sequencing. The first is known as the Maxam-Gilbert method or the chemical degradation method (figure 3.6). It uses chemicals to cleave DNA at specific bases, resulting in fragments of different lengths. These fragments, labeled at one end with ^{32}P, are then subjected to a set of four partial, but base-specific, cleavages that produce a series of subfragments that are separated by gel electrophoresis.

The second method, known as the Sanger or the dideoxy method, uses enzymatic procedures to synthesize DNA fragments of varying length in four different reactions that run in parallel. Each reaction contains one of the four labeled dideoxynucleotides in which the hydroxyl group at the position 3' is replaced by a hydrogen atom, which renders this nucleotide incapable of forming a phosphodiester bond with another nucleotide. When such a nucleotide is introduced into the DNA growing chain, the synthesis is interrupted at the point where the natural deoxynucleotide is introduced. The resulting sets of products are analyzed by gel electrophoresis.

Extent of DNA Variation

The study of DNA by recombinant DNA technologies has revealed that most of the genetic variation lies outside of the coding regions of the gene. This is not surprising, as it is well known that DNA sequences of structural or functional importance (such as genes) are found in highly conserved regions of the genome. Coding regions of active genes are therefore the most stable in terms of intraspecific individual variation and exhibit a lower degree of polymorphism compared with that found within introns and flanking regions of genes (1). Along the 3 billion base pairs that constitute the human genome, it is believed that there is variation every 100 to 300 base pairs. Even if a more conservative estimate of one mutation for every 500 base pairs is used, genetic individuality still remains the norm (1). The study of human DNA

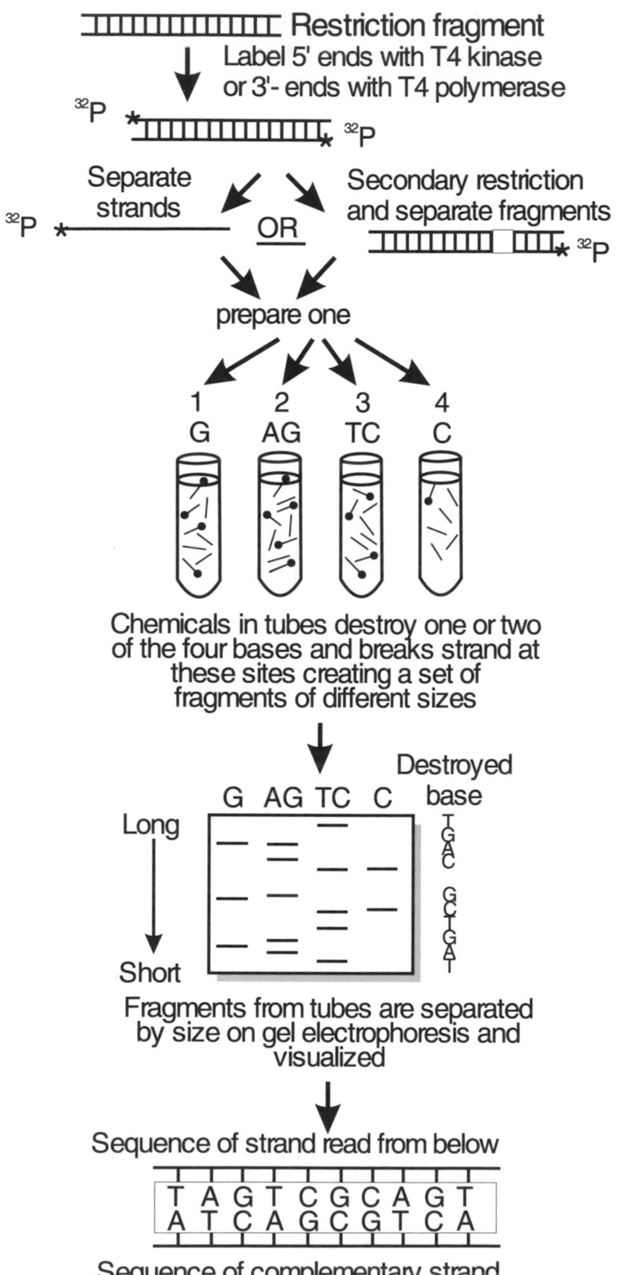

Figure 3.6 Maxam-Gilbert DNA sequencing method. Reprinted, by permission, from R. Roberts, J. Towbin, T. Parker et al., 1992, *A Primer of Molecular Biology* (New York: Chapman & Hall).

variation provides information on the evolution of the human species or on fundamental mechanisms of population genetics, such as selection, mutation gene flow, and genetic drift. Another modern application of the study of DNA variation is DNA fingerprinting, as in forensic medicine.

The average protein heterozygosity detected by two-dimensional electrophoresis (2DE) is estimated to be around 3% (6,8), and slightly more than 10% of cellular proteins and 20% of serum proteins have been estimated to be polymorphic (6). Based on the study of 42 polymorphic protein loci detected by 2DE, Goldman and Merril (7) found that 55% of these loci had PIC values below 0.3. As shown in table 3.1, the extent of DNA polymorphism is higher for RFLPs or short tandem-repeat polymorphisms (microsatellites) with PIC values generally higher than 0.6 for the latter.

Table 3.1
Comparison of Polymorphism Information Content (PIC) for Proteins, RFLPs, and Microsatellites

	PERCENTAGE		
PIC	**Proteins**	**RFLPs**	**Microsatellites**
< 0.1	31	2	
0.1 - 0.2	24	7	
0.2 - 0.3	14	15	
0.3 - 0.4	26	45	3
0.4 - 0.5	5	13	16
0.5 - 0.6		10	22
0.6 - 0.7		3	30
0.7 - 0.8		2	20
0.8 - 0.9		1	9
0.9 - 1.0		1	

Data for protein polymorphisms are derived from Goldman and Merril (7); data for RFLPs and microsatellites are from Kaplan and Delpech (12).

Summary

Genetic variation results from modifications of the genetic material either at the chromosome level (chromosomal mutations) or at the DNA level (gene mutations). The study of genetic variation at the chromosome level resulting in alterations in the structure or the number of chromosomes is usually performed for clinical purposes. Gene mutations altering the sequence of bases within the DNA are those responsible for the genetic diversity observed in human populations. The concept of genetic polymorphism is used to assess the importance of genetic variation at the DNA level. A gene is said to be polymorphic when an alternative form of this gene is observed with a frequency of at least 1% in the population. Genetic variation can be detected either at the protein level, by a technique known as electrophoresis, or at the DNA level by various methods collectively referred to as recombinant DNA technology. These methods have led to the identification of various types of polymorphisms in the human genome, including restriction fragment length polymorphisms (RFLPs) and tandem-repeat polymorphisms. The study of these polymorphisms reveals that genetic variation is more important in noncoding regions of DNA and is ubiquitous in the human genome.

References

1. Bowcock, A.; Cavalli-Sforza, L. The study of variation in the human genome. Genomics. 11:491-98; 1991.
2. Carroll, W.L. Introduction to recombinant-DNA technology. Am. J. Clin. Nutr. 58:249S-58S; 1993.
3. Cotton, G.H.; Rodrigues, N.R.; Campbell, R.D. Reactivity of cytosine and thymine in single-base-pair mismatches with hydroxylamine and osmium tetroxide and its application to the study of mutations. Proc. Natl. Acad. Sci. 85:4397-4401; 1988.
4. Cummings, M.R. Human heredity: Principles and issues. St. Paul: West Publishing Company; 1991.
5. Erlich, H.A.; Arnheim, N. Genetic analysis using the polymerase chain reaction. Annu. Rev. Genet. 26:479-506; 1992.
6. Goldman, D.; Goldin, L.G.; Rathnagiri, P.; O'Brien, S.; Merril, C.R. Twenty-seven protein polymorphisms by two-dimensional electrophoresis of serum, erythrocytes and fibroblasts in two pedigrees. Am. J. Hum. Genet. 37:130-38; 1985.
7. Goldman, D.; Merril, C.R. Protein polymorphisms detected by two-dimensional electrophoresis: An analysis of overall informativeness of a panel of linkage markers. J. Psychiatr. Res. 21:597-608; 1987.
8. Hanash, S.M.; Baier, L.J.; Welch, D.; Galteau, M. Genetic variants detected among 106 lymphocyte polypeptides observed in two-dimensional gels. Am. J. Hum. Genet. 39:317-28; 1986.

9. Hayashi, K. PCR-SSCP: A simple and sensitive method for detection of mutations in the genomic DNA. PCR Methods Appl. 1:34-38; 1991.
10. Hobbs, H.H.; Brown, M.S.; Russel, D.W.; Davignon, J.; Goldstein, J.L. Deletion in the gene for the low-density-lipoprotein receptor in a majority of French Canadians with familial hypercholesterolemia. N. Engl. J. Med. 317:734-37; 1987.
11. Housman, D. Human DNA polymorphism. N. Engl. J. Med. 332:318-20; 1995.
12. Kaplan, J.C.; Delpech, M. Biologie moleculaire. Paris: Flammarion; 1993.
13. Klug, W.S.; Cummings, M.R. Essentials of genetics. New York: Macmillan; 1993.
14. Leitersdorf, E.; Tobin, E.J.; Davignon, J.; Hobbs, H.H. Common low-density lipoprotein receptor mutations in the French Canadian population. J. Clin. Invest. 85:1014-23; 1990.
15. Ma, Y.; Betard, C.; Roy, M.; Davignon, J.; Kessling, A. Identification of a second "French Canadian" LDL receptor gene deletion and development of a rapid method to detect both deletions. Clin. Genet. 36:219-28; 1989.
16. O'Farrell, P.H. High resolution two-dimensional electrophoresis of proteins. J. Biol. Chem. 250:4007-21; 1975.
17. Orita, M.; Iwahana, H.; Kanazawa, H.; Hayashi, K.; Sekiya, T. Detection of polymorphisms of human DNA by gel electrophoresis as single-stranded conformation polymorphisms. Proc. Natl. Acad. Sci. 86:2766-70; 1989.
18. Primer on molecular genetics. Washington, DC: U.S. Department of Energy, Office of Energy Research, Office of Health and Environmental Research; 1992.
19. Roberts, R.; Towbin, J.; Parker, T.; Bies, R.D. A primer of molecular biology. New York: Elsevier; 1992.
20. Rosenthal, N. Tools of the trade: Recombinant DNA. N. Engl. J. Med. 331:315-17; 1994.
21. Southern, E.M. Detection of specific sequences among DNA fragments separated by gel electrophoresis. J. Mol. Biol. 98:503-17; 1975.

CHAPTER 4

Understanding the Methods

Advances in molecular biology and statistical methods have changed the strategies for investigating the genetic basis of complex human traits. The genetic analysis of quantitative phenotypes integrates methods from the fields of quantitative genetics, population genetics, and molecular biology.

This chapter presents an overview of the basic concepts and methods of genetic epidemiology that are currently used to address these questions. Some of the basic methodological issues have been reviewed elsewhere (27,28,57,64,69) and have also been addressed in past reviews dealing with the genetics of obesity (9) and physical fitness and performance (5,6,37). As will become evident in subsequent chapters, some of these methods have not yet been applied extensively to study of the genetic basis of fitness and performance phenotypes.

Overview of Research Strategies

An overview of the most common research strategies to investigate the genetic basis of quantitative multifactorial phenotypes is presented in figure 4.1. Approaches based on both human and animal studies can be used to identify the genes underlying quantitative phenotypes. Both approaches are increasingly used in collaborative efforts for reasons that should become clear in this chapter. In humans, statistical analyses based on the paradigms of genetic epidemiology have been traditionally used to investigate the contribution of genetic determinants to quantitative phenotypes and to test hypotheses regarding a variety of general and specific models of inheritance. More recently, molecular studies based on random genetic markers or candidate genes have been increasingly used in linkage and association studies designed to identify genes contributing to phenotypes. These complementary approaches are reviewed below.

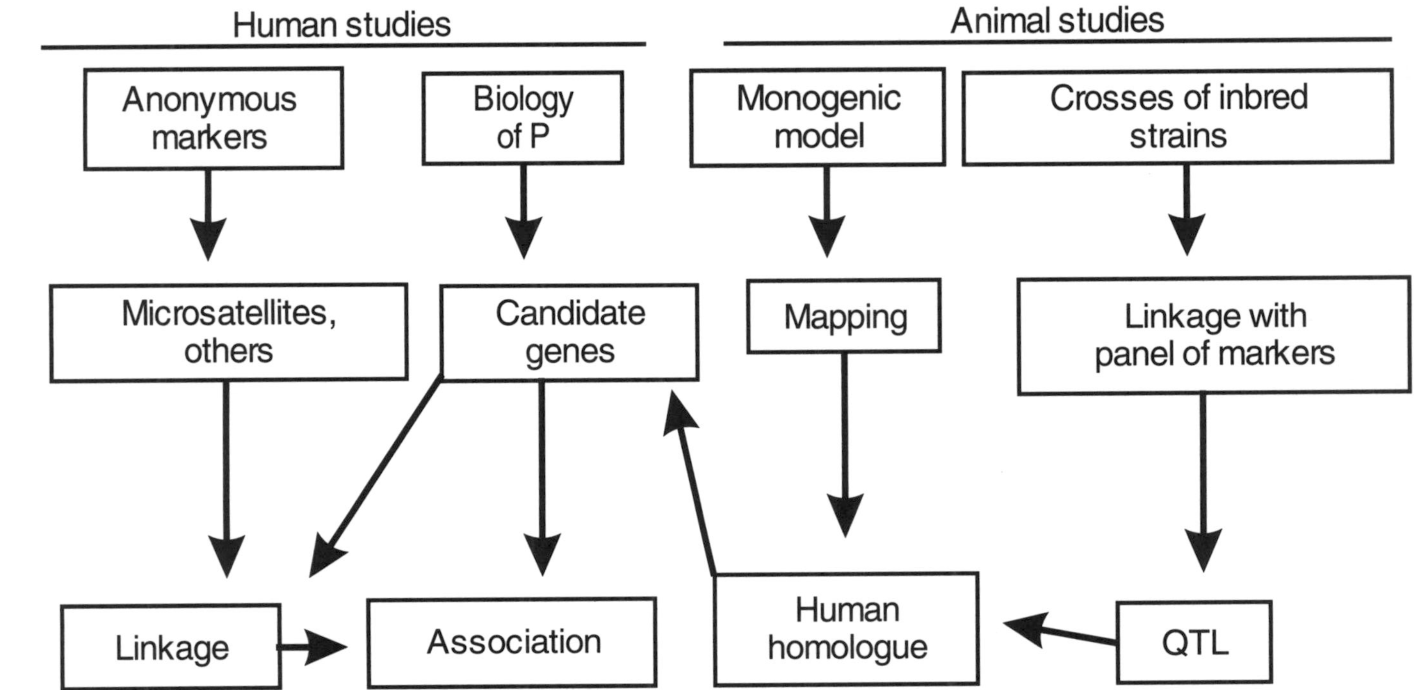

Figure 4.1 Overview of research strategies used to study the genetic and molecular basis of complex multifactorial phenotypes. P = phenotype.

Human Studies

The methods used to study the genetic basis of quantitative phenotypes can be divided into two basic approaches (figure 4.2): the unmeasured genotype and the measured genotype approaches (65). The unmeasured genotype approach is based on statistical analysis of the distribution of phenotypes in individuals and families. This approach is also referred to as the top-down approach, since inferences about the influence of genes are made from the phenotype. The measured genotype approach uses genetic variation in random genetic markers or in candidate genes and attempts to evaluate the impact of variation at the DNA level on the quantitative phenotype under study. Since inferences about the role of genes are made from the DNA to the phenotype, the approach is also known as the bottom-up approach (65,66).

The different steps in the investigation of the genetic basis of a quantitative phenotype are summarized in figure 4.3 and broadly constitute the field of genetic epidemiology (27). The first step deals with whether or not the phenotype under study aggregates in families. If there is evidence of familial aggregation, the second step is to determine

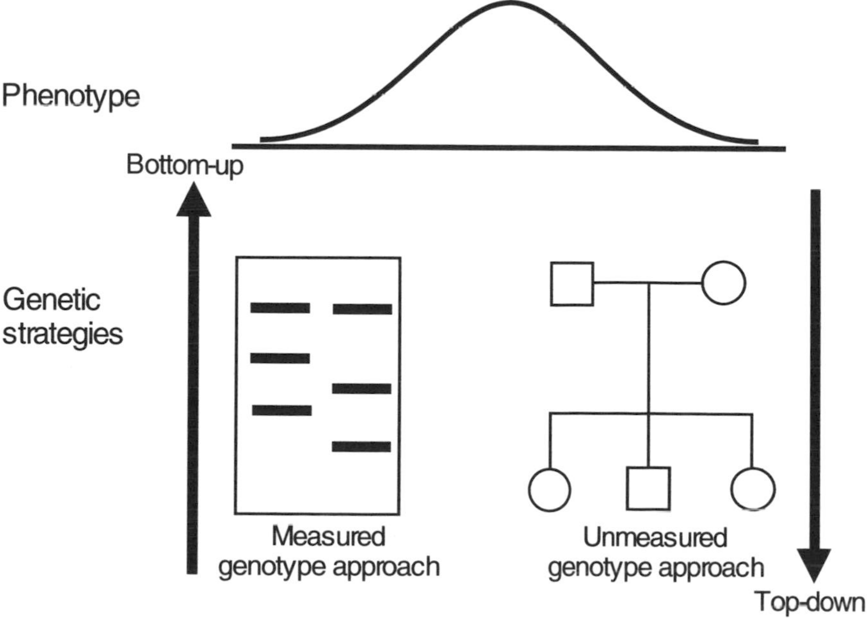

Figure 4.2 Bottom-up and top-down approaches in the study of continuously distributed phenotypes. Adapted from Sing and Boerwinkle (65).

whether or not genetic factors contribute to familial aggregation by assessing the genetic heritability of the phenotype. If there is evidence for genetic factors, the third step addresses whether or not there is a major genetic effect on the phenotype, that is, a major gene. Finally, association studies with candidate genes and linkage studies with random genetic markers or candidate genes are undertaken to identify chromosomal regions encoding the genes underlying the observed genetic effect.

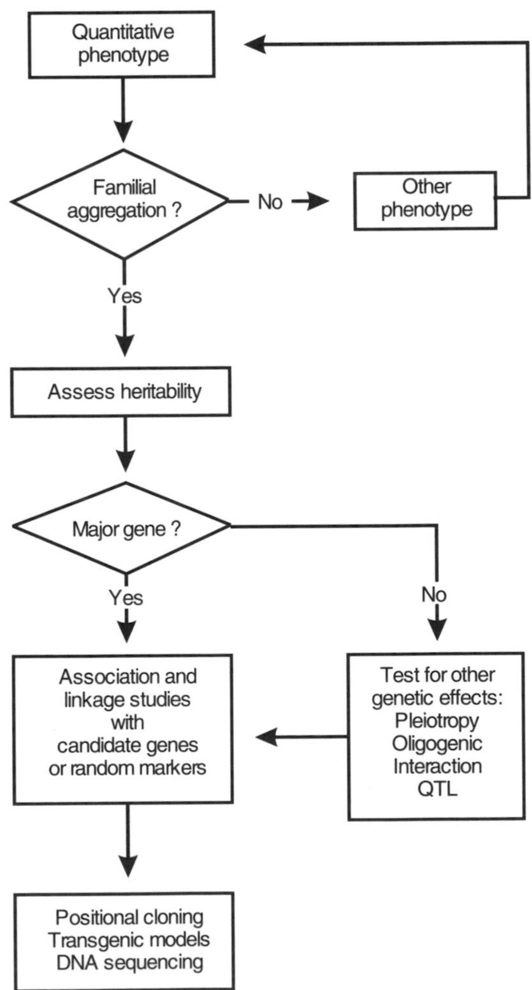

Figure 4.3 Flow chart describing the questions addressed in study of quantitative phenotypes in genetic epidemiology.

Assessment of Familial Aggregation

The first step in the study of the genetic basis of a quantitative multifactorial phenotype is to determine whether or not the phenotype aggregates in families. The presence of familial aggregation, or familiality, is demonstrated by the higher occurrence of a trait or a disease within families compared to the population at large (69). The power to detect a gene affecting a quantitative trait depends on its contribution to the increase in risk to the total variance of that trait. The power to detect linkage between a disease and a susceptibility locus depends on its contribution to the increase in risk among relatives compared to the population prevalence (62). If K denotes the population prevalence and K_R the recurrence risk for a relative of an affected individual, the lambda coefficient (λ_R) can be simply computed as:

$$\lambda_R = K_R/K$$

This coefficient can be computed from family studies provided that an appropriate estimate of the population prevalence is available. The same rationale applies to quantitative phenotypes if λ_R is computed for, say, high values or low values of a phenotype. For instance, in the case of performance, it could be useful to know λ_R for $\dot{V}O_2$max values ≥ 75 ml $\cdot$ kg^{-1} $\cdot$ min^{-1}.

For a quantitative phenotype, the presence of familial aggregation can be tested using a one-way analysis of variance in which the variance within families is compared to the variance between families. A significantly higher between-family than within-family variance (significant F-ratio) suggests that individuals of the same family are more similar than individuals of different families, which in turn suggests familial resemblance. The importance of familial resemblance can be assessed with intraclass correlations. The higher the correlation, the higher the degree of familial resemblance. In the presence of familial aggregation, correlations among family members are expected to be significantly different from zero. To provide evidence for familial aggregation, correlations between family members (i.e., spouses, parents, offspring, and siblings) provide insights about the relative importance of genetic and environmental factors. A phenotype showing no correlation between spouses but significant parent-offspring and sibling correlations is likely to be more influenced by genetic factors than one showing a spouse correlation of the same magnitude as the parent-offspring correlation.

Assessment of Heritability

After a particular trait has been shown to aggregate in families, the next step is to quantify the contribution of genetic factors to the familial aggregation. This involves the concepts and methods of quantitative genetics. Unlike the pea plants studied by Gregor Mendel, which were "tall" or "short," quantitative phenotypes cannot be sorted into distinct categories. They show continuous variation rather than the discrete variation characteristic of Mendelian phenotypes. Compared to classical Mendelian phenotypes that are determined by a single gene, quantitative phenotypes may be influenced by several genes that may contribute additively to the phenotype, each one having individually a small effect on the phenotype. Such phenotypes are said to be polygenic. Quantitative phenotypes may also be influenced by environmental (nongenetic) factors and therefore are referred to as multifactorial phenotypes.

Francis Galton was first to recognize that the inheritance of such phenotypes would be best analyzed by using statistical methods. The renowned statistician R.A. Fisher showed in a seminal paper (19) that the correlations observed between relatives for quantitative phenotypes could be explained by the actions of many individual genes, thus demonstrating the compatibility between Mendelian inheritance and polygenic inheritance.

The basic model

The genetic analysis of multifactorial phenotypes is based upon partitioning the total phenotypic variance (V_P) into genetic and environmental components as follows:

$$V_P = V_G + V_C + V_E$$

where V_G is the genetic component of the variance, V_C is the common (shared) environmental variance, and V_E is the residual or nonshared environmental component of the variance. These components of phenotypic variance can be further partitioned to include specific components such as gene-by-environment interaction (GxE), gene-by-gene interactions (epistasis), or dominance deviations. In addition to the assumptions underlying polygenic inheritance (several genes with small effects contributing additively to the phenotype), other assumptions are made in many genetic models that are used to assess genetic and environmental sources of variation. It is often assumed that there are no interactions either among genes or between the genotype and the environment. These assumptions are rarely met. Absence of genotype-environment interaction is especially unrealistic, since it implies that all genotypes behave similarly in a given environmental condition. As shown later, the existence of considerable individual differences in the response to training and the evidence that these differences are partly dependent on genotype suggests that genotype-environment interaction is probably the rule rather than the exception for health- and performance-related phenotypes.

Random mating is another basic assumption of the model. However, selection of mates among humans is rarely random. Individuals tend to select one another on the basis of some physical or behavioral characteristics. For example, an active person engaged in various sport activities is more likely to choose a partner who has a similar lifestyle. If these characteristics are influenced by genes, this nonrandom or assortative mating will increase the estimate of genetic variance.

Another assumption is the absence of genotype-environment correlation, which refers to the differential exposure of some genotypes to specific environments. Although this assumption has been discussed in detail in genetic studies of behavior (51), it has received little attention in genetic studies of biological traits. The presence of genotype-environment correlation will distort estimates of genetic and environmental variances in a way that is dependent on the direction of the correlation.

The concept of heritability is frequently used when estimating the relative contribution of genetic and environmental sources of variation to a multifactorial phenotype. A distinction is sometimes made between genetic heritability and cultural heritability. Genetic heritability (h^2) is defined as the fraction of phenotypic variance that is genetic ($h^2 = V_G / V_P$) and represents a measure of the degree of genetic determination. Since not only genes but also cultural factors can be transmitted from parents to offspring, the environmental variance can be partitioned into variance components reflecting shared family environment (V_C) and environmental factors unique to each individual (V_E). Cultural heritability (c^2) is defined as the proportion of phenotypic variance attributable to shared familial environmental effects ($c^2 = V_C/V_P$), and it measures the importance of cultural transmission across generations. Assessment of cultural heritability is not always possible. It depends on study design (see following section) and is sometimes confounded with genetic heritability. It is important to note that heritability is a ratio of variances and is therefore dependent on changes in either the numerator or the denominator. Heritability is also a population measure and does not apply to individuals. Estimates of heritability for a given phenotype are likely to vary among populations, depending on genetic and environmental characteristics of the respective populations.

Study designs used to assess heritability

Estimation of the heritability of multifactorial phenotypes requires data on various kinds of relatives with different degrees of relatedness. Several designs are available to assess the heritability of quantitative phenotypes in humans (23,27,28,57,64,69). All of the designs fall into one of the following broad categories: twin studies, adoption studies, and family studies.

Twin studies. The classical twin design has been the most widely used method to assess the heritability of a phenotype. Discovery of this method is usually credited to Francis Galton, who introduced it at the end of the 19th century, although this has been questioned (59). The aim of the twin method is to compare the resemblance of identical (monozygotic, MZ) twins to fraternal (dizygotic, DZ) twins. MZ twins are identical genetically since they originate from the division of one zygote, while DZ twins share only one-half of their genes, on average, by descent. Thus, any difference in the resemblance between MZ and DZ twin pairs is ascribed to genetic factors, assuming that both types of twin subjects are exposed to similar environmental conditions. This assumption is most critical and represents a major limitation of the twin method.

Analysis of twin data is based on analysis of variance (ANOVA). If genetic factors are involved in determining the phenotype under study, the within-pair variance will be lower for MZ twins than for DZ twins. The F-ratio of the ANOVA is thus used to test for the presence of a genetic effect. The resulting estimate of genetic variance can also be used to estimate the heritability of the phenotype. Several methods have been proposed to estimate heritability from twin data (11,12,26), and the most widely used method estimates heritability as twice the difference between the MZ and DZ intraclass correlations:

$$h^2 = 2[r_{MZ} - r_{DZ}]$$

Although it has limitations, the twin method represents a powerful design to detect the presence of a genetic effect in a quantitative phenotype. However, twins share prenatal as well as postnatal environments to a unique extent and may not be representative of the population at large. In particular, the environmental component shared by identical brothers and sisters may be quite different from that shared by members of DZ pairs. Heritability estimates derived from twin data should be interpreted with caution and should be considered as upper bound estimates of the heritability of a phenotype.

Several extensions of the classical twin method can be used to assess genetic and environmental sources of variation in a quantitative phenotype. One extension includes data on the spouses and offspring of adult twins, the twin-family method (42). In contrast to the classical twin method, the twin-family method provides more information on environmental sources of variance and on the role of assortative mating in familial resemblance and includes more degrees of genetic relatedness. It thus increases the external validity of the heritability estimates. The twin-family method has been used, for example, to estimate genetic and cultural heritabilities for HDL cholesterol, but it has been found to have reduced power for tests of hypotheses (38).

Another extension of the twin method compares twins discordant for exposure to a factor in the environment. This is the co-twin control method. It is an ideal design to control for host characteristics such as age, sex, and genetic composition, and it is a powerful way to estimate the effects of an environmental factor on a phenotype. This approach has been used recently to show that smoking has an impact on bone density (24).

Finally, data on twins reared apart can also be used in combination with twins reared together, in order to assess heritability. This design alleviates some of the limitations of the classical twin method, since twins reared apart do not share a common environment in the strict sense. Correlation between characteristics of MZ twins reared apart provides a more direct estimate of heritability. This approach has been used to assess heritability of the body mass index (52,68) and determinants of personality (10). Unfortunately, any postnatal environmental similarity among members of MZ pairs reared apart overestimates the genetic component.

Adoption studies. The study of adopted children is a powerful design to assess genetic and cultural heritabilities. Resemblance between an adopted child and members of his or her biological family is attributable largely to the genes they share in common, while resemblance between an adopted child and his or her adoptive family is due primarily to a shared family environment. In most adoption studies, data on biological parents of the adoptee are not available; this type of study is referred to as a partial adoption design. When data on both biological and adoptive relatives of the adoptee are available, the study is called a complete adoption design. Obviously, the latter design is more powerful to estimate genetic and cultural heritabilities, but as noted, biological parents of adoptees are rarely available.

In a partial adoption study, the resemblance between biological and adoptive relatives living in the same family environment is compared.

For example, the absence of a significant correlation between adoptive parents and their adopted children, combined with a significant correlation between parents and their biological children, would suggest that the phenotype under study is more influenced by genetic factors than by the family environment.

Although adoption studies offer an attractive design to assess heritability, they have limitations. One important variable to control in adoption studies is age at adoption. Adoption should ideally have taken place immediately after birth so that the resemblance between the adoptee and his or her biological parent can be defined as entirely genetic and not be confounded by any effect of familial environment. It is often advisable to adjust the correlations for age at adoption and to test whether or not they are different from the nonadjusted correlations. An assumption underlying the adoption design is the absence of selective placement of adoptees, which occurs when adoption agencies try to match adoptive parents and biological parents on a variety of characteristics (e.g., socioeconomic status, complexion, and others). In addition, it is assumed that adoptive families are an unbiased sample of the population. This is not the case, since adoptive families are not randomly selected.

Family studies. Studies of families, including nuclear families (parents and their offspring) and those with extended pedigrees, are the most widely used designs to investigate the genetic basis of quantitative phenotypes. One major advantage of this design is that families are more representative of the population at large than fixed sets of relatives like twins or adoptees. However, data on nuclear families alone do not contain sufficient information to quantify the relative contribution of genetic (V_G) and cultural (V_C) components of variance and can be used only to assess familiality, that is, the fraction of phenotypic variance attributable to the combined effects of all familial influences (57).

Two different approaches can be used to distinguish between the genetic (h^2) and cultural (c^2) components of variance using family studies. First, an index of the family environment may be constructed by regressing the phenotype on relevant indicators of the family environment in order to find the set of variables that best predicts the phenotype (55,56). With this approach, two pieces of information are available on each individual: the phenotype being measured and the index derived from the regression model. Correlations between the phenotype and the index within and among members of the nuclear family are computed and expressed in terms of a genetic model that takes into account the following: genetic and cultural inheritance, heterogeneity in heritabilities across generations, a nontransmitted common sibling environment, maternal effects in cultural transmission, and marital resemblance (54). This approach has been applied to the study of blood pressure (48).

The use of an environmental index, however, has limitations that can lead to bias in heritability estimates (58). An important assumption is the absence of common genetic factors between the phenotype and the index, which means that there should be no genetic correlation between the phenotype and the index. Ignoring such a correlation can lead to an underestimation of h^2 with a compensatory overestimation of c^2 (39). Another bias resulting in underestimation of c^2 and overestimation of h^2 may be introduced if the index is only a partial reflection of the familial environment (58).

As an alternative to the use of indices, information on additional relatives by descent (MZ and DZ twins, uncles and aunts, first-degree cousins) and adoption may be incorporated to produce multiple degrees of relatedness. With this approach, estimates of h^2 and c^2 are based on a much larger number of relative types and may be expected to be more reliable. This approach has been used in the Quebec Family Study to assess genetic and environmental sources of variance in health-related fitness phenotypes (8,46,47,49).

Statistical methods

Several statistical approaches can be used to estimate the components of phenotypic variance. A widely used approach to estimate these components in genetic epidemiology is path analysis (53,54). Path analysis is a method that decomposes familial correlations into genetic and environmental sources of variation. It is based on a path model in which any correlation between two variables can be expressed as a function of all of the paths by which the two variables are connected. The major strength of path analysis is its modeling flexibility that allows the researcher to incorporate into the path model different factors that may influence the phenotype. Different path analysis models have been developed, depending on the family structure of the data (13,14,54,60,61).

Path analysis based on model-fitting approaches and using structural equation models have also been developed to analyze twin data (43). Estimates of genetic and environmental sources of variation are obtained by fitting a model directly to the observed correlations using the maximum likelihood method. This approach permits the analysis of twin data jointly with other types of family data and is statistically more rigorous. Heritability estimates are derived from all available information in the data and are reported with associated standard errors, thus providing information about the accuracy of the estimate.

Prior to fitting a model to estimate the heritability of a quantitative phenotype, a researcher must adjust the phenotype for concomitant variables. This adjustment is performed by regressing the phenotype on the relevant concomitants (most often age and sex) and then computing the residual scores, which are then used as phenotypes in the analysis. Concomitant variables can also be modeled directly in the path model.

Detection of Major or Single-Gene Effects

Once the contribution of genetic factors to a multifactorial phenotype has been established, the next step is to identify the genes underlying the genetic effect.

Basic concepts

The first step in reaching this goal is to establish whether or not the pattern of inheritance is compatible with the segregation of a major gene. A major gene or major locus is a single gene that has a detectable effect on the level of a quantitative trait or on disease susceptibility, if it is a disease trait (34). The effect of a major locus on a quantitative phenotype (top panel) and on a disease or dichotomous trait (bottom panel) is shown in figure 4.4. A major locus with two codominant alleles, A_1 and A_2, with frequencies p and q, respectively, will give rise to three major locus genotypes, A_1A_1 (mean μ_{11}), A_1A_2 (mean μ_{12}) and A_2A_2 (mean μ_{22}). The distribution of each genotype is represented by the dashed curves, while the solid curve represents the overall distribution of the phenotype. For each major locus genotype, variation around the mean is due to polygenes and to environmental effects. It is assumed that the three genotypes are in Hardy-Weinberg proportions (p^2, $2pq$, q^2), that the major gene effect is independent of the polygenic effect, and that environmental effects are random. The impact of the major gene on the phenotype is measured in terms of its contribution to the variance of the trait. There is no general agreement on the magnitude of the effect on phenotypic variance that is necessary to define a locus as "major" (34).

The major locus model described for quantitative traits (top panel of figure 4.4) can be modified for disease traits, for which individuals are categorized as affected or unaffected. In such circumstances, the risk of developing the disease is defined as a disease liability that is influenced by genetic and environmental factors. When the liability reaches a certain threshold value T, individuals become affected (shaded area of the distribution). As shown in the bottom panel of figure 4.4, A_1A_1 individuals are at less risk of developing the disease than A_2A_2 individuals. The presence of major gene effects in quantitative phenotypes can be

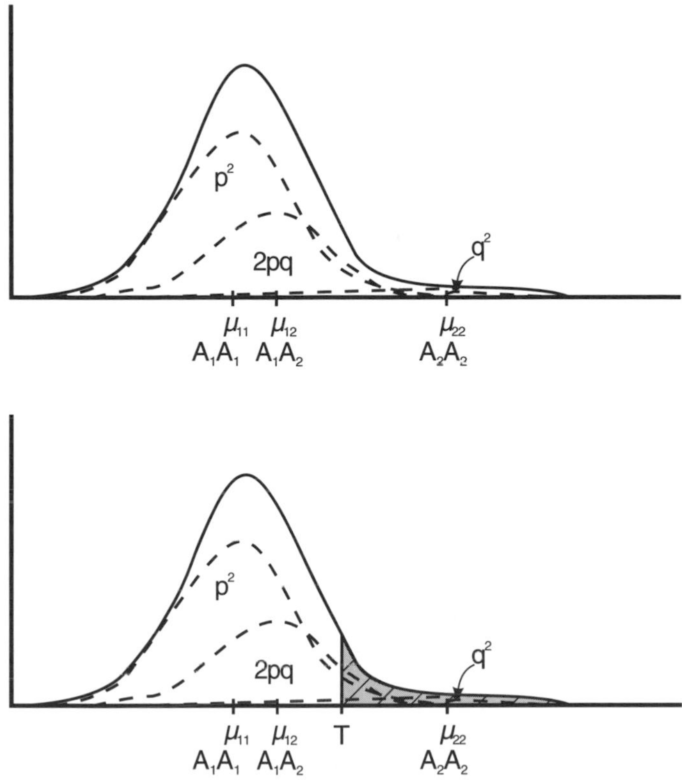

Figure 4.4 Major locus model for a quantitative trait (*top panel*) and disease (dichotomous) trait (*bottom panel*).

detected either with tools from the unmeasured genotype (hypothesis testing) or the measured genotype (with the gene at hand) approach.

Unmeasured genotype approach

Several statistical methods are available to detect major gene effects in multifactorial phenotypes using the unmeasured genotype approach (33,34). These include commingling analysis, segregation analysis, sibship variance tests, and some nonparametric methods based on comparison of phenotypic values among family members. Commingling analysis and segregation analysis are the most widely used methods.

Commingling analysis. In the presence of a major gene, distribution of a phenotype will be characterized by a mixture of two or more distributions corresponding to the genotypes at the major locus (figure 4.4). Commingling analysis is used to test whether the distribution of a quantitative phenotype is better characterized by a mixture (commingling) of distributions rather than by a single distribution (36). Although compatible with the presence of a major gene, commingling can arise as a result of nongenetic factors and should not be taken as *prima facie* evidence of a major gene effect. The advantage of commingling analysis is that it does not require family data; it is therefore useful as a preliminary method to screen for a major gene effect.

Segregation analysis. Segregation analysis is used to test the mode of inheritance of a trait within families. When applied to a complex multifactorial phenotype, complex segregation analysis allows detection of a major gene effect on a background of multifactorial inheritance. The analytical strategy relies on fitting a series of models with alternative hypotheses about the presence of a multifactorial component, the presence of a major effect, the environmental versus genetic transmission of this major effect, and the mode of transmission (dominant, recessive, or codominant) of the major gene. The model that best fits the data based on specific statistical criteria is then considered the one that explains the distribution of the trait within families. This model permits estimation of the frequency of a major gene in the population (q), and of the proportion of phenotypic variance attributable to the major gene and to the multifactorial component of inheritance.

Segregation analysis provides only statistical evidence about the contribution of a major gene to a quantitative phenotype. It indicates nothing about the specific gene or genes. Other methods based on the measured genotype approach are needed to identify the genes involved.

Measured genotype approach

With the increasing number of highly polymorphic genetic markers spanning the entire genome, the measured genotype approach is now favored by geneticists for the "genetic dissection" of complex multifactorial phenotypes (31,32,35). Three methods are particularly useful to identify the genes contributing to multifactorial phenotypes: association analysis, linkage analysis, and quantitative trait locus (QTL) mapping. Other methods are also available and are discussed briefly.

Association analysis. The objective of association studies is to test for co-occurrence of a specific allele at a marker locus and a trait in the population. The locus may be a random biochemical or DNA marker. However, the marker loci used in association studies are preferably candidate genes (i.e., genes that are likely, for biological reasons, to influence the phenotype under study). The rationale of the method is illustrated in figure 4.5. If the trait is a disease, association studies be-

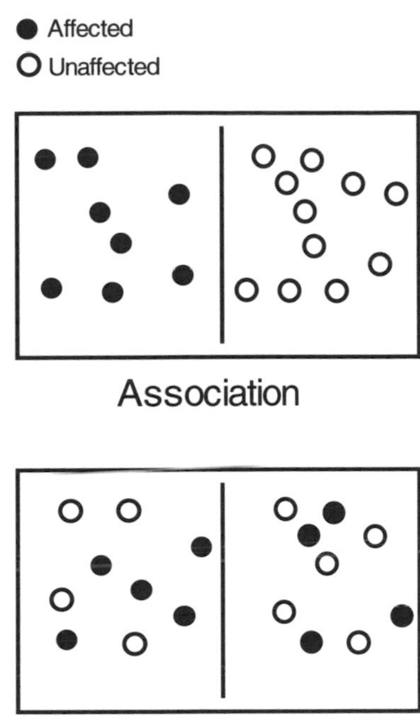

Figure 4.5 Rationale underlying association studies. In the presence of association between a specific allele at a marker locus and a disease (*upper part*), all affected subjects are expected to carry the marker allele (closed circles), while all unaffected subjects lack the allele (open circles). In the absence of association (*lower part*), the marker allele will be evenly distributed among diseased individuals and controls (22).

come a case-control design. They test whether a specific allele is observed with a higher frequency among cases (filled symbols) compared to control subjects (open symbols). If the trait is quantitative, the analysis compares mean phenotypic values among individuals of different genotypes at the marker locus, or between carriers and noncarriers of a specific allele. Association analysis can also be performed on family data. The simplest strategy analyzes only unrelated subjects, namely parents, and considers them as a random sample of the population. In another strategy, the model that incorporates familial correlations, relevant covariates and the marker locus are compared to a model without the marker locus (20). A better fit of the former compared to the latter indicates an association.

Although the association between a marker locus and a given trait suggests the presence of a major gene effect, results from association studies are difficult to interpret. Three factors can give rise to an association between a marker locus and a trait (32). First, association can occur if the marker allele is actually the allele causing the major gene effect. In such a case, one expects to find the same association in every population studied, which rarely occurs. Second, the allele itself does not cause the major gene effect, but an allele at a locus (trait locus) in linkage disequilibrium with the marker locus does. Linkage disequilibrium occurs when the alleles of the trait and the marker loci both cosegregate in families and show association in the population. Linkage does not normally lead to an association in the population because, under random mating, repeated recombination events over many generations will cause the trait locus and the marker locus to segregate independently. Linkage equilibrium is thus reached. Under linkage disequilibrium, the association observed at the population level between alleles at linked loci is preserved. This implies that the presence of an association is highly dependent on the population's natural history. Linkage disequilibrium is most likely to occur in a young isolated population (32). Third, association can also arise by chance due to population heterogeneity. In a mixed population, for example, any trait present at a higher frequency in a particular ethnic group will show association with any allele that happens to be more frequent in that group (32).

Linkage analysis. When two loci are relatively close to each other on the same chromosome, they tend to be transmitted together from parents to offspring. The probability of cosegregation (cotransmission) of the two loci depends on the distance between them on the same chromosome and on how often they are separated by recombination events during meiosis (see chapter 2). Thus, the frequency of recombination or recombination fraction, denoted by the Greek letter θ, can be used as a measure of the genetic distance or map distance between two loci. The unit of measurement is the centimorgan (cM), 1 cM corresponding to a recombination fraction of 1%. The recombination fraction ranges from 0% for loci so close to each other that recombination never occurs, to 50% ($\theta = 0.5$) for loci located on different chromosomes or far apart on the same chromosome. Two loci are said to be linked or to show genetic linkage when $\theta < 0.5$.

The objective of linkage analysis is to estimate θ and to test if it is less than 0.5. For this purpose, the likelihood (L) of obtaining the observed pattern of inheritance of the trait and the marker locus within families is computed for different values of θ. For a given value of θ, the odds of linkage are in the form of a lod score (logarithm of the odds) which is computed from the following ratio:

$$Z(\theta) = \log_{10} [L(\theta)/L(\theta = 0.5)]$$

A positive lod score indicates the presence, while a negative lod score indicates the absence, of linkage. Conventionally, a lod score ≥ 3.0 (1,000 to 1 odds in favor of linkage) is considered significant evidence of linkage, while a lod score of ≤ –2.0 (100 to 1 odds against linkage) is considered sufficient to reject the hypothesis of linkage. However, if the lod score falls between –2.0 and 3.0, no firm conclusion about the presence or absence of linkage can be reached, and data on more families are needed.

Linkage analysis using the lod score method requires knowledge of the mode of inheritance of the trait under investigation. Such methods are parametric (45) or model-based (18). Segregation analysis is sometimes performed before undertaking linkage analysis with the lod score method, in order to obtain estimates of the required parameters. However, the mode of inheritance of the trait is usually unknown, and nonparametric or model-free methods of linkage analysis are used (18,45). These methods make no attempts to estimate the recombination fraction θ, but are based on the number of alleles shared at the marker locus between pairs of relatives (most of the time sib pairs) and are referred to as allele-sharing methods (32).

The most widely used of these methods is the sib-pair linkage method of Haseman and Elston (22). The rationale underlying the method, which is shown in figure 4.6, is as follows: If a polymorphic marker is

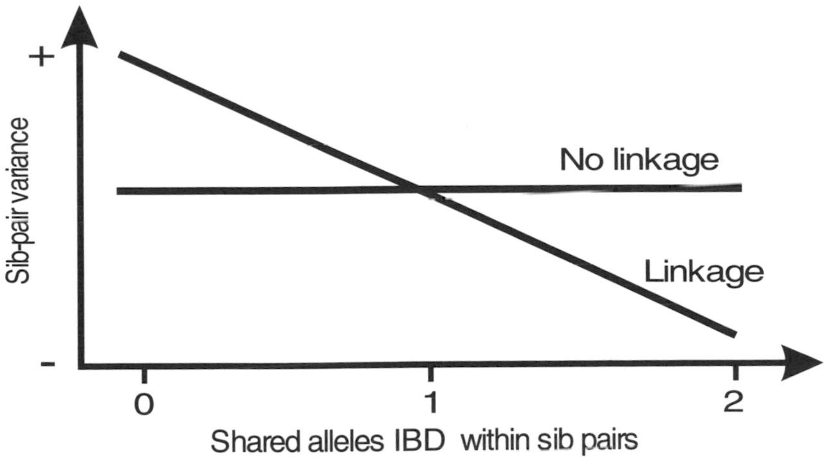

Figure 4.6 The sib-pair linkage method. In the presence of linkage between a marker locus and a trait locus, the variance within sib pairs is expected to be inversely related to the number of alleles shared identical by descent (IBD) within sib pairs.

tightly linked to a putative locus influencing a quantitative trait, sib pairs sharing a greater proportion of alleles identical by descent at the marker locus will tend to have a more similar phenotype than sibs sharing fewer alleles. Therefore, in the presence of linkage, the squared sib-pair trait difference is expected to decrease with an increase in the proportion of alleles shared identically by descent by the sib pair at the marker locus. No relationship is expected under the null hypothesis of no linkage. This method has been generalized to different pairs of relatives available in family data (44). Because the sib-pair method is a model-free method of linkage analysis, it tends to be more robust than the traditional lod score method. However, it is less powerful than the lod score method when the model is correctly specified. The sib-pair linkage method should, therefore, be considered as a good screening tool to identify regions of the genome where genes affecting a quantitative trait are located. The lambda coefficient (λ_R), described earlier in this chapter, has been shown to be a critical parameter related to the power of the sib-pair method (63).

Multipoint identical-by-descent distributions (72,73) permit linkage analysis using all available markers at one time. The sib-pair linkage method has recently been extended to include information on multiple genetic markers simultaneously (29). The possibility of performing multipoint linkage mapping in humans using sib pairs should increase the capacity to map and identify new loci affecting quantitative phenotypes such as those pertaining to fitness and performance.

Association versus linkage analyses. Association and linkage analyses are two important methods used in the identification of genes contributing to multifactorial phenotypes. However, there are fundamental differences between the methods (table 4.1). The goal of association studies is to test for the correlated occurrence between specific alleles at the marker locus and the trait locus; linkage studies test the hypothesis of cosegregation between a marker locus and the trait within families. Since linkage depends on the relative position of the marker and trait loci on the chromosome, any allele at the marker locus could be involved in the cosegregation with the trait. Linkage is, therefore, a property of loci, while association is a property of alleles. While population or family studies can be used to detect an association between marker alleles and a quantitative trait, only family and pedigree studies can be used to detect linkage. In association studies, phenotypic means among genotypes at a given marker locus are compared using methods such as analysis of variance or chi-square for the case-control design. In linkage studies, the likelihood of observing a specific pattern of marker genotypes and trait values within fami-

Table 4.1
Comparison of Association and Linkage Studies

Characteristic	Association studies	Linkage studies
Goal	Search for evidence of statistically significant association between an allele and trait	Search for evidence of cosegregation between a marker locus and a quantitative trait
Approach	Compare mean phenotypic values among individuals of different genotypes at the marker locus	Calculate the likelihood of obtaining the parent-to-offspring transmission pattern of a trait and a marker locus
Design	Population or family studies	Family studies
Methods	ANOVA Chi-square analysis Regression analysis (ASSOC)	Sibship variance tests Sib pairs Lod scores
Etiology	Susceptibility gene	Necessary gene
Property	Association is a property of alleles	Linkage is a property of loci

lies is computed using variance sibship tests, or the sib-pair or lod score method.

Finally, association studies are most likely to identify susceptibility genes (i.e., genes that increase the risk of a disease but that are not absolutely necessary for disease expression), while linkage analyses are more likely to detect necessary genes (i.e., genes that are necessary for a disease expression). Genes involved in determining health-related fitness phenotypes are more likely to be of the susceptible type. Computer simulations have shown that if the risk conferred by the associated allele is not greater than 10 times the risk associated with the normal allele, it will be very difficult to detect linkage (21). In this context, linkage analysis is likely to yield little more information than is obtained from association studies. Thus, it is possible to find association without linkage. The reverse—that is, the presence of linkage without association—is also possible in a situation of genetic heterogeneity where several alleles at the marker locus could be responsible for the major gene effect.

Combination of the unmeasured and measured approaches

A combination of the top-down (unmeasured) and bottom-up (measured) approaches offers more possibilities to understand the genetic basis of quantitative phenotypes. It allows determination of which combination of genes explains the polygenic inheritance of a trait (67).

This combination of strategies, for example, has been used to estimate the impact of measured genetic variation at the apolipoprotein E gene and the fraction of phenotypic variance attributable to polygenes on lipid and lipoprotein levels (2).

Joint use of segregation and linkage analysis is an example of the combination of the two approaches. It is essentially an extension of segregation analysis applied to a multivariate phenotype with one or more marker loci (3). The advantage of this method is that it can test whether a major gene effect found by segregation analysis is attributable to variation in specific marker loci that are known to be associated with the trait at the population level. This strategy has, for example, demonstrated that a variant of the angiotensin I-converting enzyme (ACE) contributes to plasma variation of ACE levels (71).

Detection of Genotype-Environment Interaction

The unmeasured and measured genotype approaches can be used to detect genotype-environment interaction effects (GxE) in multifactorial phenotypes. Methods for detecting GxE effects are outlined in table 4.2.

Unmeasured genotype approach

Two strategies can be used to detect GxE effects based on the unmeasured genotype approach. The first is statistical modeling, in which statistical genetic models incorporate GxE effects (1,16,17,51). Ignoring

Table 4.2
Methods for Detecting Genotype-Environment Interaction Effects

A. Unmeasured genotype approach
   1. Statistical modeling
   2. Experimental treatment

B. Measured genotype approach
   3. Compare the effect of a gene on a given phenotype between populations of different ethnic and cultural backgrounds
   4. Compare the effect of a gene on a given phenotype between individuals of the same population categorized on variables that can influence the phenotype (age,sex, etc.)
   5. Compare the phenotypic response to an environmental stimulus among individuals with different genotypes at a given locus

such interaction effects, when they do in fact exist, reduces the power of detecting major gene effects (4,70).

With the unmeasured genotype approach, another way to test for the presence of a GxE effect in humans is to challenge several genotypes in a similar manner by submitting both members of MZ twin pairs to a standardized treatment and then to compare the within- and between-pair variances in the response to the treatment (7). The presence of a significantly higher variance in the response between pairs than in that within pairs suggests that changes induced by the treatment are more heterogeneous in genetically dissimilar individuals. Experiments using either exercise training or overfeeding as treatments with MZ twins to investigate GxE effects in several health-related fitness phenotypes have been conducted over the past 10 years by the Laval University group. These experiments are summarized later in the volume.

Measured genotype approach

The measured genotype approach provides three strategies to detect GxE effects in humans. In the first, the influence of a specific gene is compared between populations with different ethnic and cultural backgrounds. In the second, the effect of a gene is studied between subgroups of individuals within the same population but categorized on the basis of variables that potentially can affect the phenotype under study (e.g., sex, age, disease status, etc.). In the third strategy, the response to an environmental stimulus (exercise training, diet, or others) is investigated among individuals with different genotypes at a given locus. An important advantage of the measured genotype approach over the unmeasured genotype approach in the study of GxE is that it makes it possible to identify the genes responsible for the interaction effect. Thus, it potentially can provide a way to screen individuals at high risk for disease or with low or high value for a multifactorial phenotype, when they are challenged by an experimental treatment or a specific environmental condition.

Animal Studies

The genetic basis of quantitative phenotypes also can be investigated using animal studies (figure 4.1). The major advantage of animal over human studies is the control that can be achieved over several factors that usually confound studies in humans. Examples include control of genetic heterogeneity by using inbred strains of animals, control of germ-line transmission in transgenic rodents, and targeted mutations for hypothesis testing (75). The advantages of animal studies are summarized in table 4.3. The high degree of homology between rodent and human genomes permits identification of quantitative trait loci in animals, whose human homologous loci can be used as potential candidate genes or chromosomal regions to be tested in association and linkage studies.

Table 4.3
Advantages of Animal Models in the Study of Polygenic Traits

Problem	Advantages of animal models
Heterogeneity	Minimize heterogeneity: Inbred animals can be kept in uniform environments on identical diets
Breeding	Can design genetic crosses as desired
Physiological studies	Can use surgery, diets, environment, and sacrifices as needed for hypothesis testing and biochemical studies
Variation	Availability of a wide variety of phenotypes in different inbred strains permits study of numerous traits
Hypothesis testing	Ability to make transgenics and targeted mutations permits hypotheses tests about roles of specific genes in complex diseases
Therapies	Can use both spontaneous and artificially produced animal models to test new therapies

From Warden and Fisler (75).

Quantitative trait locus (QTL) mapping

Quantitative trait locus (QTL) mapping has recently been developed to identify loci influencing quantitative phenotypes. This approach is described in detail by Lander and Botstein (30). The method has been used successfully to identify genes contributing to atherosclerosis (74) and obesity (75).

Figure 4.7 presents an overview of the QTL method. It is based on experimental crosses between inbred strains of animals (usually mice or rats) that differ as much as possible for the phenotype of interest. If, for example, an investigator is interested in mapping the genes responsible for genetic differences in response to endurance training, the first step would be to identify low-responder (strain A) and high-responder (strain B) strains of rodents to endurance training and then to cross them to produce F1 animals with genotype AB. The F1 animals can be crossed either to each other to produce F2 animals with genotypes AA, AB, or BB, or with one of the parental strains to produce backcrosses with

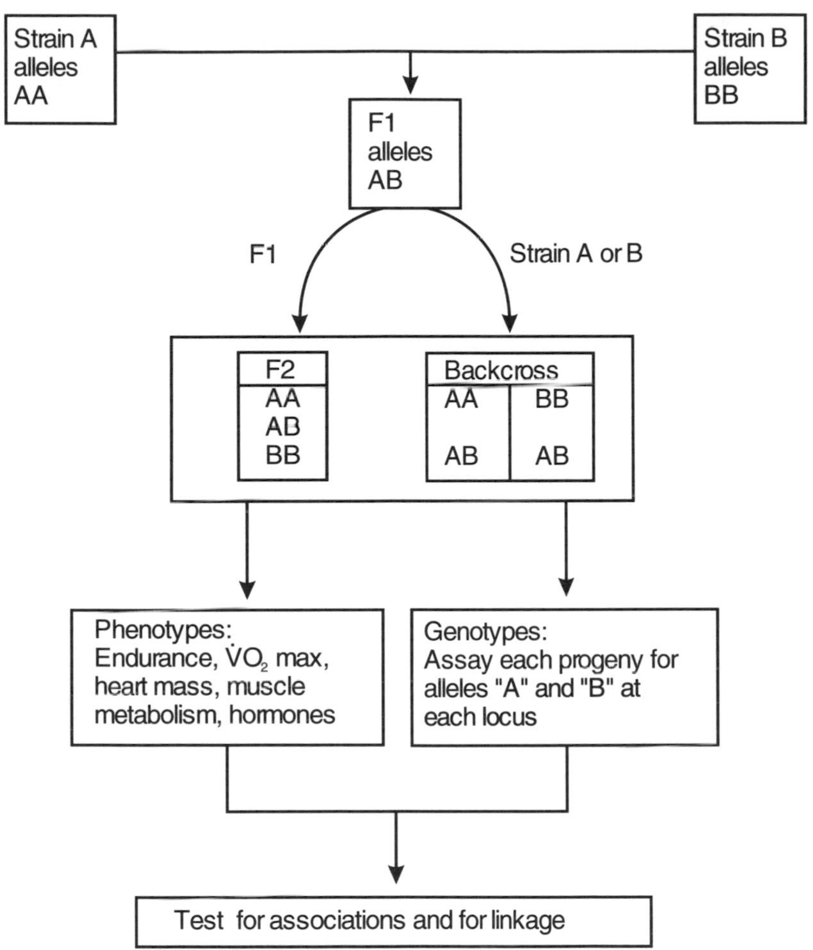

Figure 4.7 Overview of the quantitative trait locus (QTL) method. Adapted from Warden and Fisler (75).

genotypes AA and AB or BB and AB, depending on the parental strain used for the backcross. In the second step, all of the progeny are phenotyped (maximal aerobic power, cardiac or skeletal muscle characteristics, etc.). In the third step, all of the progeny are genotyped for markers spanning the genome at 10 to 20 cM intervals. Finally, statistical analyses are performed to identify and map QTLs. Because of the high degree of homology in the genomes of mammalian species, QTLs identified in animal models would then allow identification of homologous human candidate chromosomal regions that subsequently can be tested for association or linkage.

QTL mapping has become an important tool in identification of genes underlying quantitative traits in humans. It is important to emphasize that the QTL method simply identifies a chromosomal region where genes affecting the trait potentially reside. Considerable work must be done before the true nature of the genes involved can be identified. Positional cloning, DNA sequencing, development of congenic strains, and other tools are needed to identify the genes implicated. A detailed discussion of these tools is beyond the scope of this book. They are described in detail elsewhere (15,32,74,75).

Transgenic animals

Transgenic animals are an important research tool to study the functions of genes in complex phenotypes (25,40,41,50). Transgenic technology consists of transferring a gene or genes into the germ line of an animal, resulting in permanent modification of genetic information at the locus or loci involved. The gene transfer is usually accomplished *in vitro* by microinjection of a DNA fragment into the pronuclei of a one-cell embryo. The embryo is then transferred to the reproductive tract of a foster mother. After the transgene has been incorporated into one of the chromosomes of the embryo, selective breeding is undertaken until the transgene is consistently expressed in homozygotic animals. These studies are generally undertaken in mouse embryos and occasionally in rats.

Transgenic animals provide an opportunity to study *in vivo* the mechanisms of gene expression and their influence on the whole organism. Hundreds of studies are currently in progress with a variety of transgenes, and the data reported thus far indicate that much will be learned about gene expression and dysfunction for a wide spectrum of phenotypes. Indeed, it is progressively clear that dysregulation in expression of an otherwise normal gene product, even in pathways that appear to be totally unrelated to the phenotype, can cause disruption of normal phenotypic variation. This technology also has the potential to increase understanding of the molecular basis of several diseases and to offer new avenues for the treatment of common diseases in the future.

Gene knockout experiments

Another approach that can be used to investigate the function of genes in animal models is to inactivate the gene by a procedure known as gene knockout and then to observe the effect on the phenotype of interest. Knockout experiments are usually performed in mice by specifically inactivating a gene in cultured undifferentiated embryonic cells, the embryonic stem (ES) cells. These cells are then transferred to mouse embryos where they potentially can be expressed in all of the mouse cell types. The resulting animals are then bred, and those whose germ cells are derived from ES cells will transmit the knockout gene to their progeny. Gene knockouts can be accomplished specifically in selected cell types, which allows the study of the inactivated gene(s) in specific tissues.

Summary

Human and animal studies are used to study the genetic basis of complex multifactorial phenotypes. Investigation of genetic factors contributing to quantitative phenotypes in humans includes consideration of familial aggregation, quantification of the contribution of genetic factors (genetic heritability), detection of major gene effects, and studies of association and linkage.

The concept of heritability is frequently used to assess the relative contribution of genetic and environmental sources of variation in multifactorial phenotypes. Genetic heritability (h^2) refers to the fraction of phenotypic variance that is attributable to genetic factors, while cultural heritability (c^2) represents a measure of the shared familial environment. Several designs—including twin, adoption, and family studies—combined with various statistical procedures can be used to quantify the various components of phenotypic variance. The underlying genetic effect estimated from these methods is said to be polygenic, that is, resulting from the contribution of several genes having individually small effects on the phenotype. In addition to the polygenic effect, a single gene having a large impact on the phenotype could also contribute to the variation observed in multifactorial phenotypes. In the presence of such a major gene effect, the distribution of the phenotype is characterized by a mixture of distributions corresponding to the genotypes at the major locus. Complex segregation analysis is used to test whether the transmission of the major effect is compatible with Mendelian inheritance.

Among the various methods available to identify genes influencing multifactorial phenotypes, the most frequently used are association analysis, linkage analysis, and quantitative trait locus mapping. Association analysis examines the co-occurrence of a specific allele at a marker locus and a phenotype in the population; linkage analysis examines the cosegregation within families of the phenotype and the marker locus. Quantitative trait locus mapping uses experimental crosses between inbred strains of animals divergent for the trait of interest; linkage analysis then identifies and maps loci contributing to the phenotype.

Identification of genetic factors contributing to multifactorial phenotypes relies increasingly on an integrated approach that uses results from both animal and human studies. It is likely that such an integrated approach will contribute greatly to the identification of several new genetic loci contributing to complex multifactorial phenotypes in the next decade.

References

1. Blangero, J. Statistical genetic approaches to human adaptability. Hum. Biol. 65:941-66; 1993.
2. Boerwinkle, E.; Sing, C.F. The use of measured genotype information in the analysis of quantitative phenotype in man. III. Simultaneous estimation of the frequencies and effects of the apo E polymorphism and residual polygenetic effects on cholesterol, betalipoprotein and TG levels. Ann. Hum. Genet. 51:211-26; 1987.
3. Bonney, G.E.; Lathrop, G.M.; Lalouel, J.M. Combined linkage and segregation analysis using regressive models. Am. J. Hum. Genet. 43:29-37; 1988.
4. Borecki, I.B.; Bonney, G.E.; Rice, T.; Bouchard, C.; Rao, D.C. Influence of genotype-dependent effects of covariates on the outcome of segregation analysis of the body mass index. Am. J. Hum. Genet. 53:676-87; 1993.
5. Bouchard, C.; Dionne, F.T.; Simoneau, J.A.; Boulay, M.R. Genetics of aerobic and anaerobic performances. Exerc. Sport Sci. Rev. 20:27-58; 1992.
6. Bouchard, C.; Malina, R.M. Genetics for the sport scientist: Selected methodological considerations. Exerc. Sport Sci. Rev. 11:275-305; 1983.
7. Bouchard, C.; Pérusse, L.; Leblanc, C. Using MZ twins in experimental research to test for the presence of genotype-environment interaction effect. Acta Genet. Med. Gemellol. 39:85-89; 1990.
8. Bouchard, C.; Pérusse, L.; Leblanc, C.; Tremblay, A.; Thériault, G. Inheritance of the amount and distribution of human body fat. Int. J. Obes. 12:205-15; 1988.
9. Bouchard, C.; Pérusse, L.; Rice, T.; Rao, D.C. The genetics of human obesity. In: Bray, G.A.; Bouchard, C.; James, W.P.T., eds. Handbook of obesity. New York: Dekker; in press.
10. Bouchard, T.J., Jr. Genes, environment and personality. Science. 264:1700-1701; 1994.
11. Christian, J.C. Testing twin means and estimating genetic variance. Basic methodology for the analysis of quantitative twin data. Acta Genet. Med. Gemellol. 28:35-40; 1979.
12. Christian, J.C.; Kang, K.W.; Norton, J.A. Choice of an estimate of genetic variance from twin data. Am. J. Hum. Genet. 26:154-61; 1974.
13. Cloninger, C.R.; Rice, J.; Reich, T. Multifactorial inheritance with cultural transmission and assortative mating. II. A general model of

combined polygenic and cultural inheritance. Am. J. Hum. Genet. 31:176-98; 1979.

14. Cloninger, C.R.; Rice, J.; Reich, T. Multifactorial inheritance with cultural transmission and assortative mating. III. Family structure and the analysis of separation experiments. Am. J. Hum. Genet. 31:366-88; 1979.
15. Collins, F.S. Of needles and haystacks: Finding human disease genes by positional cloning. Clin. Res. 39:615-23; 1991.
16. Eaves, L.J. Including the environment in models for genetic segregation. J. Psychiatr. Res. 21:619-47; 1987.
17. Eaves, L.J. The resolution of genotype-environment interaction in segregation analysis of nuclear families. Genet. Epidemiol. 1:215-28; 1984.
18. Elston, R.C. Linkage and association to genetic markers. Exp. Clin. Immunogenet. 12: 129-40; 1995.
19. Fisher, R.A. The correlation between relatives on the supposition of Mendelian inheritance. Trans. Royal Soc. Edinburgh. 52:399-433; 1918.
20. George, V.T.; Elston, R.C. Testing the association between polymorphic markers and quantitative traits in pedigrees. Genet. Epidemiol. 4:193-201; 1987.
21. Greenberg, D.A. Linkage analysis of "necessary" disease loci versus "susceptibility" loci. Am. J. Hum. Genet. 52:135-43; 1993.
22. Haseman, J.K.; Elston, R.C. The investigation of linkage between a quantitative trait and a marker locus. Behav. Genet. 2:3-19; 1972.
23. Heath, A.C.; Kendler, K.S.; Eaves, L.J.; Markell, D. The resolution of cultural and biological inheritance: Informativeness of different relationships. Behav. Genet. 15:439-65; 1985.
24. Hopper, J.L.; Seeman, E. The bone density of female twins discordant for tobacco use. N. Engl. J. Med. 330:387-92; 1994.
25. Jaenisch, R. Transgenic animals. Science. 240:1468-74; 1988.
26. Kang, K.W.; Christian, J.C.; Norton, J.A., Jr. Heritability estimates from twin studies. I. Formula of heritability estimates. Acta Genet. Med. Gemellol. 27:39-44; 1978.
27. Khoury, M.J.; Beaty, T.H.; Cohen, B.H. Fundamentals of genetic epidemiology. New York: Oxford University Press; 1993.
28. King, M.C.; Lee, G.M.; Spinner, N.B.; Thompson, G.; Wrensch, M.R. Genetic epidemiology. Annu. Rev. Public Health. 5:1-52; 1984.
29. Kruglyak, L.; Lander, E.S. Complete multipoint sib-pair analysis of qualitative and quantitative traits. Am. J. Hum. Genet. 57:439-54; 1995.
30. Lander, E.S.; Botstein, D. Mapping Mendelian factors underlying quantitative traits using RFLP linkage maps. Genetics. 121:185-99; 1989.
31. Lander, E.S.; Kruglyak, L. Genetic dissection of complex traits: Guidelines for interpreting and reporting linkage results. Nat. Genet. 11:241-47; 1995.

32. Lander, E.S.; Schork, N.J. Genetic dissection of complex traits. Science. 265:2037-48; 1994.
33. MacCluer, J.W. Biometrical studies to detect new genes with major effects on quantitative risk factors for atherosclerosis. Curr. Opinion Lipidol. 3:114-21; 1992.
34. MacCluer, J.W. Statistical approaches to identifying major locus effects on disease susceptibility. In: Lusis, A.J.; Sparkes, R.S., eds. Genetic factors in atherosclerosis: Approaches and model systems. Basel: Karger; 1989:50-78.
35. MacCluer, J.W.; Kammerer, C.M. Dissecting the genetic contribution to coronary heart disease. Am. J. Hum. Genet. 49:1139-44; 1991.
36. MacLean, C.J.; Morton, N.E.; Elston, R.C.; Yee, S. Skewness in commingled distributions. Biometrics. 32:695-99; 1976.
37. Malina, R.M.; Bouchard, C. Genetic considerations in physical fitness. In: Assessing physical fitness and physical activity in population-based surveys. Hyattsville, MD: U.S. Department of Health and Human Services; 1989:453-73.
38. McGue, M.; Rao, D.C.; Iselius, L.; Russell, J.M. Resolution of genetic and cultural inheritance in twin families by path analysis: Applications to HDL-cholesterol. Am. J. Hum. Genet. 37:998-1014; 1985.
39. McGue, M.; Wette, R.; Rao, D.C. Path analysis under generalized marital resemblance: Evaluation of the assumptions underlying the mixed homogamy model by the Monte Carlo method. Genet. Epidemiol. 6:373-88; 1989.
40. Merlino, G.T. Transgenic animals in biomedical research. FASEB J. 5:2996-3001; 1991.
41. Metsäranta, M.; Vuorio, E. Transgenic mice as models for heritable diseases. Ann. Med. 24:117-20; 1992.
42. Nance, W.E.; Corey, L.A. Genetic models for the analysis of data from the families of identical twins. Genetics. 83:811-26; 1976.
43. Neale, M.C.; Heath, A.C.; Hewitt, J.K.; Eaves, L.J.; Fulker, D.W. Fitting genetic models with LISREL: Hypothesis testing. Behav. Genet. 19:37-49; 1989.
44. Olson, J.M.; Wijsman, E.M. Linkage between quantitative trait and marker loci: Methods using all relative pairs. Genet. Epidemiol. 10:87-102; 1993.
45. Ott, J. Analysis of human genetic linkage. Baltimore: Johns Hopkins University Press; 1991.
46. Pérusse, L.; Després, J.P.; Tremblay, A.; Leblanc, C.; Talbot, C.; Allard, C.; Bouchard, C. Genetic and environmental determinants of serum lipids and lipoproteins in French Canadian families. Arteriosclerosis. 9:308-18; 1989.
47. Pérusse, L.; Lortie, G.; Leblanc, C.; Tremblay, A.; Thériault, G.; Bouchard, C. Genetic and environmental sources of variation in physical fitness. Ann. Hum. Biol. 14:425-34; 1987.
48. Pérusse, L.; Rice, T.; Bouchard, C.; Vogler, G.P.; Rao, D.C. Cardiovascular risk factors in a French-Canadian population: Resolution of genetic and

familial environmental effects on blood pressure by using extensive information on environmental correlates. Am. J. Hum. Genet. 45:240-51; 1989.

49. Pérusse, L.; Tremblay, A.; Leblanc, C.; Bouchard, C. Genetic and environmental influences on level of habitual physical activity and exercise participation. Am. J. Epidemiol. 129:1012-22; 1989.
50. Pinkert, C.A. Introduction to transgenic animals. In: Pinkert, C.A., ed. Transgenic animal technology: A laboratory handbook. New York: Academic Press; 1994:3-12.
51. Plomin, R.; DeFries, J.C.; Loehlin, J.C. Genotype-environment interaction and correlation in the analysis of human behavior. Psychol. Bulletin. 84:309-22; 1977.
52. Price, R.A.; Gottesman, I.I. Body fat in identical twins reared apart: Roles for genes and environment. Behav. Genet. 21:1-7; 1991.
53. Rao, D.C. Statistical consideration in applications of path analysis in genetic epidemiology. In: Rao, D.C.; Chakraborty, R., eds. Handbook of statistics. Amsterdam, Netherlands: Elsevier Science; 1991:63-80.
54. Rao, D.C.; McGue, M.; Wette, R.; Glueck, C.J. Path analysis in genetic epidemiology. In: Chakravarti, A., ed. Human population genetics: The Pittsburgh symposium. New York: Van Nostrand Reinhold Co.; 1984:35-81.
55. Rao, D.C.; Morton, N.E.; Yee, S. Analysis of family resemblance. II. Linear model for familial correlation. Am. J. Hum. Genet. 26:331-59; 1974.
56. Rao, D.C.; Morton, N.E.; Yee, S. Resolution of cultural and biological inheritance by path analysis. Am. J. Hum. Genet. 28:228-42; 1976.
57. Rao, D.C.; Vogler, G.P. Assessing genetic and cultural heritabilities. In: Goldbourt, U.; de Faire, U.; Berg, K., eds. Genetic factors in coronary heart disease. Dordrecht, Netherlands: Kluwer Academic; 1994:71-81.
58. Rao, D.C.; Wette, R. Environmental index in genetic epidemiology: An investigation of its role, adequacy, and limitations. Am. J. Hum. Genet. 46:168-78; 1990.
59. Rende, R.D.; Plomin, R.; Vandenberg, S.G. Who discovered the twin method? Behav. Genet. 20:277-85; 1990.
60. Rice, J.P.; Cloninger, C.R.; Reich, T. General causal models for sex differences in the familial transmission of multifactorial traits: An application to human spatial visualizing ability. Social. Biology. 27:36-47; 1980.
61. Rice, J.P.; Cloninger, C.R.; Reich, T. Multifactorial inheritance with cultural transmission and assortative mating. I. Description and basic properties of the unitary models. Am. J. Hum. Genet. 30:618-43; 1978.
62. Risch, N. Linkage strategies for genetically complex traits. II. The power of affected relative pairs. Am. J. Hum. Genet. 46:229-41; 1990.
63. Risch, N. Linkage strategies for genetically complex traits. III. The effect of marker polymorphism on analysis of affected relative pairs. Am. J. Hum. Genet. 46:242-53; 1990.
64. Schull, W.J.; Weiss, K.M. Genetic epidemiology: Four strategies. Epidemiol. Rev. 2:1-18; 1980.

65. Sing, C.F.; Boerwinkle, E.A. Genetic architecture of inter-individual variability in apolipoprotein, lipoprotein and lipid phenotypes. In: Bock, G.; Collins, G.M., eds. Molecular approaches to human polygenic disease. New York: Wiley; 1987:99-127.

66. Sing, C.F.; Boerwinkle, E.A.; Moll, P.P.; Templeton, A.R. Characterization of genes affecting quantitative traits in humans. In: Weir, B.S.; Eisen, E.J.; Goodman, M.M.; Namkoong, G., eds. Proceedings of the 2nd International Conference on Quantitative Genetics. Sunderland: Sinauer Associates; 1988:250-69.

67. Sing, C.F.; Moll, P.P. Genetics of atherosclerosis. Annu. Rev. Genet. 24:171-87; 1990.

68. Stunkard, A.J.; Harris, J.R.; Pedersen, N.L.; McClearn, G.E. The body-mass index of twins who have been reared apart. N. Engl. J. Med. 322:1483-87; 1990.

69. Susser, M. Separating heredity and environment. Am. J. Prev. Med. 1:5-23; 1985.

70. Tiret, L.; Abel, L.; Rakotovao, R. Effect of ignoring genotype-environment interaction on segregation analysis of quantitative traits. Genet. Epidemiol. 10:581-86; 1993.

71. Tiret, L.; Rigat, B.; Visvikis, S.; Breda, C.; Corvol, P.; Cambien, F.; Soubrier, F. Evidence, from combined segregation and linkage analysis, that a variant of the angiotensin I-converting enzyme (ACE) gene controls plasma ACE levels. Am. J. Hum. Genet. 51:197-205; 1992.

72. Todorov, A.A. Multilocus IBD distributions in sibships with applications. Genet. Epidemiol. 1:A308; 1994.

73. Todorov, A.A.; Gu, C.; Elston, R.C. Multipoint IBD distribution in sibships. Genet. Epidemiol. Pending revision.

74. Warden, C.H.; Daluiski, A.; Lusis, A.J. Identification of new genes contributing to atherosclerosis: The mapping of genes contributing to complex disorders in animal models. In: Lusis, A.J.; Rotter, J.L.; Sparkes, R.S., eds. Molecular genetics of coronary artery disease: Candidate genes and processes in atherosclerosis. Basel: Karger; 1992:419-41.

75. Warden, C.H.; Fisler, J.S. Identification of gene underlying polygenic obesity in animal models. In: Bouchard, C., ed. The genetics of obesity. Boca Raton, FL: CRC Press Inc.; 1994:181-97.

CHAPTER 5

Defining Performance, Fitness, and Health

Concepts of health-related fitness, performance-related fitness, and health are discussed in this chapter. Phenotypes associated with these concepts, which are considered in subsequent chapters, are also identified. Views of health-related fitness and performance-related fitness are still evolving in the exercise science community. The intricacies and/or merits of the current debate are not considered. This chapter provides an overview of concepts and component phenotypes with little discussion of their relative merits and controversies.

The World Health Organization (14) defined fitness as "the ability to perform muscular work satisfactorily." Although this definition was quite useful in the past, it is only partially relevant to the domains of performance and health. Definitions are needed that permit the ready derivation of phenotypes and intermediate phenotypes, particularly within the paradigms of molecular biology and genetics.

Performance-Related Fitness

Performance-related fitness refers to those components of fitness that are necessary for optimal work and motor and sport performance (8,10). It is often defined in terms of motor abilities and performance on standardized tests. Performance-related fitness depends heavily upon specific motor skills, cardiorespiratory power and capacity, aerobic and anaerobic power and capacity, muscular strength, power and/or endurance, body size and composition, motivation, and nutritional status. Most of these factors can be measured in the laboratory and in the field and are amenable to genetic studies by a variety of approaches.

It is important to recognize that performance is best defined in terms of the specific requirements of an event. In this sense, performance is a heterogeneous concept. It applies to the marathon runner, the platform diver, the ski jumper, the artistic skater, the weight lifter, the decathlete, and so on. Some would argue that it is even more encompassing and includes auto racing, chess, video games, violin playing, and other similar activities. In these contexts, performance is usually viewed in terms of the highly skilled or the elite. Performance-related fitness also applies to the general population, to the child learning to run or jump, or to the aging adult whose sense of balance may be waning. Hence, the major determinants of physical performance in a variety of tasks, and not of the events per se, are important and are considered in several specific chapters.

Health-Related Fitness

Health-related fitness refers to those components of fitness that relate to the health status of the individual and that may be influenced by regular physical activity. It is defined as the state of physical and physiological characteristics that define risk levels for the premature development of several diseases or morbid conditions, where these diseases or conditions are related to a sedentary lifestyle (6). Important determinants of health-related fitness include factors such as body mass for stature, body composition, subcutaneous fat distribution, abdominal visceral fat, bone density, strength and endurance of the abdominal and dorsolumbar musculature, heart and lung functions, blood pressures, maximal aerobic power and tolerance for submaximal exercise, glucose and insulin metabolism, blood lipid and lipoprotein profile, and the ratio of lipid to carbohydrate oxidized in a variety of situations. A desirable profile for these complex factors presents an advantage in terms of health outcomes as assessed in morbidity and mortality statistics.

There are various ways to define and classify components of health-related fitness. One proposal is summarized in table 5.1 (6). It includes morphological, muscular, motor, cardiorespiratory, and metabolic components. Within each of the components, several phenotypes are indicated that have a relationship with health-related fitness and health status.

Health

Defining health is a major challenge, despite progress made in treating diseases and increasing the average life span in Western societies. Health can be defined as a human condition with physical, social, and

Table 5.1
The Components and Factors of Health-Related Fitness

Morphological component
- Body mass for height
- Body composition
- Subcutaneous fat distribution
- Abdominal visceral fat
- Bone density
- Flexibility

Muscular component
- Power
- Strength
- Endurance

Motor component
- Agility
- Balance
- Coordination
- Speed

Cardiorespiratory component
- Submaximal exercise capacity
- Maximal aerobic power
- Heart functions
- Lung functions
- Blood pressure

Metabolic component
- Glucose tolerance
- Insulin sensitivity
- Blood lipids and lipoproteins
- Substrate oxidation characteristics

Reprinted, by permission, from C. Bouchard and R.J. Shephard, 1994, Physical activity, fitness and health: The model and key concepts. In *Physical activity, fitness, and health*, edited by C. Bouchard, R.J. Shephard, and T. Stephens (Champaign, IL: Human Kinetics), 81.

psychological dimensions, each characterized on a continuum from positive to negative poles. Positive health is associated with a capacity to enjoy life and to withstand challenges; it is not merely the absence of disease. Negative health is associated with morbidity and, in the extreme, with premature mortality (6).

Morbidity can be defined as any departure, subjective or objective, from a state of physical or psychological well-being, short of death. Morbidity can be measured as the number of persons who are ill per unit population per year, the incidence of specific conditions per unit population per year, and the average duration of these conditions. On the other hand, wellness is a holistic concept, describing a state of positive health in the individual. It comprises physical, social, and psychological components of well-being (6).

Traditional illness and mortality statistics do not provide a full assessment of health. A more comprehensive approach requires that the health status of an individual or a population be established in terms of common health outcomes, information on temporary and chronic disabilities, absenteeism, overall productivity, and use of all forms of medical services, including prescribed and nonprescribed drugs. In this approach, life expectancy is assessed in light of a variety of indicators of the quality of life, including mobility, personal autonomy, and disabilities.

Individual Differences in Trainability

Repeated exposure to exercise of a given type triggers a series of responses in body tissues, organs, and systems. It is generally conceived that these responses are usually, but not always, adaptive. Similar adaptive responses occur with motor learning and adaptation to stress agents: Most, but not all, people are able to learn or cope.

In order to determine an effect of genotype on response to training, one must have evidence of individual differences in trainability. There is now considerable support for this concept (1,4). Table 5.2 illustrates the extent of individual differences in response of performance phenotypes to training. For example, the same training program may result in almost no change in $\dot{V}O_2$max for some subjects, while others gain as much as 1L of O_2 uptake.

The extent of individual differences in the response of maximal oxygen uptake to training is best illustrated in figure 5.1. After training programs lasting from 15 to 20 weeks in 47 young men, some exhibited almost no change, while others gained as much as 1 L in O_2 uptake. Such differences in response to the same training stimulus cannot be attributed to age (all subjects were young adults, 17 to 29 years old) or gender (all males). Initial pretraining $\dot{V}O_2$max accounted for about 25% of the variance in the response to training; the lower the initial level of $\dot{V}O_2$max, the greater was the increase with training. Thus, about 75% of the heterogeneity in response to systematic training was not explained.

Similar individual differences are observed in endurance performance, measured as total work output in megajoules (MJ) during a

Table 5.2
Performance Phenotypes Before Training and the Response to Training in Young Adult Males From Several Studies

	PRETRAINING PHENOTYPE		CHANGES IN PHENOTYPE			
	Mean	***SD***	**Mean**	***SD***	**Min**	**Max**
$\dot{V}O_2max$						
Quebec (*n* = 17)[a]	2.9	0.42	0.63	0.25	0.13	1.03
Arizona (*n* = 29)[b]	3.4	0.57	0.42	0.22	0.06	0.95
Max work output (MJ) in 90 min (*n* = 24)[c]	567	140	213	155	42	516
Max work output (kJ) in 90 s (*n* = 17)[d]	25.7	4.1	7.5	2.6	2.5	13.0
Max work output (kJ) in 10 s (*n* = 17)[d]	6.7	0.8	1.4	0.7	0.3	2.7

[a]Subjects were trained for 20 weeks following the procedures described earlier in Dionne et al. (7) and Lortie et al. (9).
[b]Subjects were trained for 12 weeks, three times per week, 40 min per session, at onset of blood lactate accumulation (OBLA) (about 70% to 77 % of $\dot{V}O_2max$). See Lortie et al. (9).
[c]See Lortie et al. (9).
[d]From Simoneau, Lortie, Bouchard, et al. (unpublished data) using high-intensity intermittent training as described in Simoneau et al. (13).

90-min maximal cycle ergometer test. Following a 20-week endurance training program, the mean improvement in endurance performance was about 40% of the pretraining value, but improvement in performance output ranged from 16% to 97% (9).

Human variation in trainability is also considerable for maximal performances of shorter duration (anaerobic), such as the 10-s and 90-s maximal work output on a cycle ergometer. In a study of 17 sedentary subjects who trained for 15 weeks with high-intensity intermittent exercise on a cycle ergometer, both phenotypes improved significantly (4,13). Again, individual differences in training responses were very large, with a five- to ninefold range between the responses of the low and high responders (table 5.2).

High levels of heterogeneity in response to regular exercise are also apparent for other relevant phenotypes: markers of skeletal muscle oxidative metabolism or adipose tissue metabolism, relative ratio of lipid and carbohydrate oxidization, fasting glucose and insulin levels and their response to a glucose challenge, and fasting plasma levels of lipids and lipoproteins (2). All of these phenotypes respond to regular

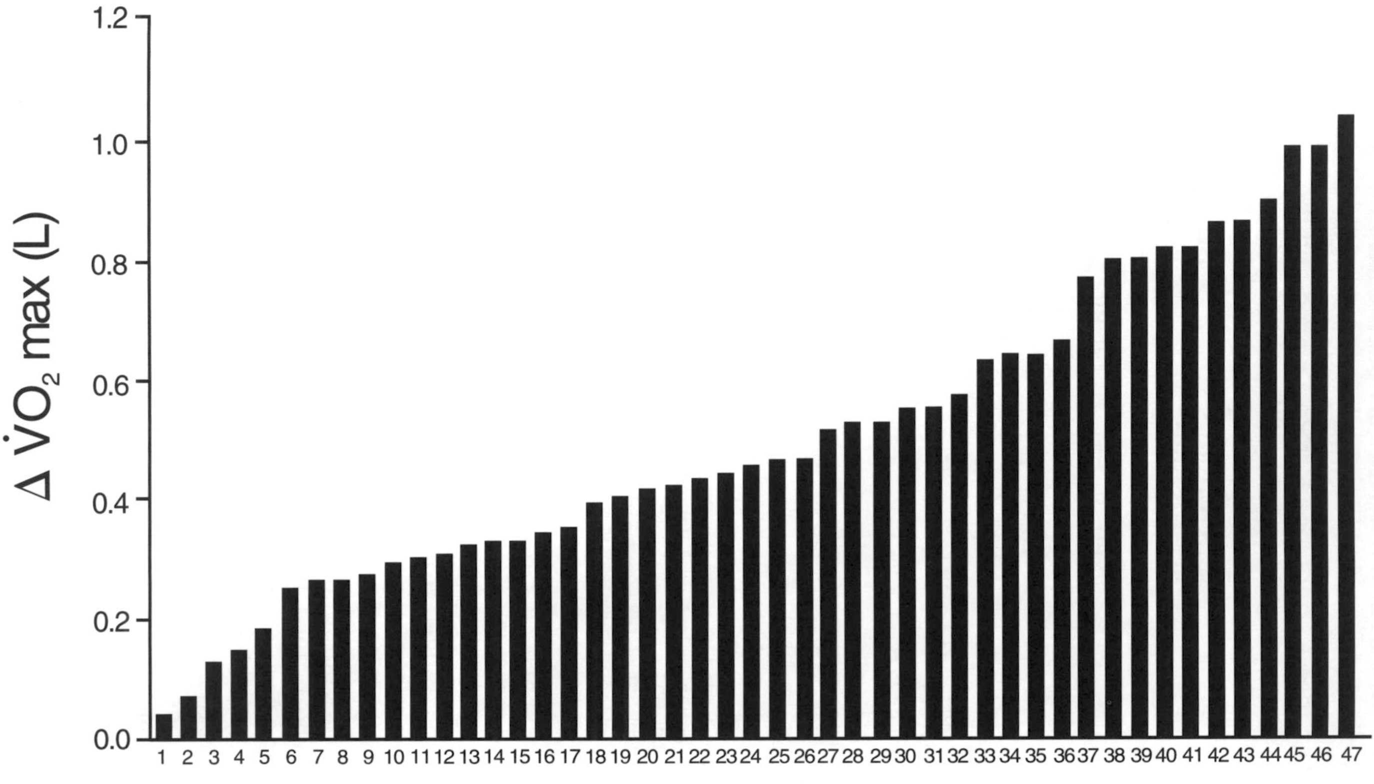

Figure 5.1 Individual differences in the response of 47 young men to training programs lasting from 15 to 20 weeks. Results are expressed as gains of V̇O_2max in liters of O_2 per minute. From Bouchard et al., unpublished data.

exercise in young adults of both sexes. However, there is a very wide range of individual differences in responses of these biological markers to training: Some individuals exhibit a high response pattern while others are almost nonresponders.

Determinants of Trainability

Figure 5.2 shows factors potentially related to variation in sensitivity to exercise training (1). The factors include age and sex, prior training experience, current phenotypic level, and genotype. Although age can be readily controlled, available evidence indicates that the response to training is quite variable even within relatively narrow age ranges. The issue of prior experience or training history has generally not been considered. However, by focusing on sedentary subjects with no history of regular participation in sport or other physical activities, researchers to some degree can control for prior experience. The remaining three factors are related in part to the individual's genotype. Sex differences in trainability may be a factor. Although maximal aerobic

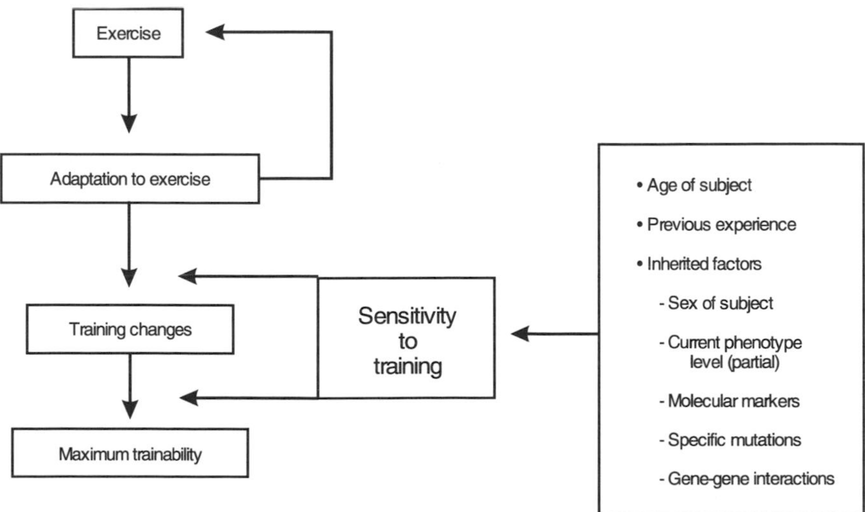

Figure 5.2 A model of factors associated with human variation in the sensitivity to exercise training. A similar sketch can be applied to motor performance. Adapted, by permission, from Ronald J. Terjung (editor), 1983, *Exercise and Sport Science Reviews* 11:297 (Indianapolis, IN: American College of Sports Medicine). ©1983 by The Franklin Institute, Philadelphia, PA.

power has the same trainability in males and females (9), endurance performance is, on the average, more trainable than maximal aerobic power, and males show a greater trainability than females under similar laboratory conditions. Hence, the sex of subjects probably accounts for some of the variation in response to training for selected phenotypes.

One of the most difficult factors to interpret in the model is the current phenotype (i.e., the pretraining level) and its relationship with the observed response to training. For instance, evidence suggests that trainability of maximal aerobic power and other aerobic output criteria is negatively related to the pretraining level (3,11,12). In a review of data from 50 published training experiments, the correlation between training-induced changes in $\dot{V}O_2max$ and the pretraining level of this attribute was negative and significant ($r = -0.5$) (5). When the frequency, intensity, and duration of training sessions, as well as the number of weeks of training were statistically controlled, the correlation was reduced only negligibly ($r = -0.4$). These observations imply that the initial phenotypic level accounts for a significant proportion of the variance in response to aerobic training. Current phenotypic levels are in part reflections of the genotypes of sedentary subjects. Under these conditions, the variance in training response associated with pretraining phenotypic level can be viewed as a component of the genotype-training interaction effect.

The genotype-training interaction effect is associated with genetic variation that is currently unknown but that is independent of pretraining phenotypic differences. The interaction component implies that sensitivity to aerobic training depends to some extent on an individual's genotype: That is, responses to training are conditioned in part by genotypic differences among individuals. This interaction component is discussed in subsequent chapters in terms of its quantitative contribution to the overall variance in response. It should be possible to describe these effects in the future in terms of genes involved, specific mutations, and eventually complex gene-gene interactions. This is an important research agenda item for those interested in the genetic and molecular bases of human fitness and performance phenotypes.

Finally, the general paradigm depicted in figure 5.2 can also be applied to motor learning and performance in motor skills. It is likely that the magnitude of genotype-learning interaction effects will vary with the motor skills in question. One can predict that the specific genes and mutations responsible for the interaction effects will be different from those involved in the trainability of aerobic and anaerobic power and capacity phenotypes.

Summary

This chapter proposes simple definitions for discussion of the various phenotypes considered later in the book. Performance-related fitness, health-related fitness, and health are defined. The notion of individual differences in trainability is introduced and illustrated with experiment results. Finally, an attempt is made to identify the determinants of trainability. The similarity between the determinants of trainability and of motor learning, seen in a genetic perspective, is emphasized.

References

1. Bouchard, C. Genetics of aerobic power and capacity. In: Malina, R.M.; Bouchard, C., eds. Sport and human genetics. Champaign, IL: Human Kinetics; 1986:59-89.
2. Bouchard, C. Genetics of the response to exercise and training. In: Fletcher, G.F., ed. Cardiovascular response to exercise. Mount Kisco, NY: Futura Publishing Co.; 1994:347-56.
3. Bouchard, C. Human adaptability may have a genetic basis. In: Landry, F., ed. Health risk estimation, risk reduction and health promotion. Proceedings of the 18th annual meeting of the Society of Prospective Medicine. Ottawa: Canadian Public Health Association; 1983:463-76.
4. Bouchard, C.; Boulay, M.R.; Simoneau, J.A.; Lortie, G.; Pérusse, L. Heredity and trainability of aerobic and anaerobic performances. Sports Med. 5:69-73; 1988.
5. Bouchard, C.; Carrier, C.; Boulay, M.R.; Thibault, M.-C.; Dulac, S. Le développement du système de transport de l'oxygène chez les jeunes. Quebec: Les Editions du Pélican; 1975.
6. Bouchard, C.; Shephard, R.J. Physical activity, fitness, and health: The model and key concepts. In: Bouchard, C.; Shephard, R.J.; Stephens, T., eds. Proceedings and consensus statement: Physical activity, fitness, and health. Champaign, IL: Human Kinetics; 1994:77-88.
7. Dionne, F.T.; Turcotte, L.; Thibault, M.-C.; Boulay, M.R.; Skinner, J.S.; Bouchard, C. Mitochondrial DNA sequence polymorphism, $\dot{V}O_2max$ and response to endurance training. Med. Sci. Sports Exerc. 23:177-85; 1991.
8. Gledhill, N. Discussion: Assessment of fitness. In: Bouchard, C.; Shephard, R.J.; Stephens, T.; Sutton, J.R.; McPherson, B.D., eds. Exercise, fitness, and health: A consensus of current knowledge. Champaign, IL: Human Kinetics; 1990:121-26.
9. Lortie, G.; Simoneau, J.A.; Hamel, P.; Boulay, M.R.; Landry, F.; Bouchard, C. Responses of maximal aerobic power and capacity to aerobic training. Int. J. Sports Med. 5:232-36; 1984.
10. Pate, R.R.; Shephard, R.J. Characteristics of physical fitness in youth. In: Gisolfi, C.V.; Lamb, D.R., eds. Perspectives in exercise science and sports

medicine. Vol. 2: Youth, exercise and sport. Indianapolis: Benchmark Press; 1989:1-45.

11. Pollock, M.L. The quantification of endurance training programs. Exerc. Sport Sci. Rev. 1:155-88; 1973.
12. Saltin, B. The effect of physical training on the oxygen transporting system in man. In: Cumming, G.R.; Snidal, D.; Taylor, A.W., eds. Environmental effects on work performance. Edmonton, AB: Canadian Association of Sport Sciences; 1972:151-62.
13. Simoneau, J.A.; Lortie, G.; Boulay, M.R.; Marcotte, M.; Thibault, M.-C.; Bouchard, C. Inheritance of human skeletal muscle and anaerobic capacity adaptation to high-intensity intermittent training. Int. J. Sports Med. 7:167-71; 1986.
14. World Health Organization. Meeting of investigators on exercise tests in relation to cardiovascular function. Geneva: WHO Technical Report. No. 388.; 1968.

CHAPTER 6

Genetics and Health Status

Good health is an essential component of well-being. The definition of health offered by the World Health Organization (43) almost 50 years ago ("a state of complete physical, mental, and social well-being and not merely the absence of disease or infirmity") highlights the multidimensionality of the concept. The concept of health has subsequently been expanded (e.g., 2,7,16). The definition offered at the 1988 Consensus Conference on Exercise, Fitness and Health, and reiterated at the 1992 Consensus Conference on Physical Activity, Fitness and Health, is as follows:

> Health is a human condition with physical, social, and psychological dimensions, each characterized on a continuum with positive and negative poles. Positive health is associated with a capacity to enjoy life and to withstand challenges; it is not merely the absence of disease. Negative health is associated with morbidity and, in the extreme, with mortality. (3, pp. 6-7)

Health status includes the presence of positive health (e.g., self-worth and physical fitness), absence of ill health (e.g., disability and mental distress), and absence of disease (e.g., communicable and chronic diseases). The state of health is thus a continuum that covers a broad range of individual and population variability.

Measuring Health Status

The concept of health is seemingly straightforward. Measurement of health status is a different matter, depending in part on the purpose of the assessment. In developing countries, for example, the growth status of children (stature and weight) is commonly accepted as an indicator of the overall health and nutritional status of a community (44). To compare living conditions in countries across the world, the United Nations

Development Program uses the Human Development Index (HDI), defined as a minimal

> measure of people's ability to live a long and healthy life, to communicate and to participate in the life of the community and to have sufficient resources to obtain a decent living. (39, p. 104)

The HDI is based on life expectancy at birth, two educational indicators (adult literacy and mean years of schooling), and per capita purchasing power parity. Table 6.1 gives the rankings of several countries according to the HDI. Resources and income are inequitably distributed in many countries, both developed and developing, so there is considerable within-country variation. The HDI for American whites (0.985) is slightly higher than that for Japan (0.983). The HDI for American blacks approximates that for Trinidad and Tobago (0.877), and that for American Hispanics is similar to that for Estonia (0.872). When the HDI is calculated by gender for American blacks and whites, it is highest for white females (0.990), followed by white males (0.975), black females (0.900), and finally black males (0.855) (39, p. 18). These results demonstrate the divergence of groups within a population relative to the national average.

The HDI is useful in evaluating aspects of health status in a global perspective. For example, globally and within specific geographic areas, countries with a high HDI characteristically have higher morbidity and mortality from diseases of the circulatory system and lower morbidity and mortality from communicable diseases. Conversely, countries with a low HDI have higher morbidity and mortality from communicable diseases.

In an attempt to estimate the burden of disease on human productivity, the World Bank developed an index of health status called the disability-adjusted life year (DALY), a statistic that

> combines the loss of life from premature death in 1990 with the loss of healthy life from disability. (42, p. 213)

The DALY statistic is weighted for potential years of life lost due to death at different ages, estimated value of a healthy year of life at different ages, and estimated value of the future relative to the present (illness and injuries can last for years or decades, and disabilities can lead to anything from nearly perfect health to the brink of death).

Estimated DALYs attributable to premature mortality and disability in several demographic regions of the world are illustrated in figure 6.1, while the estimated impact of communicable and noncommunicable diseases and injuries is summarized in figure 6.2. The burdens of DALYs

Table 6.1
Human Development Index for Selected Industrial and Developing Countries in 1990

Industrial countries		Developing countries	
Japan	0.983	Hong Kong	0.913
Canada	0.982	Uruguay	0.881
Norway	0.978	South Korea	0.872
Switzerland	0.978	Chile	0.864
Sweden	0.977	Singapore	0.849
United States	0.976	Argentina	0.832
Australia	0.972	Venezuela	0.824
France	0.971	Mexico	0.805
Netherlands	0.970	Brazil	0.730
United Kingdom	0.964	Saudi Arabia	0.688
Iceland	0.960	South Africa	0.673
Germany	0.957	Philippines	0.603
Denmark	0.955	China	0.566
Finland	0.954	Nicaragua	0.500
Austria	0.952	Morocco	0.433
Belgium	0.952	Egypt	0.389
New Zealand	0.947	Kenya	0.369
Luxembourg	0.943	Papua New Guinea	0.318
Israel	0.938	Pakistan	0.311
Ireland	0.925	India	0.309
Italy	0.924	Namibia	0.289
Spain	0.923	Haiti	0.275
Hungary	0.887	Nigeria	0.246
Russian Federation	0.862	Bangladesh	0.189
Portugal	0.853	Cambodia	0.186
Poland	0.831	Ethiopia	0.172
Romania	0.709	Sierra Leone	0.065

Adapted from United Nations Development Program (39).

lost to disability are reasonably similar in both developing and developed countries, while the burdens of DALYs lost to premature mortality are dramatically greater in developing areas, especially Sub-Saharan Africa (figure 6.1). The magnitudes of the burdens of noncommunicable

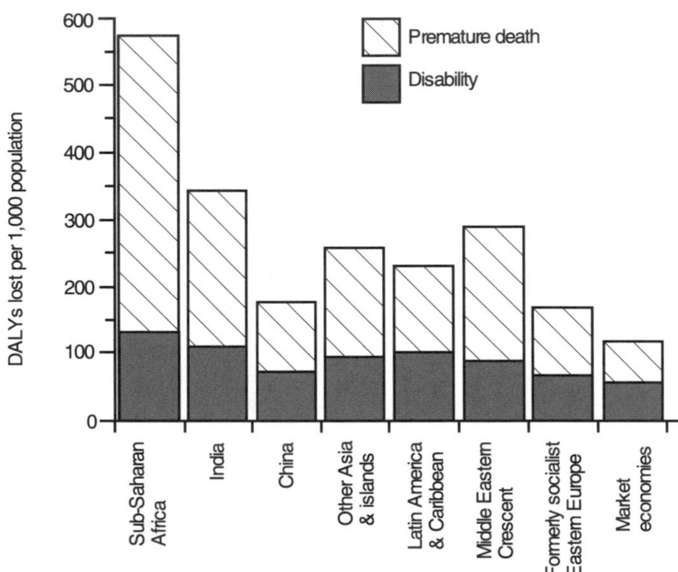

Figure 6.1 Disability-adjusted life years (DALYs) lost to premature death and disability in different demographic areas of the world. Drawn from data reported by the World Bank (42).

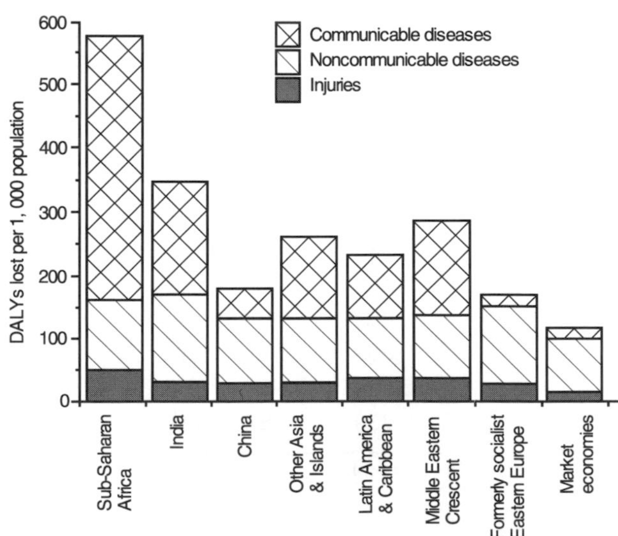

Figure 6.2 Disability-adjusted life years (DALYs) lost to communicable diseases, noncommunicable diseases, and injuries in different demographic areas of the world. Drawn from data reported by the World Bank (42).

diseases and injuries are also reasonably similar in different regions of the world. However, DALYs lost to communicable diseases vary with geographic region (figure 6.2). In developed countries, particularly the established market economies and the formerly socialist economies of eastern Europe, the burden of DALYs lost to noncommunicable diseases, a major component of which are cardiovascular, far outweighs those lost to communicable diseases and injuries. In the established market economies, cardiovascular diseases account for about 30% of DALYs lost to noncommunicable diseases (42).

The indicators developed by the United Nations and the World Bank highlight the importance of health status in the global community. The indices clearly illustrate the health and economic disparities in the world and are potentially useful in identifying correlates and determinants and in assessing needs and allocating resources on a global basis. The perspective of the World Bank (42) is largely economic and emphasizes the potential contribution of improved health to economic growth in four ways: (1) improved worker productivity as a result of reduced illnesses; (2) improved education, since healthier and adequately nourished children are more likely to enroll in school and are better able to learn; (3) enhanced alternative use of resources that would otherwise be used for medical expenses; and (4) improved utilization of natural resources, especially the land, through reduction of the burden of diseases that limit productivity.

In a recent comprehensive review of the measurement of health status and well-being, Caspersen et al. (7) use seven categories of measures ranging from the traditional (i.e., mortality and morbidity) to newer, more qualitative measures. The specific categories, subcategories, and suggested measures within each are summarized in table 6.2. The utility of specific measures varies with age. Many measures are not applicable to infancy (birth to one year) and have limited utility during preschool ages (one through four years). For individuals older than four years, data are generally available except for the heading of healthy life years. A healthy life year is defined as a

> year free from chronic, disabling diseases and conditions, from preventable infections, and from serious injury; a year with the full range of functional capacity enabling one to enter into satisfying relationships with others, to work, and to play. (7, p. 199)

The measures of health status and well-being proposed by Caspersen et al. (7) are more comprehensive than the HDI and DALY. The former were defined in the context of the United States and Canada, while the latter were developed for global comparisons. Nevertheless, the categories of measures summarized in table 6.2 have potential utility for international comparisons as data become available.

Health status can be measured for the individual and for the population. Individually, the measures are used to screen and diagnose, to plan treatment, and to evaluate progress. At the population level, the measures are used to assess needs, identify determinants and correlates, evaluate programs and services, and allocate resources.

The overall health status of the American population has changed considerably in the last century. Infant mortality has declined from 95.7/1,000 live births in 1915-1919 to 9.2/1,000 live births in 1990. However, mortality of black infants is more than double that of white infants: 17.0/1,000 versus 7.7/1,000 live births in 1990 (41). Life expectancy has also increased significantly (table 6.3). The gender- and race-related variations in life expectancy are consistent with those noted earlier for the HDI. Note that the life expectancy at birth approximately doubled for black males and more than doubled for black females between 1900 and 1990.

The improved health and morbidity status of the American population is also indicated by changes in leading causes of death since the turn of the century (table 6.4). In 1900, infectious diseases were the major cause of death and accounted for about 25% of all mortality. In 1990, infectious diseases accounted for only about 5% of the total mortality (but a new infectious disease, HIV infection, appeared among the 10 leading causes of death). Heart diseases and cancers caused more than half of all deaths in 1990, compared to only about 12% in 1900.

The changing pattern of mortality is related to generally improved living conditions, to advances in the medical and health-related sciences, and more recently to improved health awareness and changes in behaviors related to diet, smoking, and physical activity. For example, although diseases of the heart accounted for 33% of total mortality in 1990, there has been a major decline in heart disease mortality in the United States since 1970 (23,36). Improved detection and control of high blood pressure, awareness of dietary fats and cholesterol, and reduced cigarette smoking are major factors in the decline (23). The fact that people now live longer also contributes to the changing trends in causes of death. This, in conjunction with improved life expectancy, especially at 65 years, has implications for the study of longevity. (See section that follows.)

Genetic Effects in Health Status

Variable proportions of individual differences in measures of health status and well-being are dependent upon undetermined genetic

Table 6.2
Measures for the Assessment of Health Status and Well-Being

1. Mortality measures
   - A. Mortality rate
   - B. Condition-specific mortality rate
   - C. Infant mortality
   - D. Maternal mortality
   - E. Condition-specific YPPL[a]
   - F. Life expectancy at birth or other ages
2. Morbidity measures
   - A. Condition-specific incidence
   - B. Condition-specific prevalence
   - C. Impairment prevalence
   - D. Days of restricted activity
   - E. Days of seeking medical attention
   - F. Days in a hospital
   - G. Active life expectancy
3. Prevalence of risk factors
   - A. Smoking
   - B. Inactivity
   - C. Poor nutritional practices
   - D. High blood pressure
   - E. High blood cholesterol
   - F. Circumference ratios
   - G. Overweight, high body mass index
4. Use of medical care
   - A. Doctor visits per year
   - B. Proportion not seeing a physician
   - C. Interval since last doctor visit
   - D. Short-stay hospital discharge rate
   - E. Short-stay episodes per person per year
   - F. Short-stay hospital days in a year
5. Disability measures
   - A. Days of restricted activity
   - B. Mobility limitation
   - C. Activity limitation
   - D. Activities of daily living
6. Function
   - A. Physical function
     1. Overall mobility
     2. Walking
     3. Lifting
     4. Reaching overhead
     5. Going up and down stairs
     6. Rising from or sitting down in a chair

(continued)

Table 6.2 *(continued)*

- 7. Dexterity
- 8. Balancing
- 9. Flexibility
- 10. Strength

B. Mental function
- 1. Short-term memory
- 2. Intelligible speech
- 3. Alertness
- 4. Orientation in time and space
- 5. Visual and auditory perception
- 6. Confusion
- 7. Attention
- 8. Comprehension
- 9. Problem solving
- 10. Literacy
- 11. Numeracy
- 12. Overall knowledge

C. Functional activities
- 1. Basic enabling activities
  - a. Personal care activities
  - b. Skills of independence
- 2. Productivity-oriented activities
- 3. Leisure/recreational activities

7. Well-being

A. Bodily well-being
- 1. Energy
- 2. Pain
- 3. Sleep

B. Emotional well-being
- 1. Positive affect
- 2. Anger/hostility/irritability
- 3. Anxiety
- 4. Depression

C. Self-concept
- 1. Self-esteem
- 2. Sense of mastery and control

D. Global perceptions of well-being
- 1. Health outlook
- 2. Current health perceptions
- 3. Life-satisfaction

8. Healthy life years

[a]YPPL = Years of potential life lost, usually before age 65 or some other selected age.

Adapted, by permission, from C. Bouchard and R.J. Shephard, 1994, Measurement of health status and well-being. In Physical activity, fitness, and health, edited by C. Bouchard, R.J. Shephard, and T. Stephens (Champaign, IL: Human Kinetics), 182-184.

Table 6.3
Life Expectancy at Birth and at Selected Ages by Sex and Race in the United States in 1900 and 1990

	WHITE		BLACK	
	Male	**Female**	**Male**	**Female**
At birth:				
1900	48.2	51.1	32.5	35.0
1990	72.7	79.4	64.5	73.6
At 1 year:				
1900	54.6	56.4	42.5	43.5
1990	72.3	78.9	65.8	73.8
At 20 years:				
1900	42.2	43.8	35.1	36.9
1990	54.0	60.3	46.7	55.3
At 65 years:				
1900	11.5	12.2	10.4	11.4
1990	15.2	19.1	13.2	17.2

Adapted from U.S. Department of Health and Human Services (41).

Table 6.4
The 10 Leading Causes of Death in the United States as a Percentage of All Deaths in 1900 and 1990.

1900[a]	1990[b]
1. Tuberculosis, 11.3%	1. Heart diseases, 33.5%
2. Pneumonia, 10.2%	2. Cancers, 23.5%
3. Diarrhea and enteritis, 8.1%	3. Strokes, 6.7%
4. Heart diseases, 8.0%	4. Accidents, 4.3%
5. Liver disease, 5.2%	5. Chronic obstructive pulmonary diseases, 4.0%
6. Injuries, 5.1%	6. Pneumonia and influenza, 3.7%
7. Strokes, 4.5%	7. Diabetes mellitus, 2.2%
8. Cancers, 3.7%	8. Suicide, 1.4%
9. Bronchitis, 2.6%	9. Chronic liver disease and cirrhosis,1.2%
10. Diphtheria, 2.3%	10. Human immunodeficiency virus infection, 1.2%

[a]Adapted from Hinman (15).
[b]Adapted from U.S. Department of Health and Human Services (41).

factors. Of particular relevance to this volume are indicators of health status related to risk factors, physical function, and bodily well-being (table 6.2). Genetic aspects of many of the specific measures are discussed in separate chapters.

The health status of populations in developed countries has improved considerably over the past century, as apparent in reduced infant and preschool mortality, taller stature and earlier maturation of children and youth, and increased longevity (4,17,18,27,41). These improvements in health status are due not only to changes in nutrition and environment, but also to advances in medicine and public health, to social changes, and to changes in public behavior.

The improved living conditions that underlie the positive changes in health status also permit enhanced opportunity to identify the contribution of genes to human diseases. Many nutritional diseases related to specific nutrients (e.g., rickets) showed a rapid decline when the nutrient involved was identified (vitamin D in the case of rickets) and made available to the population (e.g., vitamin D-enriched milk). However, new cases of rickets have continued to appear because of an inborn error of metabolism that is not fully compensated for by nutritional supplementation in some children (31). This is an example of changing heritability of human diseases as a consequence of changes in environmental conditions.

Advances in diagnostic medical technology have contributed to the detection of other diseases inherited in a Mendelian manner. This is evidenced in the rapid growth in the number of these diseases which have been described and catalogued (21,32). Although chromosomal anomalies and single-gene Mendelian diseases have a dramatic impact on the health and well-being of individuals, however, they have a relatively minor effect on the overall health status of the population because of their low prevalence. Common polygenic and multifactorial diseases have a greater impact on the health of the community than single-gene diseases. This is especially evident in worldwide variation in DALYs lost to noncommunicable diseases (figure 6.2).

There is increasing evidence that genetic factors are important determinants in some noncommunicable diseases. For example, Sorensen and colleagues (34) compared the risks of premature death in adult adoptees whose biologic or adoptive parent died of the same cause to the risks of adoptees whose parents were both alive (table 6.5). Death of a biologic parent before the age of 50 resulted in relative risks of death in the adoptees of 2.0 for all natural causes and 4.5 for vascular causes, compared to relative risks of 1.0 and 3.0, respectively, for death of an adoptive parent. The results suggest that genetic factors play a significant role in the development of some diseases leading to premature mortality.

Table 6.5
Effect of the Death of a Biologic and Adoptive Parent Before the Age of 50 on the Risk of Premature Mortality in Adoptees From Concordant Causes

Causes of death	Biologic parent	Adoptive parent
Natural causes	2.0	1.0
Vascular causes	4.5	3.0

Values are relative risks derived from the ratio of the cause-specific mortality rate among adoptees with at least one parent who died of the same cause to that among adoptees whose parents were both alive. Adapted from Sorensen et al. (34).

Similar trends are evident for various morbid conditions. In the Framingham Heart Study, for example, having an older brother with coronary heart disease was a significant risk factor for development of this disease in males even after smoking, total serum cholesterol, and systolic blood pressure were statistically controlled (33). The relationships between genetic elements and risk factors for common diseases and morbidity indicators are discussed in subsequent chapters.

Longevity and Aging

Improvement continues to be seen in infant and preschool mortality, life expectancy, and longevity; however, the secular trend toward taller height and earlier maturation has apparently stopped in the United States, Japan, and many western European countries (11,17). Children generally appear to be nearing their genetic potential for height and timing of maturation. In contrast, longevity continues to increase. For example, the number of individuals 85 years of age and older in the United States has increased by 232% from 1960 to 1990 (6). Three projections for the increase in the number of individuals 85 years of age and older in the United States from 1991 to 2050 are shown in figure 6.3. Note, for example, that the oldest members of the baby-boomer generation, those born in 1946, reach 65 years of age in 2011 and in 2031 the surviving members of this generation will reach 85 years of age. Some of the projections are based on mortality data (for the cohort that is already in middle age), while others are based on the assumption that improvements in disease prevention and treatment will continue (6).

The increase in longevity over time and the dramatic projections for the number of individuals 85 years of age and older begs the issue of the

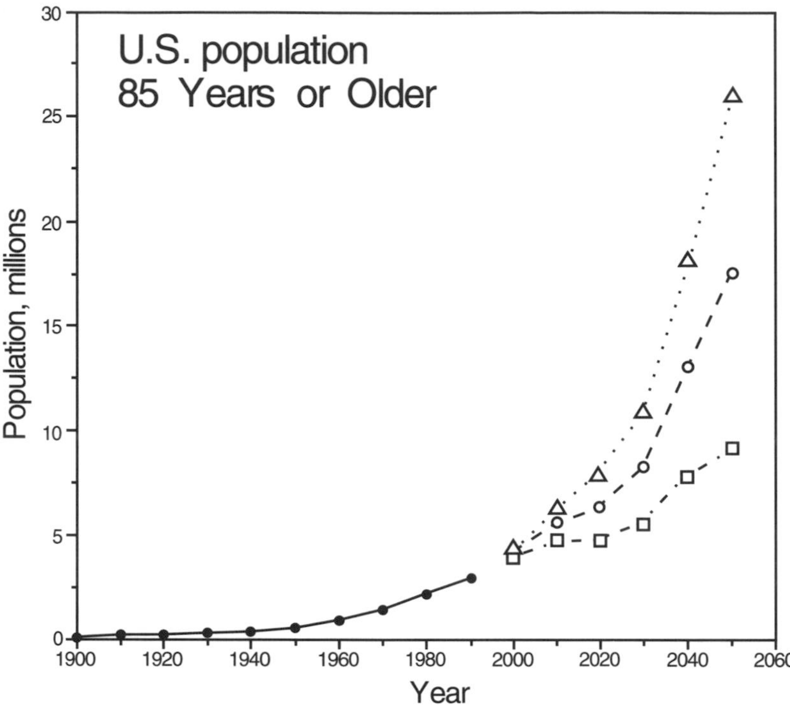

Figure 6.3 United States population 85 years of age or older. Data from 1900 to 1990 are based on census figures. Projections for 2000 to 2050 are based on the lowest (open squares), intermediate (open circles), and highest (open triangles) estimates using varying fertility, mortality, and immigration assumptions. Drawn from data reported by the U.S. Bureau of the Census (40).

inheritance of longevity. Family studies indicate a familial component to longevity, while twin studies suggest smaller intrapair life span differences between MZ than between DZ twins (4,28,38). Mean life duration differences within members of pairs of MZ and DZ twins, however, decrease with increased longevity so that at older ages (70+ in males, 80+ in females), differences in longevity between members of MZ and DZ pairs are small.

Rather than indicating genes associated with increased longevity, family and twin studies can be interpreted as emphasizing a role for deleterious genes that lead to premature death due to detrimental effects: single genes associated with lipid abnormalities, for example, or multiple genes associated with heart disease and some cancers. Brown offers the following perspective:

> People are likely to show greater homogeneity in their physical characteristics as they approach the upper limits of their natural life spans, because detrimental genes have been eliminated and they have been selected for superior health. (4, p. 918)

Although Brown deals specifically with twin data, results of family studies support similar conclusions for other genetic relationships.

Theories of aging can be classified into two categories: genetic program theories and wear and tear theories (4). Genetic theories are developmental (e.g., aging begins at conception) and emphasize modifications in gene structure and expression throughout the life span (20). Wear and tear theories, including stochastic theories, emphasize the consequences of accumulated insults over time. The insults may result in damage to DNA and subsequent alterations in cellular function. The underlying processes emphasized by all the theories of aging point to a significant role for complex gene-environment interactions at the molecular and cellular levels (10).

Syndromes of premature aging

Two rare conditions are characterized by premature or accelerated aging, the Hutchinson-Gilford and the Werner syndromes (4,19). The Hutchinson-Gilford syndrome, known as the progeria of childhood, has an incidence of about one in 8 million births and is inherited as a sporadic autosomal dominant. Individuals with the syndrome are normal at birth but show growth stunting during the first year with a low ratio of weight for length. They also present hair loss, thinning of the skin, loss of subcutaneous tissue, demineralization of bones, and severe atherosclerosis, among other features. The atherosclerosis is especially significant, for more than 80% of deaths in youngsters with the syndrome are due to myocardial infarction and congestive heart failure. These individuals do not mature sexually.

Werner syndrome is sometimes labeled as the progeria of the adult and may have its onset between 15 and 20 years, although patients are usually diagnosed in their 30s. The syndrome is inherited as an autosomal recessive. Estimated frequencies of Werner syndrome heterozygotes are about one to five per 1000 individuals and of homozygotes about 1 to 25 per one million people. Patients with Werner syndrome are characterized by short stature due to early cessation of growth, premature graying of the hair, hypogonadism, cataracts, peripheral muscular atrophy, osteoporosis, high prevalence of diabetes, and severe atherosclerosis, among others. Death usually occurs due to complications of atherosclerosis when patients are in their 40s (4,19).

Both syndromes clearly implicate the role of genes in aging processes per se and in degenerative conditions associated with early mortality. Cytogenetic analysis of MZ twins with Hutchinson-Gilford syndrome suggests that a gene for this progeria may be located on chromosome 1 (5), while several analyses (linkage, homozygosity mapping) of Werner syndrome place the locus for this condition on chromosome 8, specifically between D8S131 and D8S87 in an 8.3 cM interval containing D8S339 (14,22,30). The similarities and differences in the progress of the progeria syndromes and normal aging merit further study. If the processes are similar, they may suggest common pathways and perhaps lead to the identification of genes that regulate normal aging.

Gene associations with longevity

A multidisciplinary study of aging and longevity in a Mennonite community in Kansas and Nebraska considered the association between age and the degree of heterozygosity at eight red blood cell loci (9). Residents in the Mennonite communities are descendants of a group that immigrated to the United States from the Ukraine in 1874, although the religious sect was founded in Europe in the 1600s. Heterozygosity was high at the eight loci, 0.44 to 0.48, in the sample of 652 (360 females and 292 males). However, level of heterozygosity was not related to age and was not significantly different between younger and older subsamples.

Studies of specific gene variants in long-lived individuals may also provide insights into aging and longevity. Proust et al. (25) noted different frequencies of alleles at the HLA in long-lived individuals compared to a control sample of individuals 10 to 50 years of age in France. For example, there was an excess of the Cw1 antigen in females older than 90 years and an excess of the Cw7 in males of the same age group. Takata et al. (37), on the other hand, found a high frequency of HLA DR-1 and a low frequency of HLA DRw9 in a sample of 82 centenarians and 20 nonagenarians in Okinawa, Japan. The presence of HLA DRw9 in high frequency and HLA DR-1 in low frequency is associated with autoimmune diseases in this population, leading Takata et al. (37) to suggest that DRw9 is a risk factor and DR-1 is a favorable factor for longevity. The results, though interesting, are different in the French and Japanese samples, and emphasize the need for further study.

The frequencies of two genes related to risk for cardiovascular disease, apolipoprotein E (APOE) and the angiotensin converting enzyme (ACE), have been compared in case-control studies of centenarians and adults 20 to 70 years of age in France (29). The frequency of the APOE ε4 allele was significantly lower while that of the APOE ε2 allele was significantly higher in the centenarians than in the control subjects (table 6.6). In a prospective study of 666 elderly Finnish men, 65 to 84 years at baseline, 70 died of coronary heart disease. Compared to the survivors, there was an excess of the APOE ε4 allele among those who died, while there was no clear trend for APOE ε2 (35). The APOE ε4 allele is related to premature atherosclerosis, while the APOE ε2 allele is related to type III and IV hyperlipidemia. The results appear contradictory, although it is possible that the APOE ε2 allele has a protective effect.

At the ACE locus in humans, there is an insertion/deletion (I/D) polymorphism in intron 16. The DD genotype and in turn the D allele have significantly higher frequencies in centenarians (table 6.7), leading the authors to suggest that the D allele may have "some unknown long-term protective effect" (29). It should be emphasized that ACE has functions other than its role in the angiotensin system. ACE genotypes also vary among ethnic groups (1).

In a sample of 538 Dutch men 70 to 89 years of age, the frequency of the APOE ε4 allele was 13.9% (12), which is significantly higher than that observed in French centenarians but similar to that for French adults 20 to 70 years of age (table 6.8). Men with the APOE ε4 allele had higher concentrations of total serum cholesterol (6.31 ± 1.17 versus 6.01 ± 1.11 mmol/L) compared to men without the APOE ε4 allele. The two groups did not differ in the prevalence of hypertension, stroke, transient ischemic attack, and myocardial infarction, although men homozygous for APOE ε4 had a significantly higher prevalence (25.0%) of diabetes mellitus than men heterozygous for APOE ε4 (8.5%) and those without APOE ε4 (7.3%) (12).

In addition to its role in atherosclerosis, APOE has been associated with Alzheimer's disease (8,24). The Dutch study considered the possible relationship between alleles at the APOE locus and changes in cognitive function with age. A subsample of 378 men was followed longitudinally for three years. Results relating the APOE ε4 allele to changes in cognitive function over three years are summarized in table 6.8. Decline in cognitive function showed the largest decline in men homozygous for APOE ε4, an intermediate decline in men heterozygous for APOE ε4, and smallest decline for men without APOE ε4. The results also suggest that the presence of the APOE ε4 allele is associated with increased risk of developing impaired cognitive performance (12). More recently, in a study of 20 DZ twins 59 to 70 years of age (mean age 63 years), discordant for the APOE ε4 allele, the twin with the APOE ε4 allele scored poorer than the respective co-twin on several standard cognitive function tests (26). The results of these two studies suggest an

Table 6.6
Frequencies of APOE Genotypes and Alleles in French Centenarians ($n = 325$) and a Control Sample of Adults 20 to 70 Years of Age ($n = 161$)

Genotypes	Centenarians	Controls	Alleles	Centenarians	Controls
$\varepsilon3/\varepsilon3$	216	110	$\varepsilon2$	0.128	0.068
$\varepsilon2/\varepsilon3$	71	18	$\varepsilon3$	0.820	0.820
$\varepsilon3/\varepsilon4$	30	26	$\varepsilon4$	0.052	0.112
$\varepsilon2/\varepsilon2$	4	0			
$\varepsilon4/\varepsilon4$	0	3			
$\varepsilon2/\varepsilon4$	4	4			

Adapted from Schächter et al. (29).

Table 6.7
Frequencies of ACE Genotypes and Alleles in French Centenarians ($n = 338$) and a Control Sample of Adults 20 to 70 Years of Age ($n = 164$)

Genotypes	Centenarians		Controls		Alleles	Centenarians	Controls
	n	%	*n*	%			
DD	134	39.6	42	25.6	D	0.615	0.533
ID	148	43.8	91	55.5	I	0.385	0.467
II	56	16.6	31	18.9			

I and D, respectively, refer to insertion and deletion alleles in intron 16 of the ACE gene in humans.

Adapted from Schächter et al. (29).

Table 6.8
Changes in Cognitive Function Over Three Years in Dutch Men 70 to 89 Years According to the Presence or Absence of the APOE ε4 Allele

	Absent	Heterozygous	Homozygous
Total sample			
n	285	83	10
Changes in exam score[a]	–0.10	–0.68	–2.38
Subsample with normal cognitive function at baseline[b]			
n	207	51	7
n (%) with impaired cognitive performance	32 (15.5)	13 (25.5)	3 (42.9)
Adjusted odds ratio	1.0	2.38	12.53

[a]Based on the score on the mental examination of Folstein et al. (13) adjusted for age, occupation, cigarette smoking, alcohol use, total cholesterol, stroke, transient ischemic attack, myocardial infarction, diabetes mellitus, and hypertension.

[b]The maximum overall score on the mental list is 30. A score of 25 or less was considered indicative of reduced cognitive function.

Adapted from Feskens et al. (12).

association between presence of the APOEε4 allele and cognitive function in older men.

Summary

Although the concept of health is seemingly straightforward, measuring health status of individuals and populations is more complicated. Several indicators of health status have been described. Genetic effects in health status, though evident, are largely unspecified. Indicators of health status related to risk factors and physical function are considered in specific chapters of this volume. Of relevance to health status is the issue of longevity, which has increased considerably. Further, projections for the future point to the issue of the inheritance of longevity. At issue is whether there are specific genes for longevity, or whether deleterious genes lead to premature death. Several examples dealing with associations between genetic markers and longevity have been discussed. The markers are also associated with several pathological conditions. Association does not imply cause; nevertheless, the associations are consistent with the concept of gerontogenes.

References

1. Barley, J.; Blackwood, A.; Carter, N.D.; Crews, D.G.; Cruickshank, J.K.; Jeffery, S.; Ogunlesi, A.O.; Sagnella, G.A. Angiotensin converting enzyme insertion/deletion polymorphism: Association with ethnic origin. J. Hyper. 12:955-57; 1994.
2. Bouchard, C.; Shephard, R.J. Physical activity, fitness, and health: The model and key concepts. In: Bouchard, C.; Shephard, R.J.; Stephens, T., eds. Proceedings and consensus statement: Physical activity, fitness, and health. Champaign, IL: Human Kinetics; 1994:77-88.
3. Bouchard, C.; Shephard, R.J.; Stephens, T.; Sutton, J.R.; McPherson, B.D. Exercise, fitness, and health: The consensus statement. In: Bouchard, C.; Shephard, R.J.; Stephens, T.; Sutton, J.R.; McPherson, B.D., eds. Exercise, fitness, and health: A consensus of current knowledge. Champaign, IL: Human Kinetics; 1990:3-28.
4. Brown, W.T. Longevity and aging. In: King, R.A.; Rotter, J.I.; Motulsky, A.G., eds. The genetic basis of common diseases. Oxford: Oxford University Press; 1992:915-26.
5. Brown, W.T.; Abdenur, J.; Goonewardena, P.; Alemzadeh, R.; Smith, M.; Friedman, S.; Cervantes, C.; Bandyopadhyay, S.; Zaslav, A.; Kunaporn, S.; Serotkin, A.; Lifshitz, F. Hutchinson-Gilford progeria syndrome: Clinical, chromosomal and metabolic abnormalities. Am. J. Hum. Genet. [Abstract]. 47:A50; 1990.
6. Campion, E.W. The oldest old. N. Engl. J. Med. 330:1819-20; 1994.
7. Caspersen, C.J.; Powell, K.E.; Merritt, R.K. Measurement of health status and well-being. In: Bouchard, C.; Shephard, R.J.; Stephens, T., eds. Physical activity, fitness, and health: International proceedings and consensus statement. Champaign, IL: Human Kinetics; 1994:180-202.
8. Corder, E.H.; Saunders, A.M.; Strittmatter, W.J.; Schemchel, D.E.; Gaskell, P.C.; Small, G.W.; Roses, A.D.; Haines, J.L.; Pericak-Vance, M.A. Gene dose of apolipoprotein E type 4 allele and the risk of Alzheimer's disease in late onset families. Science. 261:921-23; 1993.
9. Crawford, M.H.; Rogers, L. Population genetic models in the study of aging and longevity in a Mennonite community. Soc. Sci. Med. 16:149-53; 1982.
10. Cristofalo, V.J.; Gerhard, G.S.; Pignolo, R.J. Molecular biology of aging. Surg. Clin. N. Am. 74:1-21; 1994.
11. Eveleth, P.B.; Tanner, J.M. Worldwide variation in human growth. 2nd ed. Cambridge: Cambridge University Press; 1990.
12. Feskens, E.J.M.; Havekes, L.M.; Kalmijn, S.; de Knijff, P.; Launer, L.J.; Kromhout, D. Apolipoprotein e4 allele and cognitive decline in elderly men. Brit. Med. J. 309:1202-6; 1994.

13. Folstein, M.F.; Folstein, S.E.; McHugh, P.R. Mini mental state. A practical method for grading the cognitive state of patients for the clinician. J. Psychiatr. Res. 12:189-98; 1975.
14. Goto, M.; Rubenstein, M.; Weber, J.; Woods, K.; Drayna, D. Genetic linkage of Werner's syndrome to five markers on chromosome 8. Nature. 355:735-38; 1992.
15. Hinman, R.A. 1889 to 1989: A century of health and disease. Public Health Rep. 105:374-80; 1990.
16. Last, J.M. A dictionary of epidemiology. 2nd ed. New York: Oxford University Press; 1988.
17. Malina, R.M. Research on secular trends in auxology. Anthropol. Anzeiger 48:209-27; 1990.
18. Malina, R.M. Secular changes in size and maturity: causes and effects. Monogr. Soc. Res. Child Dev. 44:59-102; 1979.
19. Martin, G.M. Genetics of human disease, longevity, and aging. In: Andres, R.; Bierman, E.L.; Hazzard, W.R., eds. Principles of geriatric medicine. New York: McGraw-Hill; 1985:397-412.
20. Martin, G.M. Genotypic theories of aging: An overview. Adv. Pathol. Aging Cancer Cell Membr. 7:5-20; 1980.
21. McKusick, V.A. Mendelian inheritance in man: Catalogs of autosomal dominant, autosomal recessive, and X-linked phenotypes. 10th ed. Baltimore: Johns Hopkins University Press; 1992.
22. Nakura, J.; Wijsman, E.M.; Miki, T.; et al. Homozygosity mapping of the Werner syndrome locus (WRN). Genomics. 23:600-608; 1994.
23. National Center for Health Statistics. Healthy people 2000 review, 1992. Hyattsville, MD: U.S. Department of Health and Human Services; 1993.
24. Poirier, J.; Davignon, J.; Bouthillier, D.; Kogan, S.; Bertrand, P.; Gauthier, S. Apolipoprotein E polymorphism and Alzheimer's disease. Lancet. 342:697-99; 1993.
25. Proust, J.; Moulias, R.; Fumeron, F.; Bekkhoucha, F.; Busson, M.; Schmid, M.; Hors, J. HLA and longevity. Tissue Antigens. 19:168-73; 1982.
26. Reed, T.; Carnelli, D.; Swan, G.E.; Breitner, J.C.S.; Welsh, K.A.; Jarvik, G.P.; Deeb, S.; Auwerx, J. Lower cognitive performance in normal older adult male twins carrying the apolipoprotein E e4 allele. Arch. Neurol. 51:1189-92; 1994.
27. Roche, A.F. Secular trends in stature, weight, and maturation. Monogr. Soc. Res. Child Dev. 44:3-27; 1979.
28. Schächter, F.; Cohen, D.; Kirkwood, T. Prospects for the genetics of human longevity. Hum. Genet. 91:519-26; 1993.
29. Schächter, F.; Faure-Delanef, L.; Guenot, F.; Rouger, H.; Froguel, P.; Lesueur-Ginot, L.; Cohen, D. Genetic associations with human longevity at the APO E and ACE loci. Nat. Genet. 6:29-32; 1994.
30. Schellenberg, G.D.; Martin, G.M.; Wijsman, E.M.; Nakura, J.; Miki, T.; Ogihara, T. Homozygosity mapping and Werner's syndrome. Lancet. 339:1002; 1992.

31. Scriver, C.R. Changing heritability of nutritional disease: Another explanation for clustering. World Rev. Nutr. Dietetics. 63:60-71; 1989.
32. Scriver, C.R.; Beaudet, A.L.; Sly, W.S.; Valle, D. The metabolic basis of inherited disease, volumes I and II. 6th ed. New York: McGraw-Hill; 1989.
33. Snowden, C.B.; McNamara, P.M.; Garrison, R.J.; Feinleib, M.; Kannel, W.B.; Epstein, F.H. Predicting coronary heart disease in siblings: A multivariate assessment. Am. J. Epidemiol. 115:217-22; 1982.
34. Sorensen, T.I.A.; Nielsen, G.G.; Andersen, P.K.; Teasdale, T.W. Genetic and environmental influences on premature death in adult adoptees. N. Engl. J. Med. 318:727-32; 1988.
35. Stengard, J.H.; Zerba, K.G.; Pekkanen, J.; Ehnholm, C.; Nissinen, A.; Sing, C.F. Apolipoprotein E polymorphism predicts death from coronary heart disease in a longitudinal study of elderly Finnish men. Circulation. 91:265-69; 1995.
36. Stern, M.P. The recent decline in ischemic heart disease mortality. Ann. Intern. Med. 91:630-40; 1979.
37. Takata, H.; Suzuki, M.; Ishii, T.; Sekiguchi, S.; Iri, H. Influence of major histocompatibility complex region genes on human longevity among Okinawan-Japanese centenarians and nonagenarians. Lancet. 2:824-26; 1987.
38. Turner, T.R.; Weiss, M.L. The genetics of longevity in humans. In: Crews, D.E.; Garruto, R.M., eds. Biological anthropology and aging: Perspectives on human variation over the life span. Oxford: Oxford University Press; 1994:76-100.
39. United Nations Development Program. Human development report 1993. Oxford: Oxford University Press; 1993.
40. U.S. Bureau of the Census. Sixty-five plus in America: Current population reports (special studies). Washington, DC: U.S. Government Printing Office, P23-178RV; 1993.
41. U.S. Department of Health and Human Services. Vital statistics of the United States 1990. Volume II. Mortality, Part A. Washington, DC: U.S. Government Printing Office, DHHS Publication No. (PHS) 95-1101; 1994.
42. World Bank. World development report 1993: Investing in health. New York: Oxford University Press; 1993.
43. World Health Organization. Constitution of the World Health Organization. Geneva: World Health Organization; 1948.
44. World Health Organization. New trends and approaches in the delivery of maternal and child care in health services. Sixth report of the WHO Expert Committee on Maternal and Child Health. Geneva: World Health Organization, WHO Technical Report Series, No. 600; 1976.

CHAPTER 7

Genetics of Metabolic Rate and Activity Level

The increased energy expenditure associated with regular physical activity may contribute to improved physical fitness. It may also contribute to improved functions of various systems of the human organism. Individual variation in energy expenditure related to physical activity is a factor in the etiology of several common diseases, including coronary heart disease, noninsulin dependent diabetes mellitus, hypertension, osteoporosis, and others (4). Variations in total daily energy expenditure are also correlated with the prevalence of obesity in industrialized societies.

Energy expenditure is a complex phenotype comprising several factors: basal and resting metabolic rates, thermic effect of food, and energy expenditure of specific activities. Children have the added energy cost associated with normal growth and maturation. Although only a few studies have addressed this issue, preliminary evidence indicates that genetic factors contribute to individual differences in energy expenditure.

Life Span and Gender Variability

Metabolic rate

Basal metabolic rate is commonly measured by indirect calorimetry early in the morning before the subject gets out of bed. Resting metabolic rate (RMR) is also measured in a fasted and quiet state, but in the laboratory. In most cases, the subject has walked or driven to the laboratory early in the morning for the assessment of RMR. A resting period of about 30 minutes is commonly imposed on the subject before the indirect calorimetry measurement of RMR begins.

Basal and resting metabolic rates increase as a function of body mass and are partly associated with the amount of fat-free mass

(FFM) in adults. When expressed per unit body weight, basal metabolic rate declines with age from infancy through adolescence into adulthood. It continues to decrease linearly with advancing age in adulthood (40). Longitudinal data demonstrate that basal metabolic rate per unit estimated FFM declines at a rate of about 1% to 2% per decade in men across the range from 20 to 75 years (21). Cross-sectional data for both men and women suggest that the decline in resting metabolic rate with advancing age is accelerated beyond the middle-aged years (33,34). Further, resting metabolic rate is, on average, lower in women than in men (1).

As for the thermic effect of food, available data do not permit description of possible age-related effects. No significant differences have been consistently observed between men and women for the thermic response to a standardized meal.

Activity level

Methods for estimating levels of habitual physical activity vary among studies and with age of the subjects. Activity levels of preschool children, for example, are derived from mechanical activity meters or motion sensors, parental ratings, and observations. The evidence indicates considerable variability among young children, although mean activity levels decline slightly from two to five years of age. On average, boys have higher activity levels than girls and there is some evidence that the sex difference is present before birth (12,13).

Data for activity levels of older children and youth are derived primarily from diaries, interviews, standardized questionnaires, and occasionally parental reports. Although absolute levels vary, several trends are apparent in habitual activity of North American and European youth. Estimated levels of physical activity generally increase from about five to six years of age into early adolescence and then decline. The decline in habitual physical activity is more apparent in later adolescence and is more evident in medium (7-10 metabolic equivalents [METs]) and heavy (10+ METs) activities. The late adolescent decline in activity is related, in part, to the social demands of adolescence and career choices in the transition from high school to work or college. Males are, on average, more active than females but also experience a greater decline in physical activity in late adolescence. Many surveys of activity levels of children and youth often overlook seasonal variation, which can be significant. The adolescent decline in activity occurs primarily in the summer, which for many youths is probably related to summer employment (24,25).

Among adults, cross-sectional surveys from different countries are consistent in reporting that leisure-time physical activity, on average,

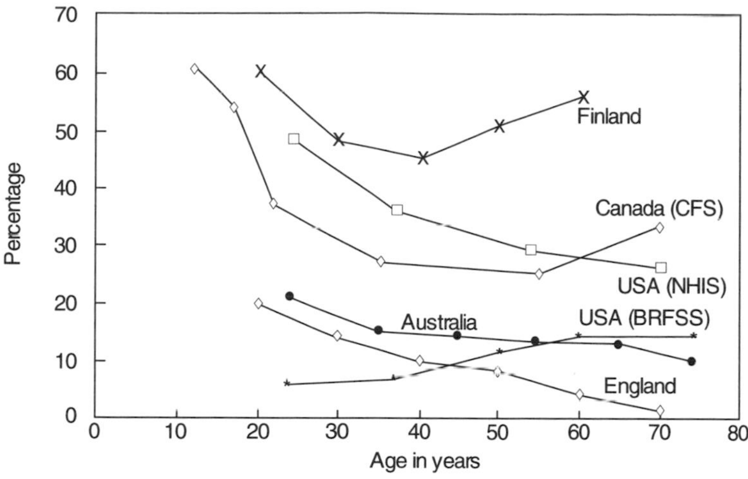

Figure 7.1 Age trends in leisure-time physical activity in different populations. The values represent percentage of active individuals (moderate and vigorous activities). Definitions vary between but are consistent within countries. CFS = Canada Fitness Survey; NHIS = National Health Interview Survey; BFRSS = Behavioral Risk Factor Surveillance System. Reprinted, by permission, from T. Stephens and C.J. Caspersen, 1994, The demography of physical activity. In *Physical activity, fitness, and health: International proceedings and concensus statement*, edited by C. Bouchard, R.J. Shephard, and T. Stephens (Champaign, IL: Human Kinetics), 208.

declines with age (41). As shown in figure 7.1, there is a reduction of about 10% to 20% in the fraction of the adult population (males and females between 20 and 40 years of age) that is moderately or vigorously active. The definition and quantification of activity levels vary among studies. The decline of activity level with age is not consistently observed throughout the life span in some populations. However, longitudinal data on age-related changes in activity during adulthood are not extensive. Those available for Canada (42) and the United States (27) are short term. Activity level also varies between the sexes; on average, males are more active than females, and the sex difference is greater for high-intensity activities than for activities of low and moderate intensity.

Genetics of Resting Metabolic Rate and Thermic Effect of Food

Resting metabolic rate (RMR) accounts, on average, for about 70% of the daily energy expenditure (36) and is influenced by age and sex (7).

Fat-free mass (FFM) is probably the single best determinant of variability in RMR (28,36). In large samples, FFM may account for as much as 50% to 80% of the total variance in RMR. Intraindividual variation in RMR is estimated at about 4% to 6%, a value that includes both normal day-to-day fluctuations in metabolic rate and technical errors associated with the assessment of RMR (3,19,36).

After adjustment for FFM, fat mass (FM), age, and sex, the residual variance in RMR remains important. The standard deviation of RMR per unit FFM over 24 hours reaches about 6 kcal in adults (5,7). After adjustment for FFM, FM, age, and sex, the standard deviation of predicted RMR reaches about ±140 kcal per day (2), which represents about ±10% of the RMR. From a review and meta-analysis of 15 samples from different studies, the standard deviation of predicted RMR adjusted for FFM is about 6 W or about 124 kcal per day (11). Thus, the 95% confidence intervals are about ±250 kcal per day. The variance in RMR that is not explained by age, sex, body mass, and body composition has implications for energy balance. These observations have been instrumental in developing the hypothesis that familial and genetic factors are associated with all or at least a fraction of the unexplained variance in RMR.

To test whether the unexplained variance in RMR aggregates in families, Bogardus et al. (2) measured RMR and body composition in 74 male and 56 female nondiabetic Pima Indians, 18 to 39 years of age, from 54 sibships. About 83% of the variance in RMR was accounted for by FFM, age, and sex. Family membership accounted for an additional 11% of the variance. These results suggest that genetic and/or shared familial factors—independent of FFM, age, and sex—contribute to individual differences in RMR.

Few attempts have been made to quantify the role of genetic factors in RMR. The heritability of RMR was estimated in the Quebec Family Study using data for parent-child pairs and MZ and DZ twins (table 7.1). Correlations within MZ twins are higher than those within DZ twins, whether RMR is expressed per kilogram of body weight or FFM (7,14). Heritability estimates based on twice the difference between MZ and DZ correlations or twice the parent-child correlations suggest that about 40% of the variance in RMR, after adjustment for age, sex, body mass, and body composition, may be inherited. More recently, familial aggregation of RMR was investigated in 121 nuclear families of the Quebec Family Study using path analysis procedures (Dériaz et al., unpublished results). After adjustment for FFM, FM, age, and sex, about 55% of the variance in RMR could be attributed to both genetic and common familial environmental factors.

Table 7.1
A Summary of Two Studies Dealing With Resting Metabolic Rate in Twin and Parent-Child Pairs

	DZ TWINS		MZ TWINS		PARENT-CHILD	
	n pairs	*r*	*n* pairs	*r*	*n* pairs	*r*
Fontaine et al. (14)						
RMR/kg	19	0.21	20	0.45*		
RMR/kg FFM	19	0.44**	20	0.64**		
Bouchard et al. (7)						
RMR/kg	21	0.38*	37	0.78***	31	0.13
RMR/kg FFM	21	0.30	37	0.77***	31	0.21
RMR adj FFM	21	0.38*	37	0.74***	31	0.23

*$p < .05$, ** $p < .01$, *** $p < .0001$

Few studies have considered the role of specific genes in the observed variation in RMR. The Alpha1, Alpha2, and Beta genes of Na,K ATPase were considered in one experiment conducted with a subset of the Quebec Family Study (10). Five genetic polymorphisms were studied, and none was associated or linked with RMR expressed variously. In another experiment conducted with members of 64 families from the Quebec Family Study, Oppert et al. (30) did not observe any association or linkage between a BclI restriction site mutation in the brown adipose tissue expressed uncoupling protein gene and RMR adjusted for the conventional concomitants.

The thermic effect of food (TEF) is the integrated increase in energy expenditure after food ingestion over the energy expended at rest before the meal. It is generally divided into an obligatory component associated with the energy cost of nutrient storage and a facultative component that has proven very difficult to estimate. Apparently only one study has dealt with the heritability of TEF (7). Energy expenditure was recorded during 4 hours after a 1,000-kcal carbohydrate meal in 21 pairs of DZ twins, 37 pairs of MZ twins, and 31 parent-offspring pairs. Correlations for TEF were 0.35, 0.52, and 0.30 in DZ, MZ, and parent-offspring pairs, respectively, suggesting a genetic effect of at least 30% and perhaps more for TEF. The biological significance of such a heritability level is highlighted by the fact that the standard deviation of TEF over 4 hours in this study reached about 20 kcal, and the 95% confidence intervals were ±40 kcal or ±4% of the energy intake.

Genetics of Activity Level

Studies on the genetic effect for activity level or the amount of energy expended for daily physical activity are not extensive, but they are more abundant than for the other components of energy expenditure. Activity level is often included as a component of personality, referring to how an individual goes about his or her daily activities. An active person is often characterized as one who is always "in a hurry" or "on the go" or one who conducts his or her activities in a vigorous manner. For example, in the theory of personality developed by Buss and Plomin (8,9), one component is activity level, since

> an active person moves around more, tends to be in motion, hurries more than others, and keeps busier than those around him. . . . His preferences in sports reflect his need for pulsating activity and a driving tempo. He likes tennis, handball, and squash best; these are followed by football, basketball, and volleyball; lower on the list are baseball, skating, swimming, and canoeing. Note that the best-liked sports involve not only a huge energy output but also bursts of exertion. Note also the sheer range of energy-depleting activities. (9, pp. 30-31)

Activity level is viewed in terms of vigor and tempo, which are perhaps alternative ways of expending energy. Twin studies and a few family studies indicate a variable pattern of genetic influence on activity level and moderate evidence for stability from early childhood into adulthood (9,15-17,23,26,35,39). A related component is fidgeting, the tendency for seemingly constant small bodily motions. Fidgeting may account for an important fraction of the 24-hour energy expenditure measured in a respiratory chamber (37). It must be emphasized, however, that activity level as defined in studies of temperament and personality is not the same as the pattern of habitual physical activity as used in studies of physical fitness and lifestyle. Nevertheless, an individual's temperament may be a significant determinant of activity interests and pursuits, which in turn influence overall daily energy expenditure.

General similarity of activity levels and patterns of children and their parents is commonly reported, but studies of familial aggregation of activity level and sports participation are relatively few. Level of habitual physical activity was investigated with a Caltrac accelerometer in 100 children, 4 to 7 years of age, and 99 mothers and 92 fathers from the Framingham Children's Study (29). Over the course of one year, data were obtained with the mechanical device for about 10 hours per day for an average of nine days in children and eight days in fathers and mothers. Active fathers (accelerometer counts per hour above the median) or active mothers were more likely to have active offspring than inactive fathers or mothers, with odds ratios of 3.5 and 2.0, respectively. When both parents were active, the children were 5.8 times more likely to be active as children of two inactive parents. These results are compatible with the notion that genetic and/or cultural factors transmitted across generations may predispose a child to be active or inactive.

Participation in sports activities may also be influenced by genetic factors. In a study based on 1,294 families including both parents and 1,587 pairs of MZ and DZ twins, Koopmans et al. (22) reported an estimated heritability of 45% for sports participation. The remaining phenotypic variance was attributed to shared familial environment (44%) and environmental factors unique to each individual (11%).

The influence of heredity on level of physical activity was also investigated in a large sample of twins from the Finnish Twin Registry (20). Leisure-time physical activity was assessed in 1,537 MZ and 3,057 DZ male twin pairs over 18 years of age. Information on intensity and duration of activity, years of participation in an activity, physical activity on the job, and subjective opinion of the subject's own activity level was obtained from questionnaires. These variables were submitted to factor analysis to generate a factor score of physical activity that was used to compute correlations within MZ and DZ twin pairs. The results indicated an estimated heritability of 62% for age-adjusted physical activity level. More recently, Heller et al. (18) reported a significantly higher concordance within 94 pairs of MZ twins than within 106 pairs of DZ twins for participation in vigorous exercise in a defined two-week period and estimated the heritability of this phenotype at about 39%.

Familial resemblance in leisure-time energy expenditure was estimated in data from the 1981 Canada Fitness Survey (31). A total of 18,073 individuals living in households across Canada completed a questionnaire on physical activity habits. Detailed information on the frequency, duration, and intensity of activities performed on a daily, weekly, monthly, and yearly basis was used to estimate average daily energy expenditure (per kilogram of body weight) for each individual. Familial correlations were 0.28, 0.12, and 0.21 for spouses (n = 1,024 pairs), parents and offspring (n = 1,622 pairs), and sibling pairs (n = 1,036), respectively. The low correlations suggest only a small contribution of genetic factors in the familial aggregation of leisure-time energy expenditure.

The relative contribution of genetic and nongenetic factors to activity level was also estimated in the Quebec Family Study, using path analysis procedures (32). Two different indicators of physical activity, habitual physical activity, and participation in moderate to vigorous physical activity were obtained from a three-day activity record completed by 1,610 members from 375 families encompassing nine types of relatives by descent or adoption. Each day of the activity record was divided into 96 periods of 15 minutes, and for each 15-minute period the individuals noted on a scale from 1 to 9 the dominant activity. Each score represented approximately a multiple of the resting metabolic rate. A score of 1 indicated that the individual was at rest, while a score of 9 indicated that the individual was doing an activity with an energy cost equivalent to about 8 times the resting metabolic rate (6). The scores 1 to 9 were summed over the 96 fifteen-minute periods of each day, and the average value for the three days was used as the indicator of habitual physical activity. Participation in moderate to vigorous physical activities was defined as the average number of periods rated 6, 7, 8, or 9 in each day. Results are summarized in figure 7.2.

Most of the variation in the two indicators of habitual physical activity level was accounted for by nontransmissible environmental factors, with values reaching 71% for habitual physical activity and 88% for exercise participation. The transmission effect across generations was also significant. The estimate for habitual physical activity was 29%, and it was entirely attributable to genetic factors. The corresponding estimate for participation in moderate to vigorous physical activity was

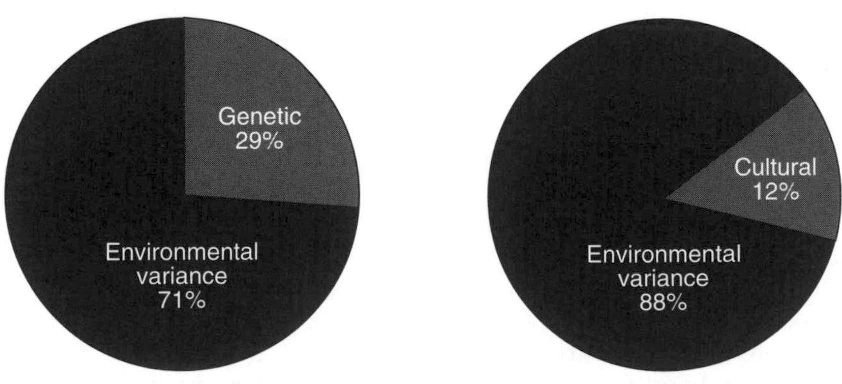

Figure 7.2 Contribution of genetic and nongenetic factors in habitual physical activity *(left)* and exercise participation *(right)*. Data from Pérusse et al. (32).

12%, and it was accounted for by cultural transmission with no genetic effect. Since habitual physical activity was computed as the sum of all scores (i.e., categories 1 through 9), and participation in moderate to vigorous activity included only activities in categories 6 through 9, activities in categories 1 through 5, that is, low-intensity activities, were characterized by the significant genetic effect. The results were thus interpreted as an indication of inherited differences in a propensity to be spontaneously active or inactive (32).

The energy cost associated with common body postures or positions (sitting or standing) and with low-intensity activities (walking, bending, reaching, etc.) repeated several times in the course of a normal day may be an important determinant of total daily energy expenditure. To test the hypothesis that such differences could be partly explained by genetic factors, the oxygen uptake required for various submaximal cycle ergometer workloads was measured in 22 pairs of male sedentary DZ twins and 31 pairs of male sedentary MZ twins aged 16 to 29 years (7). The correlations shown in table 7.2 suggest a significant genetic effect for oxygen uptake per unit body weight at low power output (50, 75, and 100 W). However, this effect became nonsignificant when energy expenditure reached about six times the resting metabolic rate. It is thus possible that there are inherited differences in the energy cost of low-intensity activities even after adjustment for variation in body mass. The significance of this phenomenon for 24-hour energy expenditure remains to be investigated.

Only one study has considered familial aggregation of 24-hour energy expenditure. Ravussin et al. (38) measured total daily energy expenditure in 94 siblings (52 males, 42 females) from 36 American

Table 7.2
Intraclass Correlations for Energy Cost of Submaximal Steady-State Cycle Ergometer Exercise

Power output (W)	DZ twins $\dot{V}O_2$/kg	MZ twins $\dot{V}O_2$/kg
50	0.22	0.67**
75	0.33*	0.72**
100	0.59**	0.82**
125	0.68**	0.79**
150	0.66**	0.75**

*$p < .05$, ** $p < .001$

Adapted from Bouchard et al. (7)

Indian families. After adjustment for age, sex, and body composition, there was significant familial aggregation of 24-hour energy expenditure with an intraclass familial correlation of 0.26.

Summary

Level of habitual physical activity is a significant predictor for several morbid conditions and even for mortality rate. Level of habitual physical activity and total daily energy expenditure are determinants of body fat content and body weight gain over time. For resting metabolic rate, thermic response to food, energy expenditure of activity, and total daily energy expenditure, there are a few family and twin studies suggesting the presence of familial aggregation for all of the phenotypes. Estimates of genetic heritability for these traits, when available, tend to be in the low range with only a few exceptions. Association and linkage studies between candidate genes and resting metabolic rate (adjusted for age, sex, body mass, and body composition) have been negative. No such studies are available for the other components of energy expenditure.

References

1. Arciero, P.H.; Goran, M.I.; Poehlman, E.T. Resting metabolic rate is lower in women than in men. J. Appl. Physiol. 75:2514-20; 1993.
2. Bogardus, C.; Lillioja, S.; Ravussin, E.; Abbott, W.; Zawadzki, J.K.; Young, A.; Knowler, W.C.; Jacobowitz, R. Familial dependence of the resting metabolic rate. N. Engl. J. Med. 315:96-100; 1986.
3. Bouchard, C. Reproducibility of body composition and adipose tissue measurements in humans. In: Roche, A.R., ed. Body composition assessment in youth and adults. Report of the Sixth Ross Conference on Medical Research. Columbus, OH: Ross Laboratories; 1985:9-14.
4. Bouchard, C.; Shephard, R.J.; Stephens, T. Physical activity, fitness, and health: International proceedings and consensus statement. Champaign, IL: Human Kinetics; 1994.
5. Bouchard, C.; Tremblay, A. Genetic effects in human energy expenditure. Int. J. Obes. 14:49-58; 1990.
6. Bouchard, C.; Tremblay, A.; Leblanc, C.; Lortie, G.; Savard, R.; Thériault, G. A method to assess energy expenditure in children and adults. Am. J. Clin. Nutr. 37:461-67; 1983.
7. Bouchard, C.; Tremblay, A.; Nadeau, A.; Després, J.P.; Thériault, G.; Boulay, M.R.; Lortie, G.; Leblanc, C.; Fournier, G. Genetic effect in resting and exercise metabolic rates. Metabolism. 38:364-70; 1989.
8. Buss, A.H.; Plomin, R. Temperament: Early developing personality traits. Hillsdale, NJ: Erlbaum; 1984.

9. Buss, A.H.; Plomin, R. A temperament theory of personality development. New York: Wiley; 1975.
10. Dériaz, O.; Dionne, F.; Pérusse, L.; Tremblay, A.; Vohl, M.C.; Côté, G.; Bouchard, C. DNA variation in the genes of the Na, K-adenosine triphosphatase and its relation with resting metabolic rate, respiratory quotient, and body fat. J. Clin. Invest. 93:838-43; 1994.
11. Dériaz, O.; Fournier, G.; Tremblay, A.; Després, J.P.; Bouchard, C. Lean-body-mass composition and resting energy expenditure before and after long-term overfeeding. Am. J. Clin. Nutr. 56:840-47; 1992.
12. Eaton, W.O.; Enns, L.R. Sex differences in human motor activity. Psychol. Bull. 100:19-29; 1983.
13. Eaton, W.O.; Yu, A.P. Are sex differences in child motor activity level a function of sex differences in maturational status? Child Dev. 60:1005-11; 1988.
14. Fontaine, E.; Savard, R.; Tremblay, A.; Després, J.P.; Poehlman, E.; Bouchard, C. Resting metabolic rate in monozygotic and dizygotic twins. Acta Genet. Med. Gemellol. 334:41-47; 1985.
15. Frischeisen-Kohler, I. The personal tempo and its inheritance. Charact. Person. 1:301-13; 1933.
16. Goldsmith, H.H. Genetic influences on personality from infancy to adulthood. Child Dev. 54:331-55; 1983.
17. Goldsmith, H.H. Studying temperament via construction of the toddler behavior assessment questionnaire. Child Dev. 67:218-35; 1996.
18. Heller, R.F.; O'Connel, D.L.; Roberts, D.C.K.; Allen, J.R.; Knapp, J.C.; Steele, P.L.; Silove, D. Lifestyle factors in monozygotic and dizygotic twins. Genet. Epidemiol. 5:311-21; 1988.
19. Henry, C.J.K.; Hayter, J.; Rees, D.G. The constancy of basal metabolic rate in free-living subjects. Eur. J. Clin. Nutr. 43:727-31; 1989.
20. Kaprio, J.; Koskenvuo, M.; Sarna, S. Cigarette smoking, use of alcohol, and leisure-time physical activity among same-sexed adult male twins. In: Progress in clinical and biological research. New York: Alan R. Liss; 1981:37-46.
21. Keys, A.; Taylor, H.L.; Grande, F. Basal metabolism and age of adult man. Metabolism. 22:579-87; 1973.
22. Koopmans, J.R.; Van Doornen, L.J.P.; Boomsma, D.I. Smoking and sports participation. In: Goldbourt, U.; de Faire, U.; Berg, K., eds. Genetic factors in coronary heart disease. Dordrecht, Netherlands: Kluwer Academic; 1994:217-35.
23. Loehlin, J.C. Heredity, environment, and the Thurstone temperament schedule. Behav. Genet. 16:61-73; 1986.
24. Malina, R.M. Growth, exercise, fitness, and later outcomes. In: Bouchard, C.; Shephard, R.J.; Stephens, T.; Sutton, J.R.; McPherson, B.D., eds. Exercise, fitness, and health: A consensus of current knowledge. Champaign, IL: Human Kinetics; 1990:637-53.
25. Malina, R.M. Physical activity and fitness of children and youth: Questions and implications. Med. Exerc. Nutr. Health. 5:125-37; 1995.

26. Matheny, A.P. A longitudinal twin study of stability of components from Bayley's Infant Behavior Record. Child Dev. 54:356-60; 1983.
27. Merritt, R.K.; Caspersen, C.J. Trends in physical activity patterns among young adults: The behavioral risk factor surveillance system, 1986-1990. Med. Sci. Sports Exerc. [Abstract]. 24:S26; 1992.
28. Miller, A.T.J.; Blyth, C.S. Lean body mass as a metabolic reference standard. J. Appl. Physiol. 5:311-16; 1953.
29. Moore, L.L.; Lombardi, D.A.; White, M.J.; Campbell, J.L.; Oliveria, S.A.; Ellison, S.A. Influence of parents' physical activity levels on young children. J. Pediatr. 118:215-19; 1991.
30. Oppert, J.M.; Vohl, M.C.; Chagnon, M.; Dionne, F.T.; Cassard-Doulcier, A.M.; Ricquier, D.; Pérusse, L.; Bouchard, C. DNA polymorphism in the uncoupling protein (UCP) gene and human body fat. Int. J. Obes. 18:526-31; 1994.
31. Pérusse, L.; Leblanc, C.; Bouchard, C. Inter-generation transmission of physical fitness in the Canadian population. Can. J. Sport Sci. 13:8-14; 1988.
32. Pérusse, L.; Tremblay, A.; Leblanc, C.; Bouchard, C. Genetic and environmental influences on level of habitual physical activity and exercise participation. Am. J. Epidemiol. 129:1012-22; 1989.
33. Poehlman, E.T.; Burke, E.M.; Joseph, J.R.; Gardner, A.W.; Katzman-Rooks, S.M.; Goran, M.I. Influence of aerobic capacity, body composition, and thyroid hormones on the age-related decline in resting metabolic rate. Metabolism. 41:915-21; 1992.
34. Poehlman, E.T.; Goran, M.I.; Gardner, A.W.; Ades, P.A.; Arciero, P.J.; Katzman-Rooks, S.M.; Montgomery, S.M.; Toth, M.J.; Sutherland, P.I. Determinants of decline in resting metabolic rate in aging females. Amer. J. Physiol. 264:E450-55; 1993.
35. Price, R.A.; Vandenberg, S.G.; Lyer, H.; Williams, J.S. Components of variation in normal personality. J. Pers. Soc. Psychol. 43:328-34; 1982.
36. Ravussin, E.; Bogardus, C. Relationship of genetics, age, and fitness to daily energy expenditure and fuel utilisation. Am. J. Clin. Nutr. 49:968-75; 1989.
37. Ravussin, E.; Lillioja, S.; Anderson, T.E.; Christin, L.; Bogardus, C. Determinants of 24-hour energy expenditure in man: Methods and results using a respiratory chamber. J. Clin. Invest. 78:1568-78; 1986.
38. Ravussin, E.; Lillioja, S.; Knowler, W.C.; Christin, L.; Freymond, D.; Abbott, G.H.; Boyce, V.; Howard, B.V.; Bogardus, C. Reduced rate of energy expenditure as a risk factor for body weight gain. N. Engl. J. Med. 318:467-72; 1988.
39. Scarr, S. Genetic factors in activity motivation. Child Dev. 37:663-73; 1966.
40. Shock, N.W. Metabolism and age. J. Chron. Dis. 2:687-703; 1955.
41. Stephens, T.; Caspersen, C.J. The demography of physical activity. In: Bouchard, C.; Shephard, R.J.; Stephens, T., eds. Physical activity, fitness, and health: International proceedings and consensus statement. Champaign, IL: Human Kinetics; 1994:204-13.
42. Stephens, T.; Craig, C.L. The well-being of Canadians: Highlights of the 1988 Campbell's Survey. Ottawa: Canadian Fitness and Lifestyle Research Institute; 1990.

CHAPTER 8

Genetics of Dietary Intake

Physical activity increases the utilization of energy and several nutrients, and the ability to perform exercise depends on adequate replacement of these nutrients in the diet. The major nutrients of interest in this context are those used as sources of energy to produce work, that is, carbohydrates, lipids, and proteins. Water, vitamins, and minerals are also important. Water is essential for thermoregulation. Vitamins and minerals are primarily regulatory in function and serve as cofactors in biochemical reactions responsible for production of energy, transport and consumption of oxygen, and maintenance of cellular integrity. Many minerals have an additional structural function, such as the mineral content of the skeleton and iron content of red blood cells.

It has been proposed that, under normal conditions, the healthy individual lives within a "nutritional window" representing minimal requirements and maximal tolerance for nutrients (41). These nutritional thresholds are affected by several factors: age, sex, body size, and growth; physiological states such as pregnancy, lactation, and illness; culturally mediated dietary habits such as vegetarianism and cult diets; environmental factors such as temperature and altitude; and level of physical activity. Genetic factors also contribute to individual differences in nutrient needs and nutrient intake. Inherited differences in the regulation of appetite and satiety, various metabolic pathways involved in food processing, and nutrient storage probably play important roles in determining individual differences in energy and in nutrient intakes. This chapter treats the contribution of genetic factors to dietary intake, taste preferences, and eating behaviors.

Life Span and Gender Variability

Daily energy and nutrient intakes vary with age and sex. Energy and macronutrient intakes based on dietary data available from Phase 1 of the Third National Health and Nutrition Examination Survey (NHANES

III) conducted between 1988 and 1991 are presented in figure 8.1 and tables 8.1 and 8.2 (8,32). The data were obtained from a 24-hour dietary recall in a population of 7,332 males and 7,479 females. Three race/ethnicity groups were included in the survey: those of European, African, and Mexican ancestry. Since energy and nutrient intakes were generally similar among the groups, data are pooled for the total population.

Mean energy intake increases from birth to adolescence (more so in males than females), reaches a plateau during late adolescence and young adulthood, and declines thereafter. Males consistently exhibit higher average intakes than females of the same age (figure 8.1). The same age and sex trends are also apparent for absolute intakes of carbohydrate, protein, and fat (tables 8.1 and 8.2). The dietary pattern remains quite stable throughout the life span in both sexes, with estimates of about 50%, 35%, and 15% of energy intake derived from carbohydrate, fat, and protein, respectively. The only exception to this pattern occurs in the first year of life, during which a slightly higher amount of energy is derived from fat (37%) and a lower amount is derived from protein (11%).

Genetics of Dietary Intake

It is widely accepted that total energy and nutrient intakes aggregate in families, with significant resemblance among spouses and between

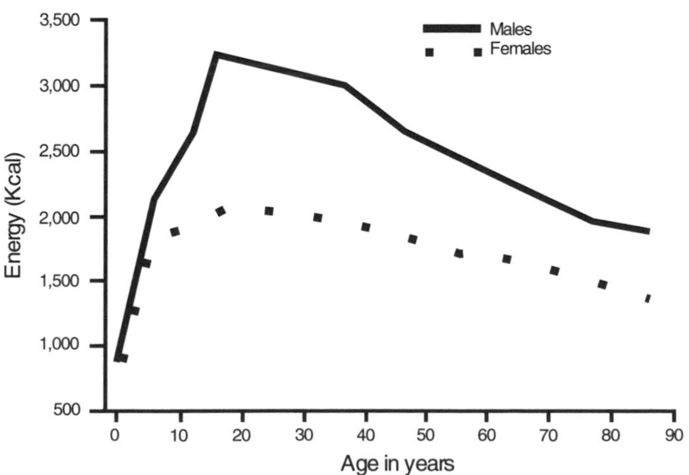

Figure 8.1 Average energy intakes in males and females by age in the U.S. population. From Briefel et al. (8).

Table 8.1
Nutrient Intakes of Males Aged Two Months and Over

	CARBOHYDRATE		PROTEIN		FAT	
Age	**g**	**%**	**g**	**%**	**g**	**%**
2-11 months	119	52.7	27	11.8	37	36.9
1-2 years	176	53.2	50	15.0	51	33.5
3-5 years	225	54.8	59	14.3	62	32.8
6-11 years	272	53.5	71	14.2	78	33.9
12-15 years	340	54.0	89	14.2	97	33.1
16-19 years	381	49.6	111	14.4	120	34.6
20-29 years	353	47.6	110	14.6	116	34.0
30-39 years	335	47.4	106	15.1	113	34.6
40-49 years	298	46.9	96	15.6	98	33.9
50-59 years	266	46.3	93	16.1	95	35.9
60-69 years	253	48.7	84	16.4	80	33.3
70-79 years	231	49.4	74	16.0	73	33.8
80 years and over	225	51.2	69	16.0	67	33.3

Data are from NHANES III (McDowell et al., 32).

parents and their children (4,17,27,29,30,34). For example, intakes of energy and several nutrients in 170 parents (mean age 33 years) and 91 children (mean age 4 years) in the Framingham Children's Study showed significant correlations between spouses and between parents and offspring (34). Correlations were generally higher for mother-child than for father-child pairs. In order to evaluate the influence of parental dietary habits on those of their offspring, odds ratios were computed for intake of a given nutrient by the child according to the level of consumption by one or both parents (table 8.3). Children with one parent eating high amounts of total fat, saturated fatty acids, and cholesterol were about two times more likely to consume a diet rich in these nutrients compared to children both of whose parents had low intakes of these nutrients (the reference group). When nutrient intake was high in both parents, the probability of having high nutrient intake in the child was about three to six times higher than the reference group. These results suggest that parents have a profound influence on the dietary habits of their children.

Table 8.2
Nutrient Intakes of Females Aged Two Months and Over

	CARBOHYDRATE		PROTEIN		FAT	
Age	**g**	**%**	**g**	**%**	**g**	**%**
2-11 months	112	52.4	25	11.2	35	37.6
1-2 years	163	53.0	45	14.9	47	34.0
3-5 years	204	54.4	54	14.3	57	33.1
6-11 years	229	52.9	63	14.5	68	34.2
12-15 years	243	54.4	62	13.5	72	33.7
16-19 years	254	52.4	67	14.1	77	34.4
20-29 years	241	49.9	69	14.5	75	34.0
30-39 years	228	49.7	70	15.3	75	34.2
40-49 years	213	49.0	67	15.8	70	34.9
50-59 years	199	49.8	64	16.1	63	33.8
60-69 years	199	51.1	64	16.6	59	32.8
70-79 years	185	52.4	58	16.6	53	32.3
80 years and over	179	54.5	52	15.9	47	31.3

Data are from NHANES III (McDowell et al., 32).

Table 8.3
Influence of High Nutrient Intake in Parents on the Intake of Their Children

Parents' intake	Odds ratio on child intake
Total fat	
One parent	2.8
Both parents	2.8
Saturated fatty acid	
One parent	2.3
Both parents	5.5
Cholesterol	
One parent	1.7
Both parents	6.3

High- and low-intake groups were defined by the median of the distribution of parent and child intakes.

Adapted from Oliveria et al. (34)

There are also familial resemblances in food preferences. In a typical study (39), ratings of 22 food items were obtained from college students and their parents. Mother-child and father-child correlations were computed for each food. The average correlation for all food items was 0.20 in mother-child pairs but only 0.04 in father-child pairs. Other studies, recently reviewed by Cavalli-Sforza (10), reveal low to moderate parent-offspring correlations for food preferences, but a stronger maternal effect in the majority of studies. The data thus suggest significant familial aggregation in energy and nutrient intakes and in food preferences. The question of interest is to what extent are the familial resemblances inherited? This question has been addressed in only few studies.

Twin studies of energy intake and food preferences (18-20,24,28,40,49) generally show greater similarities in dietary intakes and food preferences between MZ twins than between DZ twins, suggesting that genetic factors might contribute to interindividual differences in dietary habits. However, the findings must be interpreted with caution. For example, Heller et al. (24) reported higher correlations in 103 MZ twin pairs than in 94 DZ twin pairs for 12 of 13 dietary variables measured using four-day weighed food diaries. Using twice the difference between MZ and DZ twin correlations as an estimate of heritability, the resulting estimates ranged from 3% to 55%. The genetic variance was statistically significant only for intake of complex carbohydrates (h^2 = 0.55). Heller et al. (24) reported that MZ twins tended to socialize together and to eat together more regularly than DZ twins, suggesting that the greater similarity in energy intake within MZ twin pairs could be explained in part by greater environmental similarity rather than by genetic factors.

Another strategy to assess the role of heredity, independent of possible confounding effects of cultural transmission, is the use of family data that incorporate several types of biological and adopted relatives. This approach was used in the Quebec Family Study in which 1,597 individuals from 375 families (including biological and adopted siblings and MZ and DZ twins) completed a 3-day dietary record. Total energy intake and intakes of carbohydrate, lipid, and protein were estimated (37). There was significant familial aggregation for energy and nutrient intake. Familial correlations for nine different types of relatives by descent or adoption revealed that individuals sharing the same household environment, whether genetically related or not, were similar in estimated energy intake, results that are compatible with a weak genetic contribution. All of the correlations were subsequently considered in a path analysis model, which assumed that the phenotypic variance resulted from the linear and additive effects of genetic and cultural transmission as well as from a residual component that included nontransmissible environmental factors. Results are summarized in figure 8.2. Most of the observed variation in estimated total energy intake (per kilogram of weight) and for estimated absolute intakes of carbohydrate, fat, and protein was accounted for by nontransmissible environmental factors (about 70%), while transmissible variance was entirely explained by cultural inheritance (about 30%) with no significant genetic effect (37). Estimated heritabilities ranged from 0% to 3%.

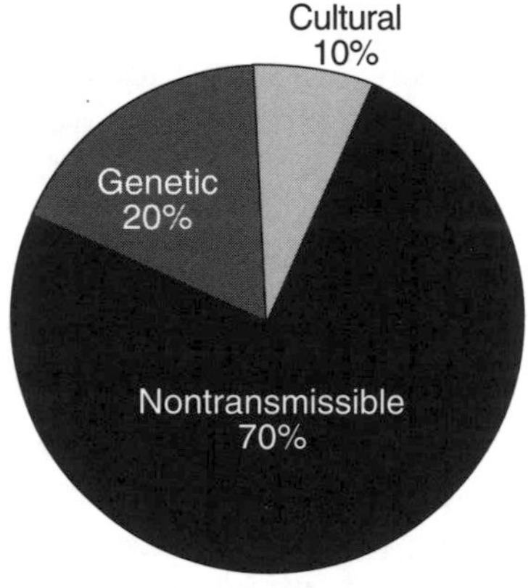

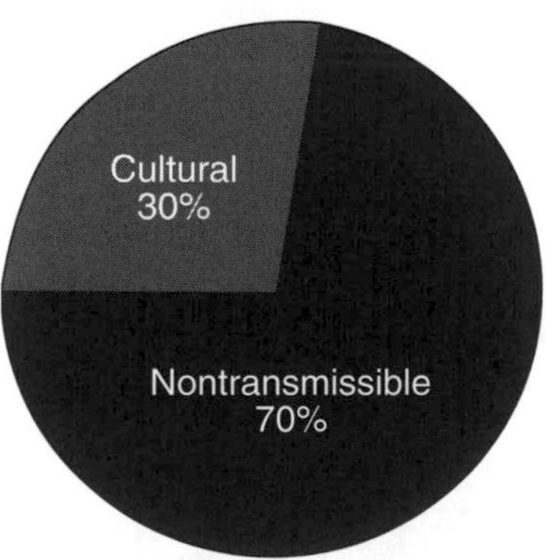

Figure 8.2 Fraction of the total phenotypic variance in energy and macronutrient intakes (*bottom*) and in percent of energy derived from carbohydrate and fat (*top*) attributable to nontransmissible, cultural, and genetic factors. Adapted from Pérusse et al. (37).

The estimated fractions of total energy intake derived from several macronutrients was also computed. When expressed as a percentage of total energy intake, intakes of carbohydrate and fat were characterized by a genetic effect of 20%, while cultural transmission and nontransmissible environmental effects explained 10% and 70% of the phenotypic variance, respectively (figure 8.2). For the percentage of energy derived from protein, the genetic effect reached 11%.

If it is hypothesized that the proportion of macronutrients in the diet reflects preferential choice for certain types of food, the results suggest that food preferences may be partly under genetic control. The existence of possible paternal or maternal effects was also tested, but there was no evidence for any specific paternal or maternal influence in the transmission of the intake of various energy components. The results of this family study thus suggest that energy and nutrient intakes and food preferences are largely influenced by nontransmissible environmental factors and cultural transmission, but that food preferences may be influenced by genetic factors to a small extent.

A very low heritability for energy intake is not surprising if the large day-to-day variation in energy intake for a given individual is considered. Total energy and nutrient intakes for seven consecutive days in 329 Hawaiian men of Japanese ancestry (33) are summarized in table 8.4. The within-individual variation in estimated energy intake reached 502 kcal for an average energy intake of 2,299 kcal. There was similar within-individual variation for protein, carbohydrate, and fat intakes and for alcohol intake. The overall energy intake of a given individual thus varied considerably during the one week of observation.

In an attempt to characterize within-individual variation in energy intake, Tarasuk and Beaton (44) recently measured the energy intakes of 13 men (22-49 years) and 16 women (20-53 years) for 365 consecutive days. There was, as expected, considerable day-to-day variation in energy intake for a given subject over the year. An example is illustrated in figure 8.3. For this subject, the cubic regression accounted for 16% of the variance in energy intake over the year with an average estimated standard deviation of 153 kcal. The standard deviations of the predicted intakes ranged from 77 to 280 kcal, illustrating the magnitude of variation in energy intake over a one-year period. Despite the apparently random day-to-day variation in energy intake, systematic long-term and weekly patterns of energy intake were detected for all but one subject. On average, these patterns explained 12.5% (0% to 37%) of the total variance in energy intake within an individual across the whole year. The shape and amplitude of the patterns were, however, specific to each individual. It would be interesting to verify whether such

Table 8.4
Mean and Within-Individual Variability of Energy and Nutrient Intakes

	Mean of 7 days	Within-person SD
Energy intake (kcal)	2,299	502
Protein (g)	94	27
Carbohydrates (g)	263	63
Fat (g)	85	29
Alcohol (g)	15	19

Data from 329 Hawaiian men of Japanese ancestry measured for seven consecutive days. Adapted from McGee et al. (33)

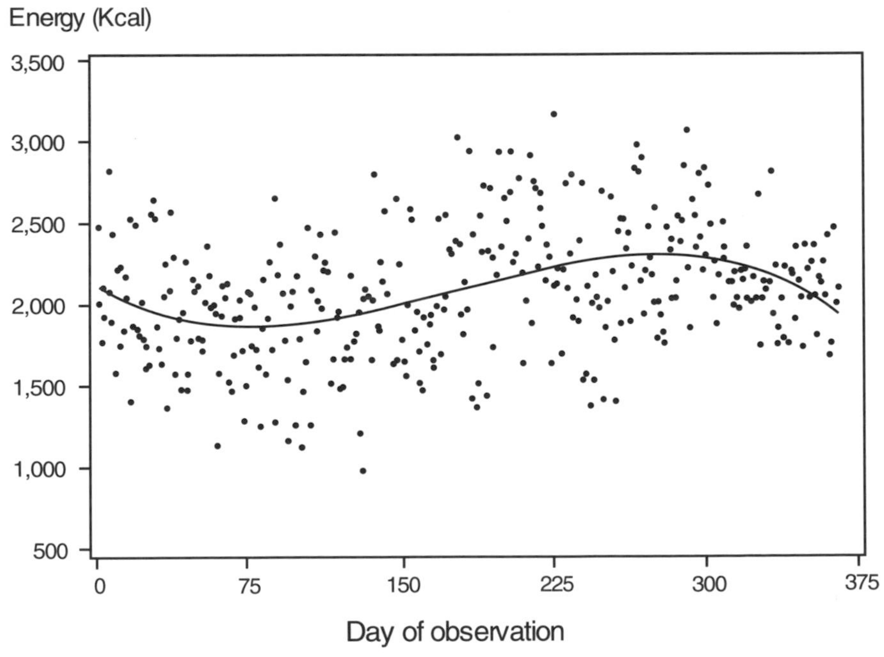

Figure 8.3 Trend in mean energy intake of one subject over one year. The cubic regression curve accounts for 16% of the subject's variance in energy intake over the year. Adapted from Tarasuk and Beaton (44).

patterns are also evident for macronutrient intakes and to test whether they are influenced by genetic factors.

Taste Preferences

Taste preferences are a major determinant of food selection and intake in humans (42) and have been linked with obesity and weight gain (15,16). Genetic differences in taste preferences may partly underlie the estimated heritabilities (about 20%) for carbohydrate and lipid food preferences (see previous discussion).

There are individual differences in responses to a variety of taste stimuli (bitterness, sourness, or sweetness). Sensitivity to phenylthiocarbamide (PTC), a bitter substance detected by some individuals (tasters) but not by others (nontasters), is inherited as an autosomal recessive trait. Studies have shown different proportions of tasters and nontasters among populations, with a frequency of nontasters of about 10% among Asians compared to about 30% among Caucasians (14).

Although taste preferences were initially believed to be restricted to bitter substances, there is now evidence that differences in taste perception include a variety of other compounds. For example, it has been shown that tasters are more sensitive to the sweet taste of sucrose and saccharin than nontasters, which suggests that PTC sensitivity may be a marker of taste preferences for several types of foods (14). This hypothesis is consistent with results of a segregation analysis of 1,152 individuals from 120 families which showed that PTC taste status was best explained by a two-locus model, one controlling PTC taste and the other a more general taste capacity (35). Although more studies are needed before a clear link between taste preferences and food preferences can be established, available data do not exclude the possibility that the genetic effect observed for food preferences may result from inherited differences in taste preferences.

Eating Behavior

Eating behavior is a broad concept that includes the process of food consumption (ingestion) and accompanying attitudes, sensations, experiences, and motivations, in addition to physiological processes (23). Dietary intake is obviously influenced by attitudes toward food. The rate at which food is eaten is an additional factor, and there is a tendency to eat faster among obese children (2) and adults (23) compared to leaner control subjects. These observations suggest that the control of appetite differs among individuals.

It is generally believed that both genetic and environmental factors influence the mechanisms regulating appetite and satiety (5). As shown in table 8.5, several peptides are known to alter feeding behavior either by stimulating or inhibiting food intake (7). It has been suggested that the peptides cholecystokinin and neuropeptide Y fit a series of criteria that qualify them as regulators of eating behavior (31). These peptides and many others represent candidate genes to study the impact of genetic variation in dietary intake and eating behavior. Moreover, it is also becoming evident that some of the peptides known to stimulate or inhibit food intake act specifically on certain nutrients such as fat or carbohydrate (7).

A peptide identified recently in mice was found to be lacking in the ob mouse model of obesity (21,51). The peptide, known as leptin, is secreted by the adipose tissue and has a profound inhibitory evidence on food intake in the ob mice and mice made obese by hyperphagia (9,21,36). Leptin is thought to exert its effects by binding to the ob-receptor in the hypothalamus (11,45). Serum leptin levels are elevated in

Table 8.5
List of Peptides That Influence Food Intake

Increase	Decrease
β-endorphin	Anorectin
Galanin	Caerulein
Growth-hormone-releasing hormone	Cholecystokinin
Growth hormone	Corticotropin-releasing hormone
Insulin	Cyclo his-pro
Neuropeptide Y	Enterostatin
Peptide YY	Insulin-like growth factor I
Somatostatin (low dose)	Glucagon
	Insulin (CNS)
	Leptin
	Neurotensin
	Neuropeptide K
	Tyrotropin-releasing hormone
	Vasoactive intestinal peptide

Adapted from Bray (7) and Levine and Billington (31).

most but not all obese subjects (12). This suggests that some individuals are insensitive to the inhibitory effects of leptin on food intake.

Data on the contribution of specific genes to individual differences in energy and nutrient intake are scarce. One recent study suggested that DNA sequence variation in the gene for the neuropeptide Y and for the Y1 receptor of NPY, identified with the HindIII and PstI restriction enzymes, were associated with daily energy intake and carbohydrate preference, particularly in women (13).

Gene-Nutrient Interactions

Individual differences in response to various dietary treatments are well documented. For example, it has long been recognized that there are considerable differences among individuals in response of serum cholesterol to dietary cholesterol (25). In some individuals (hyporesponders), serum cholesterol level is relatively insensitive to dietary cholesterol, while in others (hyperresponders), serum cholesterol is quite sensitive to dietary cholesterol (3). Although baseline dietary cholesterol is the best predictor of changes in serum cholesterol (25), three lines of evidence suggest that the response may be genetically mediated. First, the same variability in the response to dietary cholesterol has been observed in a variety of animal species, and hypo- and hyperresponsive animal strains have been obtained through selective breeding. Second, baseline serum cholesterol levels are correlated with the response to dietary cholesterol in these animal studies, and it is known that about 50% of the population variance in serum cholesterol is genetically determined. The third line of evidence comes from intervention studies in which both members of MZ twin pairs were submitted to short-term (38) and long-term (6) overfeeding. Both experiments resulted in significant changes for several morphological (see chapters 9 and 10) and metabolic (see chapter 13) fitness phenotypes. However, for the majority of these phenotypes, the F-ratio of the between to within genotype variance indicated that there was two to six times more variance between pairs of MZ twins than within pairs of MZ twins in the response to the overfeeding regimen, suggesting that the adaptation to changes in diet is genotype dependent.

Results of the overfeeding experiments also suggest that undetermined genetic characteristics specific to each individual are associated with the response to dietary changes. Use of the measured genotype approach may eventually contribute to the identification of genes involved in determining this response. Genetic variation at the apolipoprotein gene loci may be helpful in this regard. Alleles at the apolipoprotein E (APOE) gene locus are among the most extensively studied genes of lipid metabolism. Impact of this gene on plasma cholesterol levels has been measured in several populations around the world. Comparison of the APOE effect on total cholesterol levels among various populations with different dietary habits provides indirect indication of a gene-diet interaction. Figure 8.4 presents the average effects of each APOE allele on total cholesterol in four populations (22). The average effect of an allele, defined as the expected deviation of the mean of a group of individuals carrying that allele from the population mean, provides a measure of the influence of an allele on a given phenotype. For each population represented in figure 8.4, the impact of the ε2 allele is to lower while that of the ε4 allele is to raise cholesterol levels. Despite consistency in the direction of effects across populations, the magnitude of the effects varies. The cholesterol-raising effect of the ε4 allele, for example, tends to be highest in populations on high-fat diets like those of Tirol and Finland, and lowest in populations on low-fat diets like those of Japan and Sudan. These results have lead to the proposition that the effect of the APOE gene on plasma cholesterol is mediated by diet (48).

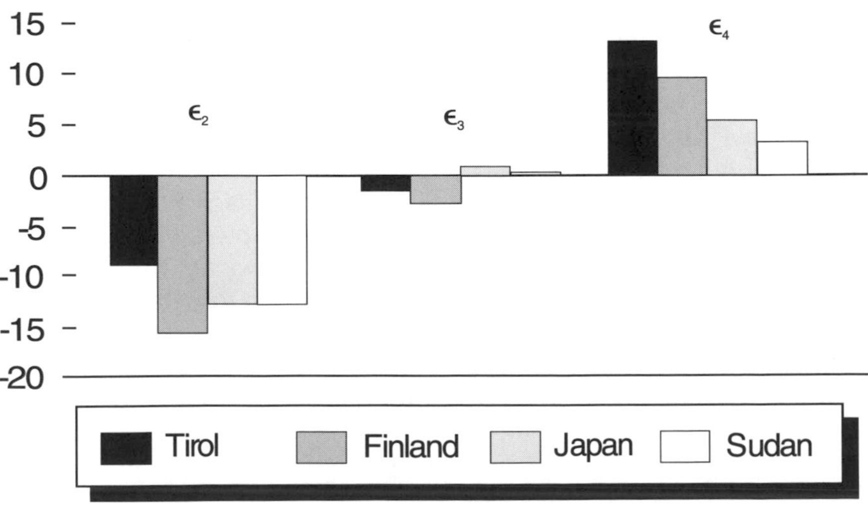

Figure 8.4 Average effects of the APOE alleles on fasting plasma total cholesterol (mg/dl) in populations with different diets. See text for explanations. Adapted from Hallman et al. (22).

Genetic variation at apolipoprotein gene loci has also been used to study the role of genetic factors in the response to dietary changes. Intervention studies were conducted on individuals from North Karelia, Finland. When subjects were switched from their traditional diet (high in fat and low in P/S, or ratio of polyunsaturated to saturated fatty acids) to a low-fat, high-P/S diet for six weeks and then switched back to the basal diet, there were considerable individual differences in the dietary response of serum lipids, lipoproteins, and apolipoproteins. When the results were analyzed in the context of genetic polymorphisms at apolipoprotein gene loci (43,46,47,50), significant gene-nutrient interactions were evident. In the case of APOE polymorphisms, results indicated that subjects homozygous for the ε4 allele exhibited greater reductions in total cholesterol compared to subjects with other genotypes when switched to the low-fat, high-P/S diet, and they showed greater increases when brought back to the basal diet (46). Differences in intestinal absorption of cholesterol were also observed between individuals with different APOE genotypes (26).

In addition to the APOE polymorphism, sequence variation detected with the XbaI restriction enzyme on the APOB gene was associated with the response to dietary intervention; subjects carrying the allele associated with the presence of the XbaI cutting site showed greater reductions in total cholesterol, LDL cholesterol, HDL cholesterol, and APOB levels (47). The effects of genetic variation at the APOA II, APOA I-CIII-AIV gene cluster, APOB, APOE, and LDL receptor genes on the response were also investigated in the same cohort (50). An effect on the response to dietary changes was observed for an MspI RFLP at the APOB gene locus, accounting for 6.3% of the variation in changes of APOA I levels. Results of these studies and those of others were recently summarized by Abbey (1). Although there are some inconsistencies among studies, it appears that genetic variation at the APOB and APOE gene loci are involved in mediating plasma lipid responses to dietary manipulations.

Summary

There are familial resemblances in dietary intake and food preferences. Although twin studies suggest that genetic factors contribute to this aggregation, nontransmissible environmental factors and cultural transmission account for most (about 80%) of the variation in energy intake and food preferences. Despite low heritability levels, there is increasing evidence that genes are important determinants of individual metabolic responses to variation in food intake. Although few candidate genes have been investigated thus far, some data suggest that markers of the neuropeptide Y and of its Y1 receptor are associated with individual differences in energy and carbohydrate intake. The most promising research avenue to pursue at this time appears, however, to be that of the *ob* gene and its leptin product, a strong inhibitor of food intake in rodents. Genetic variability in response to nutritional stresses has the potential to place some individuals more at risk of developing obesity as well as vascular and other diseases for which nutrition is a risk factor. In this context, genetic polymorphisms in the APOE and APOB loci are involved in modulating plasma lipids and lipoproteins in response to dietary interventions.

References

1. Abbey, M. The influence of apolipoprotein polymorphism on the response to dietary fat and cholesterol. Curr. Opin. Lipidol. 3:12-16; 1992.
2. Barkeling, B.; Ekman, S.; Rossner, S. Eating behavior in obese and normal weight 11-year-old children. Int. J. Obes. 16:355-60; 1992.
3. Beynen, A.C.; Katan, M.B.; Van Zutphen, L.F. Hypo- and hyperresponders: Individual differences in the response of serum cholesterol concentration to changes in diet. Adv. Lipid. Res. 22:115-71; 1987.
4. Birch, L.L. The relationship between children's food preferences and those of their parents. J. Nutr. Educ. 12:14-18; 1980.
5. Blundell, J.E. Appetite disturbance and the problem of overweight. Drugs. 39:1-19; 1990.
6. Bouchard, C.; Tremblay, A.; Després, J.P.; Nadeau, A.; Lupien, P.J.; Thériault, G.; Dussault, J.; Moorjani, S. The response to long-term overfeeding in identical twins. N. Engl. J. Med. 322:1477-82; 1990.
7. Bray, G.A. The effect of peptides on nutrient intake and the sympathetic nervous system. In: Bray, G.A.; Ryan, D.H., eds. The science of food regulation: Food intake, taste, nutrient partitioning and energy expenditure. Baton Rouge: Louisiana State University Press; 1992:257-76.
8. Briefel, R.R.; McDowell, M.A.; Alaimo, K.; Caughman, C.R.; Bischof, A.L.; Caroll, M.D.; Johnson, C.L. Total energy intake in the US population: The Third National Health and Nutrition Examination Survey, 1988-1991. Am. J. Clin. Nutr. 62:1072S-80S; 1995.
9. Campfield, L.A.; Smith, F.J.; Guisez, Y.; Devos, R.; Burn, P. Recombinant mouse OB protein: Evidence for a peripheral signal linking adiposity and central neural networks. Science. 269:546-49; 1995.
10. Cavalli-Sforza, L.L. Cultural transmission and nutrition. World Rev. Nutr. Diet. 63:35-48; 1990.
11. Chen, H.; Charlat, O.; Tartaglia, L.A.; et al. Evidence that the diabetes gene encodes the leptin receptor: Identification of a mutation in the leptin receptor gene in db/db mice. Cell. 84:491-95; 1996.

12. Considine, R.V.; Sinha, M.K.; Heiman, M.L.; Kriauciunas, A.; Stephens, T.W.; Nyce, M.R.; Ohannesian, J.P.; Marco, C.C.; McKee, L.J.; Bauer, T.L.; Caro, J.F. Serum immunoreactive-leptin concentrations in normal-weight and obese humans. N. Engl. J. Med. 334:292-95; 1996.

13. Côté G.; Tremblay, A.; Dionne, F.T.; Bouchard, C. DNA sequence variation at the NPY and NPY Y1 receptor loci and human food intake. Obes. Res. [Abstract]. 3: 3545; 1995.

14. Drewnowski, A. Genetics of taste and smell. World Rev. Nutr. Diet. 63:194-208; 1990.

15. Drewnowski, A.; Holden-Wiltse, J. Taste responses and food preferences in obese women: Effects of weight cycling. Int. J. Obes. 16:639-64; 1992.

16. Drewnowski, A.; Kurth, C.L.; Rahaim, J.E. Taste preferences in human obesity: Environmental and familial factors. Am. J. Clin. Nutr. 54:635-41; 1991.

17. Eastwood, M.A.; Brydon, W.G.; Smith, D.M.; Smith, J.H. A study of diet, serum lipids, and fecal constituents in spouses. Am. J. Clin. Nutr. 36:290-93; 1982.

18. Fabsitz, R.R.; Garrison, R.J.; Feinleib, M.; Hjortland, M. A twin analysis of dietary intake: Evidence for a need to control for possible environmental influences in MZ and DZ twins. Behav. Genet. 8:15-25; 1978.

19. Faust, J. A twin study of personal preferences. J. Biosociol. Sci. 6:75-91; 1976.

20. Greene, L.S.; Desor, J.A.; Maller, O. Heredity and experience: Their relative importance in the development of taste preference in man. J. Comp. Physiol. Psychol. 89:279-84; 1975.

21. Halaas, J.L.; Gajiwala, K.S.; Maffei, M.; Cohen, S.L.; Chait, B.T.; Rabinowitz, D.; Lallone, R.L.; Burley, S.K.; Friedman, J.M. Weight reducing effects of the plasma protein encoded by the obese gene. Science. 269:543-46; 1995.

22. Hallman, D.M.; Boerwinkle, E.; Saha, N.; Sandholzer, C.; Menzel, H.J.; Csazar, A.; Utermann, G. The apolipoprotein E polymorphism: A comparison of allele frequencies and effects in nine populations. Am. J. Hum. Genet. 49:338-49; 1991.

23. Hammer, L.D. The development of eating behavior in childhood. Ped. Clin. N. Am. 39:379-94; 1992.

24. Heller, R.F.; O'Connel, D.L.; Roberts, D.C.K.; Allen, J.R.; Knapp, J.C.; Steele, P.L.; Silove, D. Lifestyle factors in monozygotic and dizygotic twins. Genet. Epidemiol. 5:311-21; 1988.

25. Hopkins, P.N. Effects of dietary cholesterol on serum cholesterol: A meta-analysis and review. Am. J. Clin. Nutr. 55:1060-70; 1992.

26. Kesaniemi, M.A.; Ehnholm, C.; Miettinen, T.A. Intestinal cholesterol absorption efficiency in man is related to apoprotein E phenotype. J. Clin. Invest. 80:578-81; 1987.

27. Kolonel, L.N.; Lee, J. Husband-wife correspondence in smoking, drinking, and dietary habits. Am. J. Clin. Nutr. 34:99-104; 1981.

28. Krondl, M.; Coleman, P.; Wade, J.; Milner, J. A twin study examining the genetic influence on food selection. Hum. Nutr. 37A:189-98; 1983.
29. Laskarzewski, P.; Morrison, J.A.; Khoury, P.; Kelly, K.; Glatfeller, L.; Larsen, R.; Glueck, C.J. Parent-child nutrient intake interrelationships in school children ages 6 to 19: The Princeton School District Study. Am. J. Clin. Nutr. 33:2350-55; 1980.
30. Lee, J.; Kolonel, L.N. Nutrient intakes of husbands and wives: Implications for epidemiologic research. Am. J. Epidemiol. 115:515-25; 1982.
31. Levine, A.S.; Billington, C.J. Selected criteria for peptides as regulators of feeding: An overview. In: Bray, G.A.; Ryan, D.H., eds. The science of food regulation, food intake, taste, nutrient partitioning and energy expenditure. Baton Rouge: Louisiana State University Press; 1992:210-23.
32. McDowell, M.A.; Briefel, R.R.; Alaimo, K.; Bischof, A.M.; Caughman, C.L.; Carroll, M.D.; Loria, C.M.; Johnson, C.L. Energy and macronutrient intakes of persons aged 2 months and over in the United States: Third National Health and Nutrition Examination Survey, Phase 1, 1988-1991. Advanced data from vital and health statistics, no. 255. Hyattsville, MD: National Center for Health Statistics. DHHS publication (PHS) 95-1250; 1994.
33. McGee, D.; Rhoads, G.; Hankin, J.; Yano, K.; Tillotson, J. Within-person variability of nutrient intake in a group of Hawaiian men of Japanese ancestry. Am. J. Clin. Nutr. 36:657-63; 1982.
34. Oliveria, S.A.; Ellison, R.C.; Moore, L.L.; Gillman, M.W.; Garrahie, E.J.; Singer, M. Parent-child relationships in nutrient intake: The Framingham Children's Study. Am. J. Clin. Nutr. 56:593-98; 1992.
35. Olson, J.M.; Boehnke, M.; Neiswanger, K.; Roche, A.F.; Siervogel, R.M. Alternative genetic models for the inheritance of the phenylthiocarbamide (PTC) taste deficiency. Genet. Epidemiol. 6:423-34; 1989.
36. Pelleymounter, M.A.; Cullen, M.J.; Baker, M.S.; Hecht, R.; Winters, D.; Boone, T.; Collins, F. Effects of the obese gene product on body regulation in ob/ob mice. Science. 269:540-43; 1995.
37. Pérusse, L.; Tremblay, A.; Leblanc, C.; Cloninger, C.R.; Reich, T.; Rice, J.; Bouchard, C. Familial resemblance in energy intake: Contribution of genetic and environmental factors. Am. J. Clin. Nutr. 47:629-35; 1988.
38. Poehlman, E.T.; Tremblay, A.; Després, J.P.; Fontaine, E.; Pérusse, L.; Thériault, G.; Bouchard, C. Genotype-controlled changes in body composition and fat morphology following overfeeding in twins. Am. J. Clin. Nutr. 43:723-31; 1986.
39. Rozin, P.; Fallon, A.E.; Mandell, R. Family resemblance in attitudes to foods. Dev. Psychobiol. 20:309-14; 1986.
40. Rozin, P.; Millman, L. Family environment, not heredity, account for family resemblances in food preferences and attitudes: A twin study. Appetite. 8:125-34; 1987.

41. Rudman, D.; Nagraj, H.S.; Caindec, N. Genetic influences on nutritional thresholds. World Rev. Nutr. Diet. 63:161-74; 1990.
42. Schiffman, S.S.; Warwick, Z.S. The biology of taste and food intake. In: Bray, G.A.; Ryan, D.H., eds. The science of food regulation: Food intake, taste, nutrient partitioning and energy expenditure. Baton Rouge: Louisiana State University Press; 1992:293-312.
43. Talmud, P.J.; Boerwinkle, E.; Xu, C.F.; Tikkanen, M.J.; Pietinen, P.; Huttunen, J.K.; Humphries, S. Dietary intake and gene variation influence the response of plasma lipids to dietary intervention. Genet. Epidemiol. 9:249-60; 1992.
44. Tarasuk, V.; Beaton, G.H. The nature and individuality of within-subject variation in energy intake. Am. J. Clin. Nutr. 54:464-70; 1991.
45. Tartaglia, L.; Dembski, M.; Weng, X.; et al. Identification and expression cloning of a leptin receptor OB-R. Cell. 83:1263-71; 1995.
46. Tikkanen, M.J.; Huttunen, J.K.; Ehnholm, C.; Pietinen, P. Apolipoprotein E4 homozygosity predisposes to serum cholesterol elevation during high fat diet. Arteriosclerosis. 10:285-88; 1990.
47. Tikkanen, M.J.; Xu, C.F.; Hamalainen, T.; Talmud, P.; Sarna, S.; Huttunen, J.K.; Pietinen, P.; Humphries, S. XbaI polymorphism of the apolipoprotein B gene influences plasma lipid response to diet intervention. Clin. Genet. 37:327-34; 1990.
48. Utermann, G. Apolipoprotein E polymorphism in health and disease. Am. Heart J. 113:433-40; 1987.
49. Wade, J.; Milner, J.; Krondl, M. Evidence for a physiological regulation of food selection and nutrient intake in twins. Am. J. Clin. Nutr. 34:143-47; 1981.
50. Xu, C.F.; Boerwinkle, E.; Tikkanen, M.J.; Huttunen, J.K.; Humphries, S.E.; Talmud, P.J. Genetic variation at the apolipoprotein gene loci contribute to response of plasma lipids to dietary change. Genet. Epidemiol. 7:261-75; 1990.
51. Zhang, Y.; Proenca, R.; Maffei, M.; Barone, M. Leopold, L; Friedman, J.M. Positional cloning of the mouse obese gene and its human homologue. Nature. 372:425-32; 1994.

CHAPTER 9

Genetics of Body Size and Physique Phenotypes

Interest in the genotypic basis of variability in human morphology is widespread. For example, it is common for people to comment on the similarity of size or build of parents and their offspring; if you would like an estimate of how tall a child will be, check the stature of his or her parents. Although there is clear familial resemblance in many morphological traits, quantifying the resemblance in terms of genetic or environmental sources of variation is a different matter. This chapter considers the genetic influence on several morphological phenotypes, including stature and weight, segment lengths and breadths, physique or body build, and skeletal (bone) tissue. Adipose tissue is the component of body composition that currently receives most attention because of its association with excess fatness and because of the strong correlation of adipose tissue with several chronic diseases of adulthood. Given the volume of recent information dealing with adipose tissue, a separate chapter is devoted to this tissue (chapter 10). Although data are not as extensive as for adipose tissue, information on genetic aspects of skeletal muscle is accumulating rapidly so that a separate chapter is devoted to this tissue (chapter 11).

Life Span and Gender Variability

Changes in size of the body as a whole and specific segments, as well as body build with age and by gender, are discussed subsequently. The trends are based on group means or medians, but there is considerable variation and overlap.

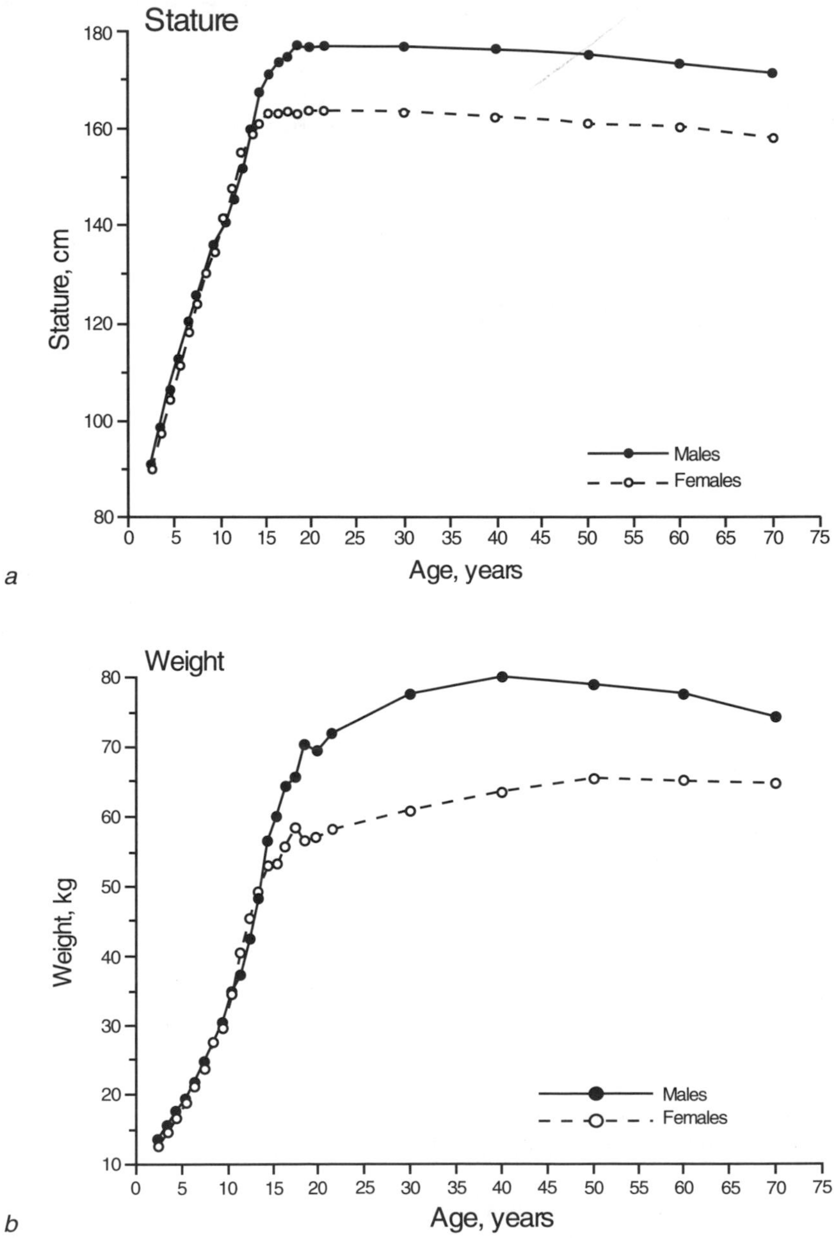

Figure 9.1 Changes in (*a*) stature and (*b*) body weight through the life span. Data are medians. Drawn from data reported by Najjar and Rowland (32).

Body size

Changes in stature and body weight over time are shown in figure 9.1, *a* and *b*. Sex differences are minor during childhood but are clearly apparent in later adolescence. There is a sex difference in the timing of the adolescent spurt, which occurs, on average, about two years earlier in girls than in boys (27).

Adult stature is attained in late adolescence, but some individuals continue to grow in stature through the mid-20s (42). Body weight continues to increase gradually from late adolescence into adulthood. Beginning in the 30s, stature tends to decline, on average, and the decline increases with advancing age. The loss occurs primarily in the trunk, reflecting the loss of compressibility of the intervertebral disks, changes in the mineral content of the vertebrae, and changes in the maintenance of posture. Body weight, however, is more variable with age, especially in cross-sectional studies.

The preceding trends are based on cross-sectional group means or medians. Longitudinal data for aging are not extensive, and there are discrepancies relative to cross-sectional trends, especially for stature. The age-associated decline in stature is not nearly as great in longitudinal data as in cross-sectional surveys (52). It is important to note that only longitudinal data permit an estimate of age changes; cross-sectional data describe only differences between age groups. Differences between age groups or cohorts can also be influenced by secular factors: that is, the older individuals were born and raised under different environmental conditions than younger adults.

Segment lengths, breadths, and circumferences

The segment lengths most often considered during growth and adulthood are sitting height and estimated leg length. The latter is derived by subtracting sitting height from stature, thus providing an estimate of subischial length. Both segments show growth patterns and sex differences similar to those of stature (figure 9.2, *a* and *b*). Leg length attains its growth spurt earlier than sitting height, and the latter continues to increase into late adolescence or early adulthood (27).

The ratio of sitting height to stature provides an estimate of relative trunk or relative leg length, and thus information on relative body proportions. There is no sex difference in the ratio during childhood and early adolescence; however, in later adolescence the ratio is higher in females and remains so through adulthood. Thus, women have, for the same stature, relatively shorter legs than men. The ratio changes during adulthood and aging largely as a function of sitting height.

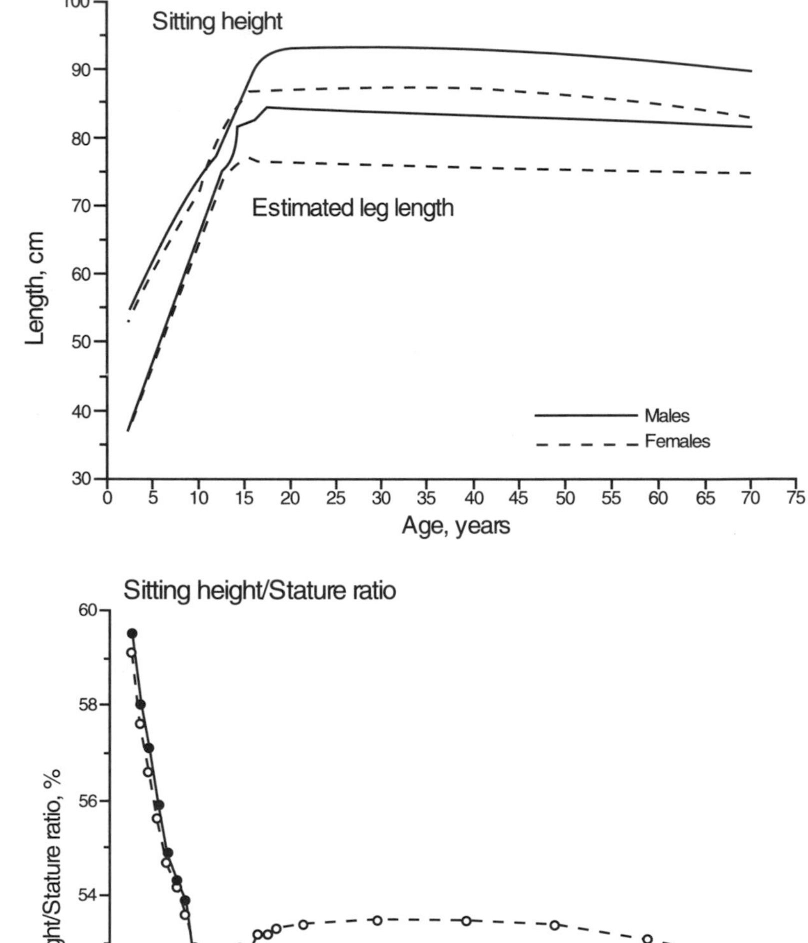

Figure 9.2 Changes in (*a*) sitting height and estimated leg length (stature minus sitting height) and (*b*) the sitting height/stature ratio through the life span. Data are for American whites. Data from 2 to 17 years are means; data for subsequent ages are calculated from mean sitting heights and statures. Drawn from data reported by Martorell et al. (28) and Najjar and Rowland (32).

Breadths across specific bone landmarks are often used to indicate the size and robustness of different segments of the body. Skeletal breadths follow a pattern of growth and sex difference similar to that of body weight. Ratios of breadth measurements provide another indication of body proportions. One of the more common ratios is that between breadths of the hips (bicristal breadth) and shoulders (biacromial breadth). The ratio does not differ during childhood, but with the onset of adolescence, males develop relatively broader shoulders compared to their hips, while females develop relatively broader hips compared to their shoulders (27).

Circumferences of the extremities are indicators of relative muscular development, while those of the trunk are more difficult to interpret given the heterogeneous composition of the contents of the trunk. Limb and trunk circumferences follow a pattern of change during growth and adulthood similar to that for body weight. In adulthood and aging, the ratio of waist to hip circumferences is often used as a surrogate indicator of upper body fat relative to lower body fat.

Physique (body build)

Physique refers to the overall shape of the body, commonly referred to as body build. Interest in physique has a very long tradition. Many approaches to describe and classify individuals on the basis of body build are apparent in the literature (6,10,40). The method most commonly used is somatotyping, an established technique for describing total body form. Three methods for estimating somatotype have been used in estimating genotypic influences: the original photoscopic procedures of Sheldon and colleagues (47,48), the anthropometric protocol of Parnell (36), and the anthroposcopic and anthropometric procedures of Heath and Carter (6). The latter method is most often used in its anthropometric form and is labeled the Heath-Carter method.

The concept of somatotype quantifies an individual's physique in terms of the contribution of three components: (1) Endomorphy is characterized by the predominance of the digestive organs and by roundness and softness of contours throughout the body; it is estimated from skinfolds in the Parnell and Heath-Carter methods. (2) Mesomorphy is characterized by the predominance of muscle, bone, and connective tissues, so that muscles are prominent with sharp definition. Mesomorphy is estimated from stature, calf and flexed arm circumferences (corrected for the medial calf and triceps skinfolds, respectively), and bicondylar and biepicondylar breadths of the femur and humerus, respectively, in the Parnell and Heath-Carter methods. (3) Ectomorphy is characterized by linearity and fragility, along with poor muscle development and a predominance of surface area over body mass; it is estimated as stature divided by the cube root of weight in all three methods. Each component is assigned a numerical score, with 0.1 the lowest possible score. The upper limit of the scoring system is 7 in the Sheldon and Parnell methods, but is open in the Heath-Carter method. The three components together define the individual's somatotype, and the ratings for each component are always reported in the following sequence: endomorphy, mesomorphy, and ectomorphy.

Physique assessments derived from the three methods and their modifications all use the terms endomorphy, mesomorphy, and ectomorphy, as well as the term somatotype. Hence, rather than being used in the sense originally intended by Sheldon, somatotype has become a more or less generic term referring to an individual's physique.

During childhood, changes in mean somatotype component ratings are relatively small. Boys show an increase in mesomorphy, especially

Table 9.1
Mean Somatotypes of Canadian Participants in the YMCA Lifestyle Inventory–Fitness Evolution Program

Age group	*n*	Endomorphy	Mesomorphy	Ectomorphy
Males				
15-19	161	3.0	4.7	2.7
20-29	2,259	3.6	5.0	2.1
30-39	2,985	4.0	5.2	1.8
40-49	2,031	4.1	5.3	1.7
50-59	1,159	4.1	5.4	1.6
60 +	375	4.0	5.1	1.8
Females				
15-19	235	4.3	3.7	2.4
20-29	1,752	4.4	3.7	2.4
30-39	1,201	4.5	3.9	2.3
40-49	787	5.1	4.4	1.8
50-59	498	5.4	4.5	1.8
60 +	156	5.3	4.7	1.7

Age-specific standard deviations range from 0.9 to 1.4.

Adapted from Bailey et al. (2).

in later adolescence, and this increase continues into the mid-20s (table 9.1). Ectomorphy tends to decline slightly in later adolescence, while trends in endomorphy are more variable. Data for girls, on the other hand, are less extensive and indicate an increase primarily in endomorphy with a decline in ectomorphy (27). Changes in somatotype with age during adulthood suggest a continued slight increase in mesomorphy and endomorphy in males and a decline in ectomorphy, with continued increase in endomorphy, in females (6). Note, however, that somatotype data for women indicate social class differences: Upper social class women are less endomorphic than those in lower social strata, which is consistent with observations for body weight and the body mass index (BMI).

Genetics and Morphological Phenotypes

Stature and Weight

Current knowledge concerning genetic regulation of stature and weight through early adulthood suggests three generalizations: (1) Genes associated with length and weight of the newborn appear to be different from those responsible for adult stature and weight; (2) there is a set of genes which is associated with adult stature and weight; and (3) there is another independent set of genes which regulates the rate of growth in stature and weight (41,53,55).

Twin, sibling, and parent-child studies have been used to document familial resemblance and in turn genetic effects on growth in stature and weight. Within-pair correlations for stature and weight of MZ and DZ twins change with age (table 9.2). Resemblance between DZ twins decreases with age, while that between MZ twins increases from birth onward. The adult pattern of twin resemblance appears to be established early in life, by about 2 to 3 years of age.

Reanalysis of reported variances, correlations, and F-ratios for stature in 19 samples of twins indicates heritabilities ranging from 0.69 to 0.96 with a mean of 0.85 and a standard deviation of 0.07 (5). Hence, stature is apparently determined mainly by genotype. This generalization applies to adequately nourished, healthy individuals; it is well established that chronic undernutrition influences attained stature.

Parent-child correlations for stature are rather stable throughout childhood, decline somewhat around puberty, and then increase during the latter phases of growth (14,15,55,58). Parent-child correlations must be interpreted in light of the fact that the correlation decreases as the difference in age between parents and offspring increases (13).

Table 9.2
Twin Correlations for Stature and Body Weight

	STATURE		BODY WEIGHT	
Age	**MZ**	**DZ**	**MZ**	**DZ**
Birth	0.66	0.77	0.64	0.71
3 months	0.77	0.74	0.78	0.66
6	0.81	0.70	0.82	0.62
12	0.86	0.69	0.89	0.58
2 years	0.88	0.59	0.88	0.55
3	0.93	0.59	0.89	0.52
4	0.94	0.59	0.85	0.50
5	0.94	0.57	0.86	0.54
6	0.94	0.56	0.87	0.57
7	0.94	0.51	0.88	0.54
8	0.95	0.49	0.88	0.54
9	0.93	0.49	0.88	0.62

Data are from the Louisville Twin Study with an original sample size of 952 children.
Adapted from Wilson (59).

On the other hand, when the stature of parents and children is correlated at the same chronological age, variation in coefficients with time is not apparent (13). Although full-sibling correlations for stature with both sibs at the same chronological age are more easily obtained than corresponding parent-offspring data, the pattern of age-associated variation is not clearly established. Thus, some studies report significant changes from birth to maturity (16), while others report only a very slight and almost linear increase with age (12,13). A summary of data for stature in siblings, as well as in parents and offspring over 15 years of age, yields a mean weighted parent-child correlation of 0.49 (6,344 pairs) and mean weighted full-sibling correlation of 0.53 (3,412 pairs) (5).

Path analysis has been applied to stature in order to detect total transmission across generations, and, if appropriate data are available, to partition genetic and cultural transmission (1,3,24,37). Results of path analysis applied to a large data set, including nine types of relatives by descent or by adoption, indicate a total transmission effect of 60% of the age and sex adjusted phenotypic variance in stature, with a heritability of about 40% (3). These results suggest a much lower genetic effect than reported in studies limited to either twins or nuclear families.

The contribution of sex-linked genes to the covariation between biological relatives is not clearly established. It has been suggested that the major fraction of variance in stature is determined by autosomal genes and that the contribution of sex-linked genes, if any, is small (16). In a recent critical evaluation of the statures of patients with sex chromosome abnormalities (table 9.3), Ogata and Matsuo (33) suggest a dosage effect of pseudoautosomal and Y-specific growth genes, in combination with a degree of growth stunting related to change in the quantity of the euchromatic or nonactivated regions as determinants of the adult stature. The results thus suggest the presence of a gene on the Y chromosome contributing to growth in stature. More recently, Salo

Table 9.3
Adult Statures of Patients With Sex Chromosome Aberrations and Appropriate Reference Data

Karyotype	*n*	Population	STATURE *M* in cm	STATURE SD	Difference in cm
47, XYY	14	American	188.6	7.6	+12.1
47, XYY	20	British	181.5	5.7	+6.8
47, XXY	17	Dutch	182.3	4.6	+4.3
47, XXY	52	British	177.7	7.0	+3.0
46, XY-reference		Dutch	178.0	6.4	
46, XY-reference		American	176.5	7.0	
46, XY-reference		British	174.7	6.7	
47, XXX	14	British	167.5	8.6	+5.3
46, XX-reference		Swiss	164.6	5.9	
46, XX-reference		American	163.6	7.0	
46, XX-reference		British	162.2	6.0	
45, X	54	American	143.5	6.1	–20.1
45, X	18	Swiss	143.3	4.8	–21.3
45, X	18	British	140.1	6.5	–22.1
46, X, del(X)(p22.32)	13	American	154.1	4.9	–9.5
46, X, del(X)(q13-21)	6	American	153.7	8.5	–9.9
46, X, i(Xq)	8	American	140.4	5.1	–23.2
46, X, t(X;autosome)	7	American	164.1	7.9	+0.5
46, X, Xp+ = Y(+)XX male	11	British	167.2	5.8	–7.5

Difference between the mean and population-specific reference data.
Adapted from Ogata and Matsuo (33).

et al. (43) related molecular markers on the Y chromosome to statures in 15 patients with Y chromosome deletions. Correlations between stature and the deletion breakpoints suggested that a growth control Y (*GCY*) gene (17) tentatively can be assigned to the proximal arm of the Y chromosome close to the centromere (43). In contrast, X-associated effects on the stature of normal individuals have been postulated more often, but the data are not conclusive.

Parent-child correlations for body weight are far less consistent than those observed for stature and suggest a lower heritability (31). Body weight by itself is of limited utility, since it is related to stature. Further, weight for stature ratios, specifically the BMI, are commonly used in epidemiological and clinical studies. Since the BMI is difficult to interpret in children and young adolescents, a good deal of the data are derived from adults. The BMI is often used as a proxy for obesity when no direct assessment of body fat is available. Hence, discussion of genetic influences on the BMI is ordinarily set within the context of overweight and/or obesity. Note, however, that the BMI is more strictly a measure of heaviness relative to stature, and is influenced by physique, FFM, and FM. Genetic analyses of the BMI are summarized in chapter 10.

Skeletal lengths and breadths

Segmental and bone lengths are apparently under a significant degree of genotypic control, probably more so than other morphological dimensions (18,45,54). Skeletal breadths also have a substantial genetic variance, although it is commonly stated that there is more stringent genetic regulation of linear dimensions (35,45).

Body segments—defined, for example, by measuring sitting height, leg length, and arm length—are composites of several bones. Skeletal breadths are often measured across single bones (e.g., across the epicondyles of the humerus) or across several bones (e.g., the distance between the acromial processes of each scapula, which also includes the clavicles and sternum). Measures of various skeletal lengths and breadths are rather common in anthropometric surveys, and available evidence from twins, siblings, and parents and offspring indicate substantial genetic variance (18,35,45,54,57).

Using published variances, correlations, and F-ratios from twin samples, heritability coefficients were computed for skeletal lengths and breadths (3,5). Estimated heritabilities for segment lengths were similar to those for stature: 0.84 ± 0.10 for suprasternal height, 0.71 ± 0.29 for sitting height, 0.84 ± 0.04 for total arm length (acromiale to dactylion), and 0.82 ± 0.02 for foot length. Those for specific bone lengths were consistent with the preceding: 0.62 ± 0.01 for arm length (acromiale to olecranon) and 0.71 ± 0.09 for forearm length (olecranon to stylion, ulna length). Corresponding estimates for skeletal breadths were reasonably similar but in some cases a bit more variable: 0.64 ± 0.22 for biacromial breadth, 0.60 ± 0.13 for bicristal breadth, and 0.34 to 0.52 for chest breadths. The latter are, of course, affected by when during the respiratory cycle the bony breadth measurements of the chest are taken. Data for other breadths—for example, wrist (bistyloid), ankle (bimalleolar), humerus (biepicondylar), and femur (bicondylar)—are less extensive, but a mean of 31 coefficients gave a heritability of 0.60 ± 0.18.

Full-sib and parent-offspring data for skeletal lengths and breadths are consistent with those of twin studies. Weighted full-sib correlations for several anthropometric dimensions were as follows: sitting height, 0.42; suprasternal height, 0.56; estimated leg (subischial) length, 0.56; total arm length, 0.47; biacromial breadth, 0.44; bicristal breadth, 0.46; biepicondylar breadth, 0.35; and bicondylar breadth, 0.52. Corresponding parent-child correlations were slightly lower: estimated leg length, 0.30; suprasternal height, 0.53; biacromial breadth, 0.34; bicristal breadth, 0.41; biepicondylar breadth, 0.32; and bicondylar breadth, 0.34 (5). The sibling and parent-child correlations did not indicate a sex-linked effect.

The genetic effect in skeletal lengths and breadths is supported by parent-child data for natural and adoptive children (table 9.4). The data are from 622 sets of midparent-natural child pairs and 154 sets of foster midparent–adopted child pairs in the Quebec Family Study. Midparent-child correlations were about twice as large for individuals sharing genes by descent and the home environment than for those sharing only the home environment. For example, the correlations were essentially zero for four of the six skeletal breadths in the foster midparent–adoptive child pairs, but consistently above 0.4 in the biological midparent–natural child pairs. The results thus suggest the presence of a rather substantial genetic effect and a somewhat stronger effect for breadths than for lengths.

Evidence from a study of 10-year-old siblings based on principal component analysis suggests genetic pleiotropic effects in skeletal lengths and breadths. One factor accounted for a major share of the variation in skeletal lengths, and another for breadth measurements, after environmental and familial factors were statistically controlled (4). Results of a bivariate path analysis of twin data for stature and bicristal breadth suggested that the same genes were perhaps contributing to both skeletal dimensions (23).

Ratios of length and/or breadth dimensions provide information on relative body proportions. An early twin study indicated a moderately high heritability (0.67) for the sitting height/stature ratio (i.e., relative trunk length), and, by inference, relative leg length (9). The estimated

Table 9.4
Midparent-Child Interclass Correlations for Selected Bone Dimensions

Variable	Foster midparent–adopted child (*N* = 154 sets)	Midparent–natural child (*N* = 622 sets)
Upper-arm length	0.20**	0.39**
Tibial height	0.29**	0.54**
Femur (bicondylar) breadth	0.15*	0.53**
Ankle (bimalleolar) breadth	0.08	0.49**
Humerus (biepicondylar) breadth	0.25**	0.62**
Wrist (bistyloidal) breadth	–0.03	0.44**
Biacromial breadth	0.01	0.43**
Biiliac breadth	0.13	0.42**

Scores were adjusted for age and gender by generation and normalized.
*$p < .05$; ** $p < .01$ Bouchard (3).

heritability for the ratio of biacromial breadth to stature was lower (0.33). The midparent-child correlation for the sitting height/stature ratio in the Quebec Family Study was about five times higher for individuals sharing genes by descent and the home environment (0.41) than for those sharing only the home environment (0.08). In contrast, the corresponding correlations for the ratio of total arm length to stature were similar, 0.36 and 0.31, respectively (Bouchard, unpublished data).

Differences in skeletal proportions among population groups are well documented and suggest that the lengths and breadths and their relationships are under significant genetic control. Populations of black African ancestry, for example, have relatively longer extremities compared to populations of European, Mexican, and Asian ancestry (26,28).

Circumferences

Circumferences are heterogeneous measurements. Those of the extremities include, for example, an outer perimeter of skin and subcutaneous tissue, muscle, and one or two bones. Trunk circumferences include—in addition to skin—subcutaneous tissue, muscle and bone, and various visceral tissues, depending on the level of measurement. Data on twin and familial resemblance in circumferences are not extensive.

Familial correlations for four body circumferences in the Nancy (France) Family Study are summarized in table 9.5. The parent-offspring and sibling correlations were similar. Mother-child and father-child correlations did not differ, except for a slightly higher mother-child correlation for waist circumference. Sex-specific sibling correlations also did not significantly differ (56). In the Quebec Family Study, midparent-natural child and foster midparent-adopted child correlations for arm and calf circumferences did not differ. The correlations were 0.26 and 0.22, respectively, for arm circumference, and 0.25 and 0.21, respectively, for calf circumference (3). In general, familial correlations for circumferences were thus slightly lower than those for skeletal lengths and breadths.

Table 9.5
Familial Correlations for Circumferences in the Nancy Family Study

Circumferences	Father-child	Mother-child	Full siblings
Suprailiac	0.33	0.32	0.32
Waist	0.38	0.47	0.49
Thigh	0.33	0.32	0.33
Arm	0.27	0.26	0.33

All correlations are adjusted for age and stature.
Adapted from Tiret et al. (56).

Physique

One of the earliest studies of familial resemblance in body build was that of Davenport (11). Using the ratio of weight/stature^2, Davenport compared builds of parents and offspring using basic Mendelian principles. The results generally indicated parent-offspring similarities in body build defined by the weight/stature ratio. However, the results also indicated that mating of two heterogeneous individuals for the weight/stature ratio can produce an exceptionally variable progeny.

Within-pair correlations for somatotype components in MZ and DZ twins are summarized in table 9.6. Correlations were higher for MZ than for DZ twins, with the exception of the female twins reported by Chovanova et al. (7). However, in a larger sample of twins from the same series, the correlations for MZ twins were higher (34). In the twins from the Quebec Family Study, intraclass correlations were also computed after statistically removing the effects of the other two components. The resulting correlations and variances were altered (51): They

Table 9.6
Intraclass Correlations of Somatotype Components in Several Samples of Twins of Similar Age

		MZ TWINS				DZ TWINS			
Study	Age range	Pairs	Endomorphy	Mesomorphy	Ectomorphy	Pairs	Endomorphy	Mesomorphy	Ectomorphy
Males									
Quebec[a]	9-23 yr	28	0.90	0.86	0.94	19	0.58	0.46	0.55
			0.75	0.51	0.74		0.64	0.22	0.60
Prague[b]	11-23 yr	14	0.83	0.90	0.90	10	0.44	0.15	0.22
Wroclaw[c]	12-19 yr	29	0.86	0.89	0.88	28	0.38	0.11	0.14
Wroclaw[d]	PrePHV	41	0.86	0.85	0.85	41	0.56	0.57	0.50
	SklMat		0.86	0.95	0.89		0.65	0.46	0.64
	Adult		0.79	0.93	0.91		0.59	0.54	0.53
Females									
Quebec[a]	9-23 yr	34	0.83	0.81	0.85	21	0.69	0.41	0.62
			0.83	0.74	0.75		0.38	0.27	0.51
Wroclaw[c]	11-19 yr	24	0.85	0.77	0.95	15	0.83	0.79	0.77
Wroclaw[d]	PrePHV	41	0.90	0.74	0.87	36	0.44	0.39	0.40
	SklMat		0.86	0.85	0.84		0.46	0.57	0.52
	Adult		0.87	0.79	0.91		0.23	0.37	0.35

[a]Song et al. (51), Heath-Carter anthropometric method. The second row of correlations is based on the residuals of the somatotype component after the effects of the other two components had been statistically controlled.
[b]Kovar (21), Heath-Carter anthropometric method.
[c]Chovanova et al. (7), Heath-Carter anthropometric method.
[d]Orczykowska-Swiatkowska et al. (34), modified Parnell anthropometric method. The twins were seen longitudinally at three ages: PrePHV, two years before peak height velocity (mean ages: male MZ, 11.9 yr; male DZ, 11.5 yr; female MZ, 10.4 yr; female DZ, 11.0 yr); SklMat, at the age of attaining skeletal maturity of the hand and wrist (mean ages: male MZ, 16.9 yr; male DZ, 16.6 yr; female MZ,14.9 yr; female DZ, 15.0 yr); Adult, young adulthood (mean ages: male MZ, 18.1 yr; male DZ, 18.2 yr; female MZ,17.7 yr; female DZ, 17.7 yr).

were lower in MZ twins of both sexes (except endomorphy in female MZ twins, no change) and in female DZ twins; in male DZ twins, they were slightly higher for endomorphy and ectomorphy and considerably lower for mesomorphy. When the effects of the other two somatotype components were statistically removed, within-pair variances of male MZ and DZ twins no longer differed, while those of female MZ and DZ twins remained significantly different. This suggests that in this sample of twins, the three components of somatotype are more closely related in male twins than in female twins. Using maximum-likelihood procedures in 10-year-old twins, Maes (25) reported higher correlations for each somatotype component within pairs of MZ twins compared with DZ twins.

Parnell (36) was apparently the first to compare somatotypes of children and their parents. There was about 75% concordance between parent-child somatotype ratings. Sibling and parent-offspring correlations for somatotype components in three studies are summarized in table 9.7. Sibling correlations are higher than parent-offspring correlations and are similar in magnitude to corresponding correlations for body size and various anthropometric dimensions described earlier. Correlations for specific sibling and parent-child pairings are generally homogeneous (4,50). Statistically controlling for seven indicators of socioeconomic status significantly reduced sibling correlations for somatotype components but did not appreciably alter the parent-child correlation for ectomorphy (4). On the other hand, statistically controlling for activity level, energy intake, and the other two somatotype components in the Quebec Family Study (50) did not alter either sibling or parent-child correlations for somatotype components. More recently, evidence from a relatively large family study in Madrid, Spain, also indicated significant familial aggregation for somatotype (44). The study used maximum-likelihood procedures and adjusted for age, sex, and the other two somatotype components (see 50,51). In contrast to the studies summarized in table 9.7, the Spanish study found that familial correlations for mesomorphy were influenced by sex of the parents and offspring. Sex of siblings did not influence the sibling correlations, but sisters resembled each other in each somatotype component more than brothers (44).

Spouse correlations for somatotype components are generally low. In the Canada Fitness Survey (37), the Quebec Family Study (50), the Leuven Longitudinal Twin Study (25), and the Family Study in Madrid (44), spouse correlations ranged from +0.15 to +0.19 for endomorphy, –0.08 to 0.13 for mesomorphy, and 0.06 to 0.14 for ectomorphy.

Twin, sibling, and parent-offspring correlations thus indicate a significant genetic effect on somatotype. The twin data show no consistent

Table 9.7
Familial Correlations for Somatotype Components

	ENDOMORPHY		MESOMORPHY		ECTOMORPHY	
	Sibling correlations	**Parent-offspring correlations**	**Sibling correlations**	**Parent-offspring correlations**	**Sibling correlations**	**Parent-offspring correlations**
Bouchard et al. (4),	0.40	—	0.30	—	0.38	0.22
10 years of age	0.25[a]	—	0.21	—	0.27	0.18
Pérusse et al. (37)	0.25[b]	0.21	0.29	0.24	0.29	0.24
Song et al. (50)	0.32[c]	0.28	0.55	0.33	0.38	0.23
	0.36[d]	0.24	0.56	0.33	0.36	0.23

[a]The second row of correlations controls for seven indicators of socioeconomic status.
[b]Adjusted for age and gender.
[c]Adjusted for age and gender, and the other two somatotype components.
[d]Adjusted for age, gender, activity level and energy intake, and the other two somatotype components.

pattern of sex differences (table 9.6), while sibling and parent-offspring correlations suggest a stronger familial aggregation for mesomorphy than for endomorphy and ectomorphy (table 9.7). The latter results are consistent with a path analysis (TAU) of the transmissibility of somatotype components in a nationally representative sample from Canada (the Canada Fitness Survey). Transmissibility was 45% for mesomorphy, 42% for ectomorphy, and 36% for endomorphy (37). The transmissible component (including genetic and nongenetic transmission) did not exceed 50% of the phenotypic variance, suggesting that nontransmissible factors may be more important than biological and cultural factors. And, for a given phenotype, the contribution of environmental factors may be quite different by age and sex and between populations. Clearly, the roles of age and of genetic and nongenetic factors in the development of somatotype need further study with larger samples and preferably with longitudinal data.

Bone mass

Familial resemblance and estimated heritability of bone mass have been approached primarily with the classical twin study design and with study of nuclear families. Two relatively recent reviews (20,39) have summarized a good deal of the information derived from radiographic and single- and dual-photon absorptiometric (SPA, DPA) methods. Most of the data are limited to the radius (forearm) and metacarpals (non-weight-bearing) with less extensive data for the vertebral column and femur (weight-bearing). The evidence suggests that the heritability of bone mass at the commonly measured sites is moderately high.

Smith et al. (49) measured bone mass and width of the radius at midshaft with SPA in 71 juvenile and 80 adult twin pairs. The intrapair variance of MZ twins was significantly less than that of DZ twins for both measures of bone tissue. Among the younger twins, the intrapair variance for 23 DZ pairs reached 0.0052 g/cm^2 for radial bone mass compared to only 0.0013 g/cm^2 in 48 MZ pairs. Corresponding intrapair variances for radial bone width at midshaft were 0.0011 cm for DZ twins and 0.0006 cm for MZ twins. The genetic effect appeared to be more pronounced in the juvenile twins (i.e., during growth) compared to the adult twins. For example, the intrapair variance in bone mass was four times greater in juvenile DZ than in MZ twins, while it was only two times greater in adult DZ than in MZ twins. Christian et al. (8) followed up a sample of the adult male twins (25 MZ and 21 DZ pairs) studied by Smith et al. (49) when they ranged in age from 60 to 71 years. After 16 years, the twins showed a loss of bone mineral content at midshaft averaging 0.098 g and 0.94 g in MZ and DZ twins respectively. The within-pair correlations for bone mass did not differ between MZ (r = 0.52) and DZ (r = 0.49) twins at follow-up, while bone width correlations were higher in MZ (r = 0.69) than DZ (r = 0.31) twins. These results suggest that the environmental component of phenotypic variance in bone mass (mineral) but not bone width appears to increase with age; that is, environmental factors apparently contribute substantially to bone mineral loss with age. This conclusion is also suggested by the lack of differences in intrapair correlations for the decrease in the ratio of mass to width (an estimate of density) over 16 years: 0.35 in MZ and 0.43 in DZ twins (8).

The transition through menopause has a significant effect on the bone mass of women. Although it is suggested that environmental factors affect bone mineral mass more during adulthood than during the years of active growth, daughters of postmenopausal women with osteoporosis have reduced bone mass themselves even before they reach menopause (46).

In a study of adult twins (primarily female), Pocock et al. (38) used SPA and DPA measurements of bone mass at the radius, lumbar spine, and three proximal femoral sites. The pattern of intrapair correlations indicated evidence for genetic determination of bone mass. The authors concluded that a single gene or a set of genes determines bone mass at all of the sites measured in the study.

McKay et al. (29) have recently reported a three-generation study of bone mineral density at the proximal femur and lumbar spine using dual-energy x-ray absorptiometry (DEXA). Mother-offspring correlations showed a clear pattern of familial resemblance in bone mineral density, although numbers of pairs were rather small. Mother–adult daughter correlations were higher than those for mother–growing daughter correlations; however, when the latter correlations were adjusted for sexual maturity status of the girls, the correlations increased considerably and approached those for the mother–adult daughter pairs. In contrast, mother–growing son correlations for bone mineral density were lower than those for girls and were increased only slightly when corrected for sexual maturity status.

The DEXA technique also permits estimates of total-body bone mineral in additional to regional sites. In a study of 160 adults from 40 families using DEXA and SPA, Krall and Dawson-Hughes (22) estimated broad and narrow heritabilities for total-body bone mineral density and bone mineral density at the femoral neck, lumbar spine, radius, and heel (table 9.8). Heritabilities were based on sibling correlations (broad sense, twice the intraclass correlations) and on the regression of midparent on offspring values (narrow sense), after adjusting for age, stature, and weight. All estimates ranged from moderate to high,

Table 9.8
Estimated Heritabilities of Total-Body and Regional Bone Density Based on Parents and Adult Offspring

	Siblings[a]	Midparent-offspring[b]	Midparent-offspring adjusted for lifestyle[c]
Total-body	0.80	0.69	0.60
Radius	0.57	0.51	0.47
Os calcis	0.86	0.64	0.56
Femoral neck	0.63	0.70	0.62
Spine (L2-L4)	0.35	0.50	0.46

[a]Broad heritabilities based on twice the intraclass correlations, adjusted for age, stature, and weight.
[b]Narrow heritabilities based on the regression of midparent on offspring values, adjusted for age, stature, and weight.
[c]Lifestyle indicators are listed in the text.
Adapted from Krall and Dawson-Hughes (22).

but those based on sibling correlations were higher for total-body, radial, and heel bone density. In contrast, those based on regression analysis were higher for femoral neck and lumbar bone density. After adjusting for several lifestyle variables (calcium and caffeine intakes, alcohol and cigarette use, and physical activity), estimated narrow heritabilities declined by 3% to 9%. The results suggest that a significant portion of the variance in bone mineral density, 46% to 62%, is related to heredity.

At present, there does not appear to be evidence for a major gene for bone mass identified through segregation analysis. However, Morrison et al. (30) have suggested that allelic variation at the gene encoding the vitamin D receptor (VDR) was predictive of up to 75% of the total genetic effect on bone mineral density at the spine in healthy women and in adult MZ and DZ twins from Australia. The relationship was weaker for bone mineral density at the hip. Comparison of groups of pre- and postmenopausal women homozygous for VDR alleles suggested 10- and 8-year differences in bone mineral density at the spine and hip, respectively; that is, the allele for low bone mineral density was associated with an estimated decrease in bone density equivalent to about 10 years of aging in the spine and 8 years in the hip (30). Another study of adult female MZ and DZ twins considered three polymorphisms at the VDR gene locus relative to bone mineral density at the spine, upper femur, and forearm (19). As expected, the estimated heritability of bone mineral density was high, greater than 70%. However, there was no relationship between VDR sequence variation and bone mineral density at any of the skeletal sites. Thus, a major role for genetic variation of VDR on bone mineral density, which would have major implications for osteoporosis, is not yet clearly established.

Summary

The available evidence indicates that a significant portion of the variation in body size and proportions, physique, skeletal lengths and breadths, limb circumferences, and bone mass is genetically determined. With the exception of the measurement of bone mass, which has benefited from recent advances in technology, the other aspects of human morphology are derived largely from anthropometry. Genetic inferences are based primarily on twin, family, and sibling studies; all studies indicate significant familial aggregation. Stature is the dimension that has received most study in both normal individuals and those with genetic anomalies. Evidence from patients with Y chromosome deletions suggests a growth control gene on the proximal arm of the Y chromosome. Segregation analysis of bone mineral has not shown evidence for a major gene effect. Studies attempting to relate morphological characteristics summarized in this chapter to specific genes and other genetic markers are extremely limited.

References

1. Annest, J.L.; Sing, C.F.; Biron, P.; Mongeau, J.G. Familial aggregation of blood pressure and weight in adoptive families. III. Analysis of the role of shared genes and shared household environment in explaining family resemblance for height, weight and selected weight/height indices. Am. J. Epidemiol. 117:492-506; 1983.
2. Bailey, D.A.; Carter, J.E.L.; Mirwald, R.L. Somatotypes of Canadian men and women. Hum. Biol. 54:813-28; 1982.
3. Bouchard, C. Genetic aspects of anthropometric dimensions relevant to assessment of nutritional status. In: Himes, J., ed. Anthropometric assessment of nutritional status. New York: Alan R. Liss; 1991:213-31.
4. Bouchard, C.; Demirjian, A.; Malina, R.M. Genetic pleiotropism in skeletal lengths and breadths. In: Ostyn, M.; Beunen, G.; Simons, J., eds. Kinanthropometry II. Vol. 9, International series on sport sciences. Baltimore: University Park Press; 1980:78-87.
5. Bouchard, C.; Lortie, G. Heredity and endurance performance. Sports Med. 1:38-64; 1984.

6. Carter, J.E.L.; Heath, B.H. Somatotyping: Development and applications. Cambridge: Cambridge University Press; 1990.
7. Chovanova, E.; Bergman, P.; Stukovsky, R. Genetic aspects of somatotypes in twins. Anthropos. 22:5-12; 1982.
8. Christian, J.C.; Slemenda, C.; Johnston, C.C. Heritability of bone mass: A longitudinal study in aging male twins. Am. J. Hum. Genet. 44:429-33; 1989.
9. Clarke, P.J. The heritability of certain anthropometric characters as ascertained from measurements of twins. Am. J. Hum. Genet. 8:49-54; 1956.
10. Comas, J. Manual of physical anthropology. Springfield, IL: Charles C. Thomas; 1957.
11. Davenport, C.B. Body build and its inheritance. Washington, DC: Carnegie Institution of Washington; 1923.
12. Furusho, T. Genetic study on growth of stature. Jpn. J. Hum. Genet. 17:249-72; 1972.
13. Furusho, T. Genetic study on stature. Jpn. J. Hum. Genet. 19:1-25; 1974.
14. Furusho, T. On the manifestation of genotypes responsible for stature. Hum. Biol. 40:437-55; 1968.
15. Garn, S.M. Body size and its implications. In: Hoffman, L.W.; Hoffman, M.L., eds. Review of child development research, 2. New York: R. Sage Foundation; 1966:529-61.
16. Garn, S.M.; Rohmann, C. Interaction of nutrition and genetics in the timing of growth and development. Pediatr. Clin. N. Am. 13:353-79; 1966.
17. Goodfellow, P.N.; Davies, K.E.; Ropers, H.H. Report of the committee on the genetic constitution of the X and Y chromosomes. Cytogenet. Cell Genet. 40:296-352; 1985.
18. Howells, W.W. Correlations of brothers in factor scores. Am. J. Phys. Anthropol. 11:121-40; 1953.
19. Hustmyer, F.G.; Peacock, M.; Hui, S.; Johnston, C.C.; Christian, J. Bone mineral density in relation to polymorphism at the vitamin D receptor gene locus. J. Clin. Invest. 94:2130-34; 1994.
20. Kelly, P.J.; Eisman, J.A.; Sambrook, P.N. Interaction of genetic and environmental influences on peak bone density. Osteoporosis Int. 1:56-60; 1990.
21. Kovar, R. Somatotypes of twins. Acta Univ. Carol. Gymn. 13:49-59; 1977.
22. Krall, E.A.; Dawson-Hughes, B. Heritable and life-style determinants of bone mineral density. J. Bone Miner. Res. 8:1-9; 1993.
23. Kramer, A.A.; Green, L.J.; Croghan, I.T.; Buck, G.M.; Ferer, R. Bivariate path analysis of twin children for stature and biiliac diameter: Estimation of genetic variation and co-variation. Hum. Biol. 58:517-25; 1986.
24. Longini, I.M.; Higgins, M.W.; Hinton, P.C.; Moll, P.P.; Keller, J.B. Genetic and environmental sources of familial aggregation of body mass in Tecumseh, Michigan. Hum. Biol. 56:733-57; 1984.

25. Maes, H.H.M. Univariate and multivariate genetic analysis of physical characteristics of twins and parents. Unpublished doctoral dissertation, Catholic University of Leuven, Leuven, Belgium, 1992.
26. Malina, R.M. Regional body composition: Age, sex, and ethnic variation. In: Roche, A.F.; Heymsfield, S.B.; Lohman, T.G., eds. Human body composition. Champaign, IL: Human Kinetics; 1996:217-55.
27. Malina, R.M.; Bouchard, C. Growth, maturation and physical activity. Champaign, IL: Human Kinetics; 1991.
28. Martorell, R.; Malina, R.M.; Castillo, R.O.; Mendoza, F.S.; Pawson, I.G. Body proportions in three ethnic groups: Children and youths 2-17 years of age in NHANES II and NHANES. Hum. Biol. 60:205-22; 1988.
29. McKay, H.A.; Bailey, D.A.; Wilkinson, A.A.; Houston, C.S. Familial comparison of bone mineral density at the proximal femur and lumbar spine. Bone Miner. 24:95-107; 1994.
30. Morrison, N.A.; Cheng, J.; Tokita, A.; Kelly, P.J.; Crofts, L.; Nguyen, T.V.; Sambrook, P.N.; Eisman, J.A. Prediction of bone density from vitamin D receptor alleles. Nature. 367:284-87; 1994.
31. Mueller, W.H. The genetics of size and shape in children and adults. In: Falkner, F.; Tanner, J.M., eds. Human growth: A comprehensive treatise. Vol. 3. Methodology, ecological, genetic, and nutritional effects on growth. New York: Plenum Press; 1985:145-68.
32. Najjar, M.F.; Rowland, M. Anthropometric reference data and prevalence of overweight. Vital Health Stat. Ser. 11, No. 238; 1987.
33. Ogata, T.; Matsuo, N. Sex chromosome aberrations and stature: Deduction of the principal factors involved in the determination of adult height. Hum. Genet. 91:551-62; 1993.
34. Orczykowska-Swiatkowska, Z.; Hulanicka, B.; Kotlarz, K. Intrapair differences in somatotype in three phases of ontogenetic development of twins. Stud. Phys. Anthropol. (Wroclaw). 9:39-59; 1988.
35. Osborne, R.H.; DeGeorge, F.V. Genetic basis of morphological variation. Harvard: Harvard University Press; 1959.
36. Parnell, R.W. Behaviour and physique: An introduction to practical and applied somatometry. London: Edward Arnold; 1958.
37. Pérusse, L.; Leblanc, C.; Bouchard, C. Inter-generation transmission of physical fitness in the Canadian population. Can. J. Sport. Sci. 13:8-14; 1988.
38. Pocock, N.A.; Eisman, J.A.; Hopper, J.L.; Yeates, M.G.; Sambrook, P.N.; Eben, S. Genetic determinants of bone mass in adults: A twin study. J. Clin. Invest. 80:706-10; 1987.
39. Pollitzer, W.S.; Anderson, J.J.B. Ethnic and genetic differences in bone mass: A review with a hereditary vs environmental perspective. Am. J. Clin. Nutr. 50:1244-59; 1989.
40. Rees, L. Constitutional factors and abnormal behaviour. In: Eysenck, H.J., ed. Handbook of abnormal psychology: An experimental approach. New York: Basic Books; 1960:344-92.

41. Robson, E.B. The genetics of birth weight. In: Falkner, F.; Tanners, J.M., eds. Human growth. Vol. 1. Principles and prenatal growth. New York: Plenum Press; 1978:285-97.

42. Roche, A.F.; Davila, G.H. Late adolescent growth in stature. Pediatrics. 50:874-80; 1972.

43. Salo, P.; Kääriäinen, H.; Page, D.C.; de la Chapelle, A. Deletion mapping of stature determinants on the long arm of the Y chromosome. Hum. Genet. 95:283-86; 1995.

44. Sanchez-Andres, A. Genetic and environmental influences on somatotype components: Family study in a Spanish population. Hum. Biol. 67:727-38; 1995.

45. Schreider, E. Biométrie et génétique. Biométrie Humaine. 4:65-86; 1969.

46. Seeman, E.; Hopper, J.L.; Bach, L.A.; Copper, M.E.; Parkinson, E.; McKay, J.; Jerunes, G. Reduced bone mass in daughters of women with osteoporosis. N. Engl. J. Med. 320:554-58; 1989.

47. Sheldon, W.H.; Dupertuis, C.W.; McDermott, E. Atlas of men: A guide for somatotyping the adult male at all ages. New York: Harper & Brothers; 1954.

48. Sheldon, W.H.; Stevens, S.S.; Tucker, W.B. The varieties of human physique. New York: Harper & Brothers; 1940.

49. Smith, D.M.; Nance, W.E.; Kang, K.W.; Christian, J.C.; Johnston, C.C. Genetic factors in determining bone mass. J. Clin. Invest. 52:2800-2808; 1973.

50. Song, T.M.K.; Malina, R.M.; Bouchard, C. Familial resemblance in somatotype. Am. J. Hum. Biol. 5:265-72; 1993.

51. Song, T.M.K.; Pérusse, L.; Malina, R.M.; Bouchard, C. Twin resemblance in somatotype and comparisons with other twin studies. Hum. Biol. 66:453-64; 1994.

52. Svanborg, A.; Eden, S.; Mellstrom, D. Metabolic changes in aging as predictors of disease: The Swedish experience. In: Ingram, D.K.; Baker, G.T.; Shock, N.N., eds. The potential for nutritional modulation of aging processes. Trumbull, CT: Food & Nutrition Press; 1991:81-90.

53. Tanner, J.M. Growth at adolescence. 2nd ed. Oxford: Blackwell Scientific; 1962.

54. Tanner, J.M. Inheritance of morphological and physiological traits. In: Sorsby, A., ed. Clinical genetics. St. Louis: CV Mosby; 1953:155-74.

55. Tanner, J.M.; Israelsohn, W.J. Parent-child correlations for body measurements of children between the ages of one month and seven years. Ann. Hum. Genet. 26:245-59; 1963.

56. Tiret, L.; Ducimetière, P.; André, J.L.; Gueguen, R.; Herbeth, B.; Spycherelle, Y.; Rakotovao, R.; Cambien, F. Family resemblance in body circumferences and their ratios: The Nancy Family Study. Ann. Hum. Biol. 18:259-71; 1991.

57. Vandenburg, S.G. How stable are heritability estimates from six anthropometric studies. Am. J. Phys. Anthropol. 20:331-38; 1962.
58. Welon, Z.; Bielicki, T. Further investigations of parent-child similarity in stature as assessed from longitudinal data. Hum. Biol. 43:517-25; 1971.
59. Wilson, R.S. Twins: Genetic influence on growth. In: Malina, R.M.; Bouchard, C., eds. Sport and human genetics. Champaign, IL: Human Kinetics; 1986:1-21.

CHAPTER 10

Genetics of Body Fat and Fat Distribution

Interest in body fat and relative fat distribution has grown over the last few decades. An excessive amount of fat, particularly when it leads to abdominal obesity, is associated with greater risk of a variety of morbid conditions as well as premature death. It is also well recognized that a low absolute and relative body fat content is associated with better performance in most performance activities. Because of these observations, and because the body of knowledge regarding the genetics of body fat and fat distribution has expanded remarkably over the last few years, this chapter is devoted entirely to these topics.

The Phenotypes

Body composition is most often viewed in the context of the two-compartment model: body mass = fat-free mass + fat mass. Newer methods permit use of a multicomponent model that attempts to partition fat-free mass (FFM) into its skeletal muscle and nonmuscle lean components, and to quantify the bone mineral compartment of FFM (105).

Total body fat is defined here as the absolute amount of energy stored in the form of triacylglycerol in the body (mostly in adipose tissue). Total body fat content is generally estimated in clinical settings, using the body mass index (BMI). The correlation between the BMI and total body fat or percentage body fat is reasonable in large and heterogeneous samples. The predictive value of the BMI is, however, much less impressive in a given individual, especially when the BMI is below about 30 kg/m^2. Thus, the BMI is primarily an indicator of heaviness and only indirectly of body fat (11,49). Any estimate of genetic effects on BMI is influenced in unknown proportions by the contribution of the genotype to fat mass, muscle mass, skeletal mass, and other components.

Although one cannot ignore BMI because of its widespread clinical and epidemiological use as a proxy for fatness, the most appropriate indicator of overall fatness or of obesity remains the percentage of body fat.

A properly assessed phenotype is a sine qua non for productive genetic studies. The data summarized in table 10.1 indicate why the BMI is only a partially acceptable surrogate measure of body fat content (11). The common variance between the BMI and percentage body fat derived from underwater weighing in large samples of adult men and women, 35 to 54 years of age, reaches only about 40%. At the extremes of the distribution of body fat content, the BMI is more closely related to percentage body fat: The common variance may reach 60% and more. This is not entirely satisfactory in genetic studies that deal with individual differences in the phenotype of interest. For such studies to be successful, the phenotype of a complex multifactorial trait must be measured with a reasonable degree of precision.

Table 10.1
Common Variance ($r^2 \times 100$) Between the BMI and Estimated Body Composition in Adults, 35 to 54 Years of Age

	BMI in 342 males	BMI in 356 females
Percent fat	41	40
Fat-free mass	37	25
Sum of 6 skinfolds	58	67
Trunk/limb skinfolds	10	8

Percentage fat and fat-free mass were derived from underwater weighing assessment of body density.

Adapted from Bouchard (11).

Percentage body fat remains very heterogeneous at any level of the BMI. This is illustrated in table 10.2, which is based on data from young and middle-aged adult males (9). For instance, in 27 men with a BMI of 28 to 30, the mean percentage body fat reached 28%, but the range was from 15% to 41%. The same phenomenon has also been observed in women.

Thus far few genetic studies have been reported on total or relative body fat content estimated with some of the more direct methods. The main exceptions are the Quebec Family Study and experiments with positive and negative energy balance, also conducted in Quebec, with

Table 10.2
Heterogeneity of Body Fat Content for a Given Class of the BMI in Adult Males

		% BODY FAT[a]		
N	**BMI**	**Mean**	**Minimum**	**Maximum**
27	20 to 22	17	8	32
76	23 to 25	22	11	35
46	26 to 27	26	16	40
27	28 to 30	28	15	41

[a]Estimated from underwater weighing assessment of body density.
Reproduced from Bouchard (9) with permission of CRC Press.

pairs of MZ twins. In these studies, body fat phenotypes were estimated from body density derived from underwater weighing.

Fat distribution characteristics are critical for the assessment of risk profiles for cardiovascular disease and noninsulin dependent diabetes mellitus, especially but not exclusively in adult men and postmenopausal women. Fat topography may also have relevance to physical performance, but the association is not clearly established.

The phenotypes of subcutaneous fat distribution that have been studied most extensively are the amount of abdominal subcutaneous fat, the amount of upper-body fat, the relative amount of upper-body fat (e.g., trunk to extremity skinfold ratios), and the amount of subcutaneous fat relative to total body fat. There is also considerable interest in abdominal visceral fat, the fat depot within the abdominal cavity, and in particular the portion of that depot that drains into the hepatic portal vein. Since abdominal visceral fat can be estimated only by imaging techniques, such as computerized tomography (CT) scanning or magnetic resonance, few genetic studies have been reported thus far for this phenotype.

Even though the covariation between total body fat and abdominal visceral fat is statistically significant, the relationship is also characterized by a high degree of heterogeneity. Table 10.3 demonstrates that, when BMI and percentage body fat are constrained to narrow ranges, there is generally a threefold range for the amount of abdominal visceral fat measured with computerized tomography in adult males (14). Thus, in 16 men with BMI values of 30 or 31 kg/m^2 and a percentage of body fat ranging from 30% to 33%, mean abdominal visceral fat was 153 cm^2 with a range from 77 cm^2 to 261 cm^2. The lack of a tight coupling among the BMI, percentage body fat, and abdominal visceral fat is also observed in adult women.

Table 10.3
Variation in Amount of Abdominal Fat Measured by CT Scans at L4-L5 for Given BMI and Percentage Body Fat Classes in Adult Males

			VISCERAL FAT IN CM²		
N	BMI	% fat[a] (range)	Mean	Minimum	Maximum
15	21 to 22	14 to 18	58	31	84
19	24 to 25	19 to 24	89	50	140
18	27 to 28	25 to 29	133	63	199
16	30 to 31	30 to 33	153	77	261

[a]% fat derived from underwater weighing.
Adapted from Bouchard et al. (14).

Because of the covariation observed between total body fat and various indicators of fat topography, it is necessary to statistically adjust or to experimentally control fat distribution phenotypes for total body fat content before undertaking genetic analyses. As shown in figure 10.1, the level of covariation among these phenotypes ranges from about 30% to 50% (9). A lack of proper control will result in unspecified and confounded influences relating to both total body fat and fat topography phenotypes.

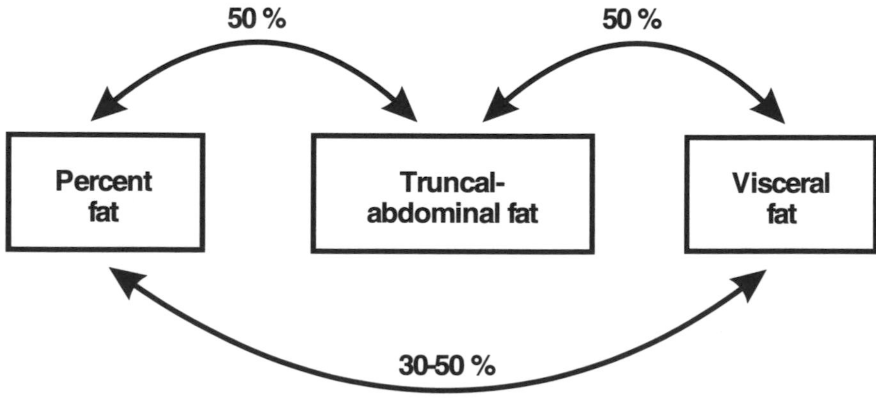

Figure 10.1 Common variance between three body fat phenotypes. Fat mass estimated from underwater weighing; truncal-abdominal fat assessed from skinfolds or CT scans; abdominal visceral fat estimated by CT scan at the L4-L5 vertebrae. From Bouchard (9).

Life Span, Gender, and Population Variability

BMI increases during childhood and adolescence. It continues to increase gradually into adulthood, on the average more in men than in women. The difference between men and women lessens after menopause. These trends are illustrated in figure 10.2. The lower BMIs in the elderly may reflect differential survival of lighter individuals. There is also social class variation in BMI, more so among women than among men. Women in the upper social strata tend to be lighter and have a lower BMI than women in the lower social strata.

Fat mass (FM) increases gradually during childhood and adolescence, more so in females who have a larger estimated fat mass. Percentage body fat increases during childhood and adolescence in girls, but temporarily declines during adolescence in males due to the rapid accumulation of fat-free mass at this time (72). On average, fat mass and percentage body fat increase gradually into the 20s or 30s in both sexes. During the 30s and 40s, FFM begins to decline, on average, in both sexes, while FM and percentage fat continue to increase. This pattern is sustained throughout the decades of life until senescence and death (71). The trends in life span mean values of males and females for body fat mass, fat-free mass, and percentage fat are summarized in figure 10.3, *a-d*.

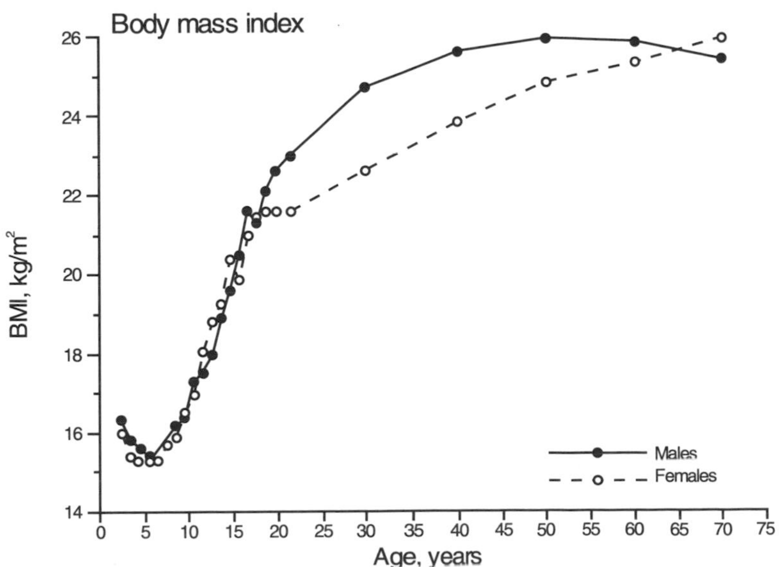

Figure 10.2 Changes in the body mass index through the life span. Data are medians. Drawn from data reported by Najjar and Rowland (80).

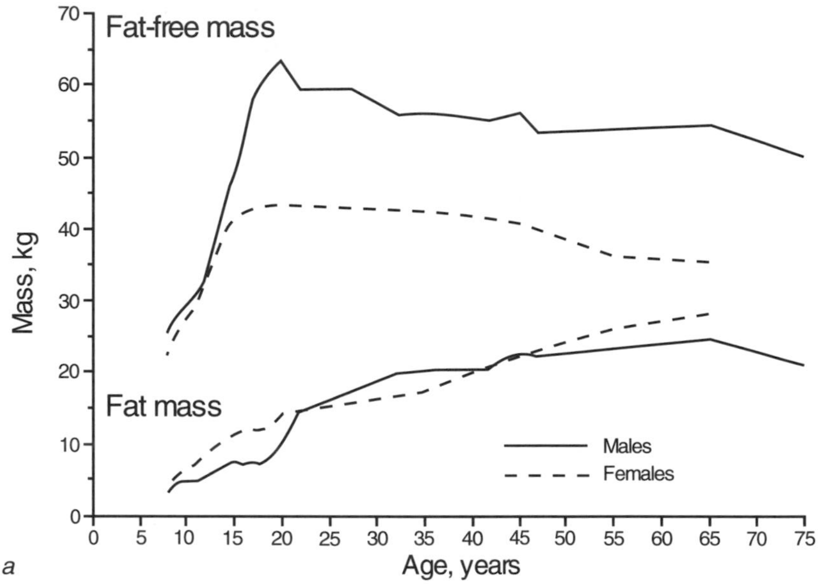

(continued)

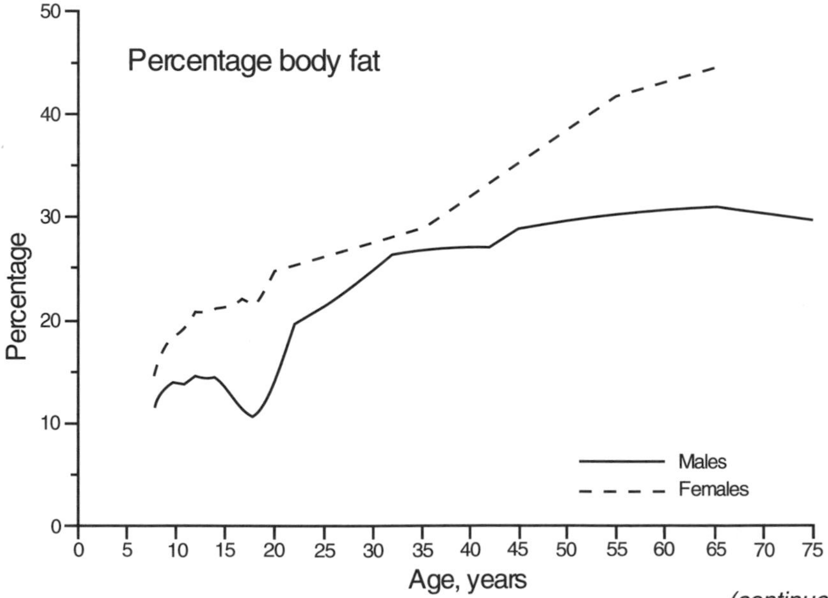

Figure 10.3 Changes in estimated fat-free mass, fat mass, and percentage body fat through the life span: *a* and *b* are based on body density; *c* and *d* are based on total body water. Values from infancy to about 20 years are composites calculated from the literature (see 72,74). Values for adulthood are estimated from data reported by Friis-Hansen (46), Krzywicki and Chinn (61), Norris et al. (84), and Young et al. (138).

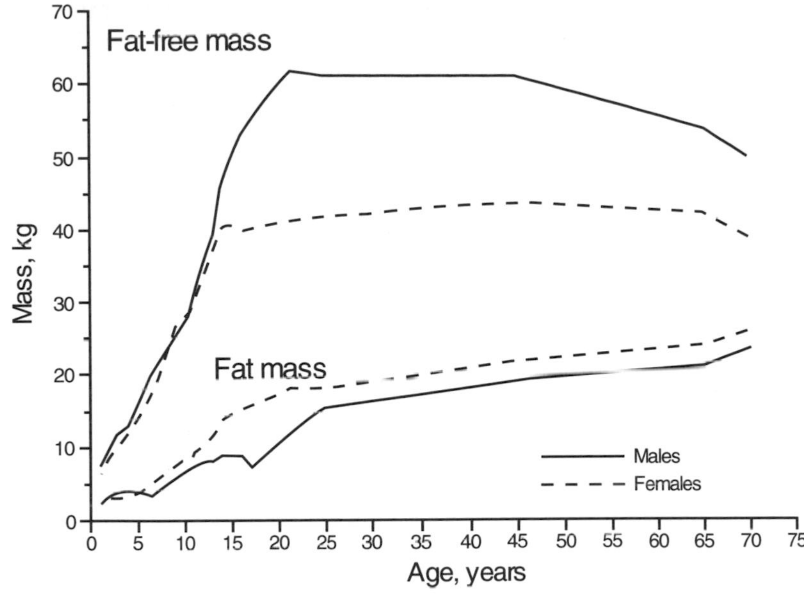

c

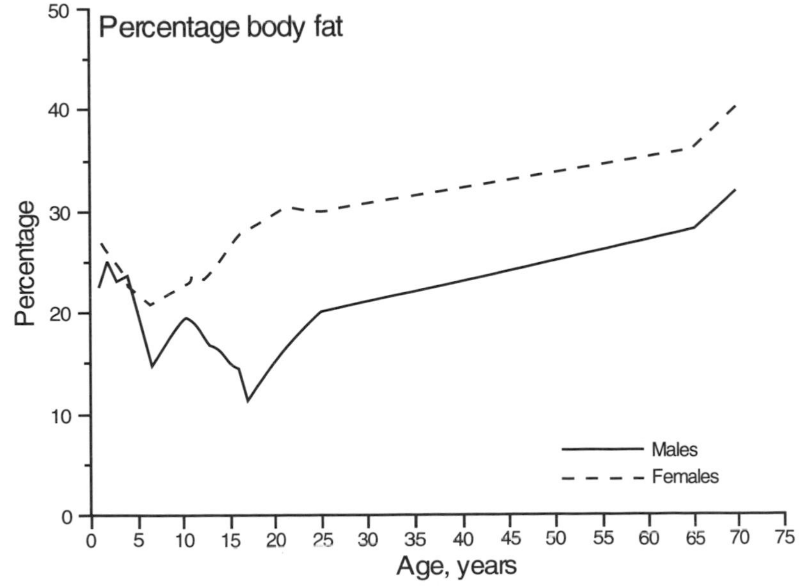

d

Figure 10.3 *(continued)*

Changes in subcutaneous fatness and relative subcutaneous fat distribution from childhood through adulthood are illustrated in figure 10.4, *a-c*. Girls average more subcutaneous fat than boys at all ages, especially during adolescence, and the sex difference is more apparent in subcutaneous fat on the extremities than on the trunk. As a result, there is a clear sex difference in fat distribution during adolescence, i.e., with sexual maturation. Relative to girls, boys have more subcutaneous fat on the trunk (truncal-abdominal fat) than on the extremities. This sex difference persists into adulthood. The absolute and relative amount of subcutaneous fat on the trunk increases gradually with age until about 60 to 70 years in both sexes, but more so in males than in females (71,73).

Information on changes in visceral fat during childhood and puberty are limited. Sex differences are minimal in late childhood, and there are no differences in visceral fat between early and late pubertal girls. Early adolescent girls have less visceral fat than adults, which suggests that abdominal visceral fat accumulates in later adolescence and/or young adulthood. Data on visceral fat in adolescent males are not currently

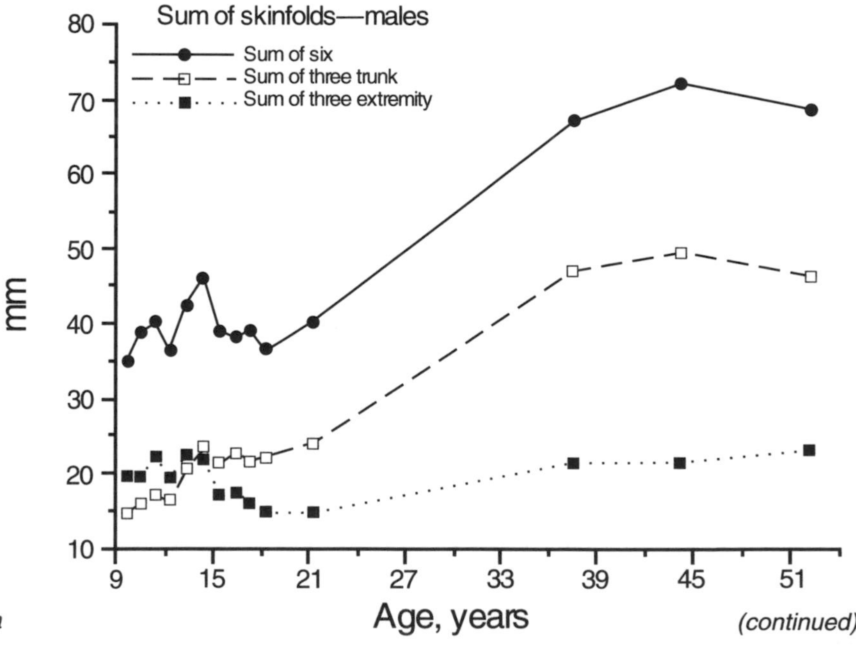

(continued)

Figure 10.4 Changes in trunk and extremity skinfolds in (*a*) males and (*b*) females and (*c*) in the ratio of trunk to extremity skinfolds from late childhood through adulthood. Redrawn from Malina (71).

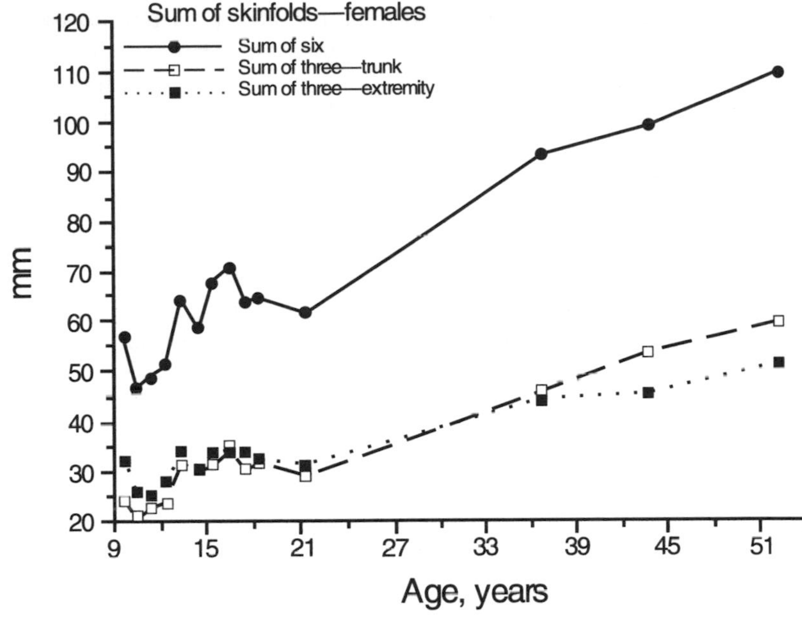

b

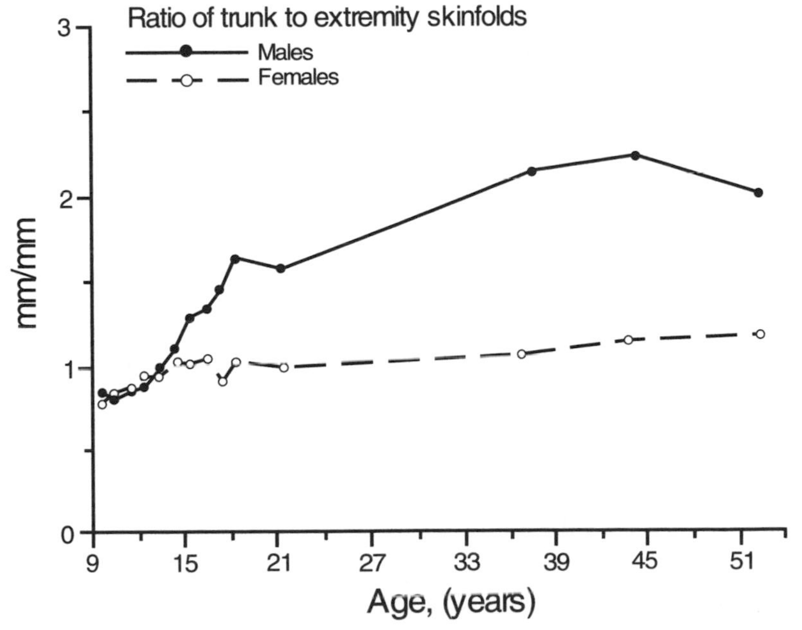

c

Figure 10.4 *(continued)*

available, but males have more visceral fat than females in the 20s, which suggests that the sex difference has its origin during puberty. Both sexes increase in visceral fat with age. Males have proportionally more abdominal visceral fat than females, and the sex difference increases with age (9).

The sex difference in the amount of abdominal visceral fat for any level of total body fat is particularly striking. Sjöström (112) estimated that the amount of CT abdominal visceral fat reached, on the average, 8% of the total body adipose tissue volume in women, but 20% in men. Men thus have a larger proportion of body fat content as visceral fat. Women will, therefore, attain a given amount of abdominal visceral fat, on average, at a much higher total body fat content than men (67).

Age- and sex-related trends in the relative distribution of body fat among various body regions are illustrated in figure 10.5, which is based on CT scans of 96 females and 66 males of Japanese ancestry ranging in age from 10 to 79 years (60). The relative contribution of abdominal visceral fat increases with age in both sexes, and more so in males than in females. The cross-sectional trends also suggest that the age-associated increase in visceral fat begins earlier in males than in females. In addition, subjects with a higher BMI tend to have a larger visceral fat component (42).

Genetic Epidemiology

Heritability of body fat content

There is still some disagreement among researchers regarding the importance of genetic factors in the familial resemblance observed for body fat (19,50). Most studies have used the BMI or skinfold thicknesses at a few sites as indicators of body fat. Heritability estimates range from almost zero to values as high as 90% for the BMI (19,117). With the use of different designs (twin, family, and adoption studies), large variation in ages of subjects, only a few types of relatives, and very often small sample sizes, such wide variation in estimated heritabilities within and between populations is not unexpected. With few exceptions, these studies could not separate the effects of genes from those of the environment shared by relatives living together in the same household. Moreover, only a handful of studies have included a wide range of BMI values to ensure that the phenotype was adequately represented.

Twin and adoption data. Comparison of MZ twins reared apart with MZ twins reared together is useful in assessing the role of heredity, since it provides some control over potentially confounding influ-

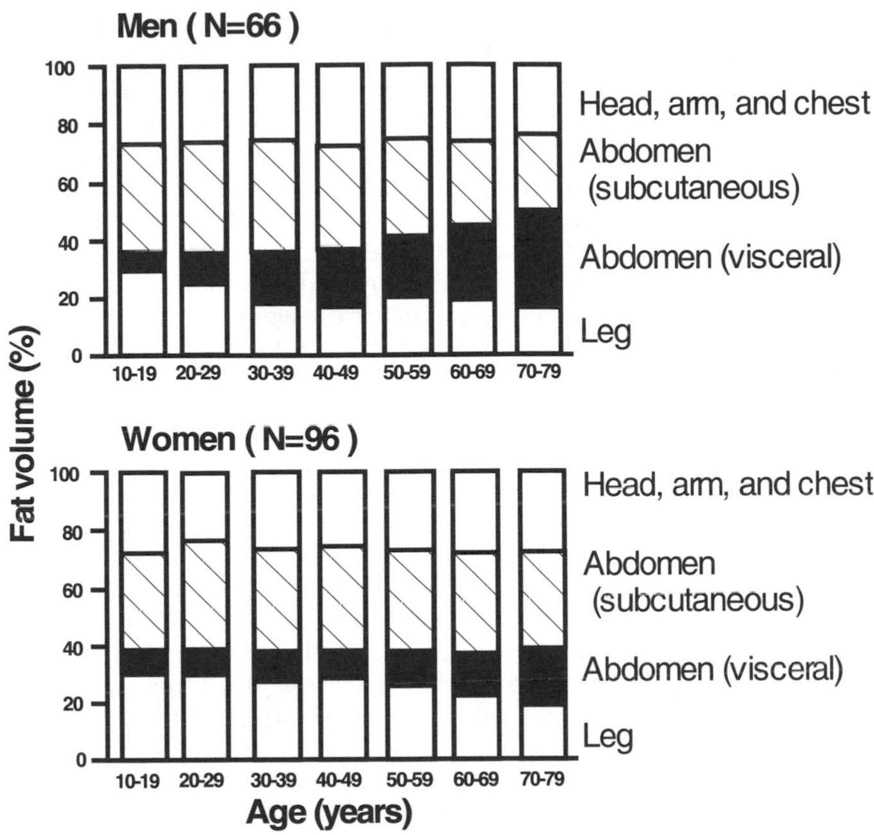

Figure 10.5 Age-related changes in whole-body fat distribution. Reprinted, by permission, from K. Kotani, K. Tokunaga, S. Fujioka et al., 1994, "Sexual dimorphism of age-related changes in whole-body fat distribution in the obese," *International Journal of Obesity* 18:207-212.

ences of a shared environment. Within-pair correlations for the BMI of MZ twins reared apart are generally similar to those of MZ twins reared together (70,95,118). The results suggest that shared familial environment did not contribute much to the variation in the BMI of MZ twins. The correlations of MZ twins reared apart provide a direct estimate of the genetic effect, if it is assumed that members of the same pair were not placed in similar environments, that the twins were not—for some unspecified reasons—behaving similarly despite the fact they were living apart, and that intrauterine factors did not influence long-term variation in the BMI. According to these studies, the heritability of the BMI would be in the range of 40% to 70%. Other twin study designs generate similar or even higher heritability estimates.

Five recent adoption studies, in which BMI data were available for both the biologic as well as adoptive relatives of the adoptees, have reported negligible effects of shared family environment on the BMI (94,114-116,119). In a recent review, Grilo and Pogue-Geile (51) also concluded that experiences shared among family members appeared largely irrelevant in determining individual differences in body weight and obesity. These findings are somewhat at odds with the strong familiality of energy expenditure and energy intake (see chapters 7 and 8). Heritability estimates for the BMI derived from adoption studies tend to cluster around 30% or less, which are lower than those based on twin studies. The entire issue needs further investigation by genetic epidemiologists.

Family studies. During the last 60 years or so, a large number of authors have reported that obese parents had a higher risk of having obese children than lean parents (27). These investigations were complemented by studies comparing resemblance in pairs of spouses, parents and children, and brothers and sisters for body weight, the BMI, and selected skinfold thicknesses (17,19,77). Table 10.4 summarizes the correlations reported in four rather large family-based studies of the BMI: the Framingham Heart Study (54), the Canada Fitness Survey (90), the Quebec Family Study (20), and the Nord-Trøndelag Norwegian National Health Screening Service Family Study (120).

In the Framingham Heart Study, adult levels of the BMI were used in both the parental and offspring generations, and the results were interpreted as providing little support for a genetic effect on the BMI (54). On the other hand, in the more recent study of 74,994 persons of both sexes, 20 years of age and older, from the population of Nord-Trøndelag, Norway, correlations were available for many types of relatives. Some of the correlations are summarized in table 10.4. A broad heritability of about 40% was obtained by fitting a path model for genetic and environmental transmission to the data (120). The authors rejected a simple model with only an additive genetic effect and an individual environmental effect.

The two studies in Canada considered transmission effects as well as heritability of both BMI and subcutaneous fat as assessed by the sum of several skinfolds. The first study was based on a stratified sample of the Canadian population and included the BMI and the sum of five skinfolds on 18,073 subjects. This large sample comprised 4,825 pairs of spouses, 8,881 parent-child pairs, 3,929 pairs of siblings, 43 uncle/aunt–nephew/niece pairs, and 85 grandparent-grandchild pairs (90). The total transmission effect across generations for the age- and sex-adjusted phenotypes reached about 35%. The second study was based on 1,698 mem-

Table 10.4
Familial Correlations for the BMI Derived From Four Large Family Studies

Relationship	Framingham Heart Study	Canada Fitness Survey	Quebec Family Study	Norway
Spouses	0.19 (1,163)	0.12 (3,183)	0.10 (248)	0.12 (23,936)
Parent-offspring	0.23 (4,027)	0.20 (7,194)	0.23 (1,239)	0.20 (43,586)
Siblings	0.28 (992)	0.34 (3,924)	0.26 (370)	0.24 (19,157)
Uncle/aunt–nephew/niece	0.08 (1,970)	–0.11 (34)	0.14 (88)	0 (1,146)
Grandparent-grandchild	NA	0.05 (32)	NA	0.07 (1,251)
DZ twins	NA	NA	0.34 (69)	0.20 (90)
MZ twins	NA	NA	0.88 (87)	0.58 (79)

Data derived from Heller et al. (54) for the Framingham Heart Study, Pérusse et al. (90) for the Canada Fitness Survey, Bouchard et al. (20) for the Quebec Family Study, and Tambs et al. (120) for Norway. Number of pairs are given in parentheses. NA = No data available.

Reproduced from Bouchard and Pérusse (18), with permission of CRC Press.

bers of 409 families and included nine types of relatives by descent or adoption (20). There was a total transmissible variance across generations for the BMI and the sum of six skinfolds of about 35%, but a genetic effect of only 5%.

Only one report has considered densitometrically estimated fat mass and percentage body fat (20). In a sample comprising nine different types of relatives, about one-half of the variance, after adjustment for age and sex, in fat mass or percentage body fat was associated with a transmissible effect, and 25% of the variance was compatible with an additive genetic effect (figure 10.6).

Trends in heritability estimates. An overview of genetic epidemiological aspects of body fat content is given in table 10.5, which describes trends in terms of the various designs used to generate the data. Heritability levels are highest with twin studies, intermediate with nuclear family data, and lowest with adoption data. When several types of relatives are used jointly in the same design, heritability

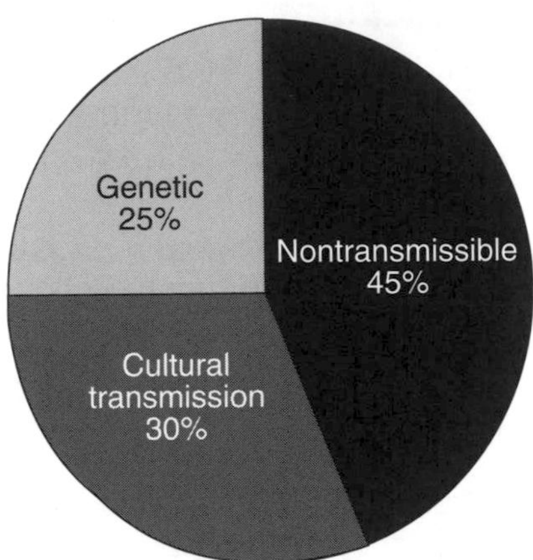

Figure 10.6 Total transmissible variance and its genetic component for percentage body fat and fat mass. Reprinted, by permission, from C. Bouchard and L. Pérusse, 1996, "Current status of the human obesity gene map," *International Journal of Obesity* 4:81-90.

Table 10.5
Overview of the Genetic Epidemiology of Human Body Fat Content

	Heritability/ transmission	Maternal/ paternal	Familial environment
Nuclear families	30 to 50	No	Minor
Adoption studies	10 to 30	Mixed results	Minor
Twin studies	50 to 80	No	No
Combined strategies	25 to 40	No	Minor

Reproduced from Bouchard (10), with permission from CRC Press.

estimates typically cluster around 25% to 40% of the age- and sex-adjusted phenotypic variance. There is no clear evidence for a specific maternal or paternal effect, and the common familial environmental effect is only marginal. The presence of a nonadditive genetic effect is often suggested from the genetic epidemiological research.

Evidence for a major gene

A common observation in familial studies of obesity is that an obese person is, on the average, 10 to 15 BMI units heavier than his or her mother, father, brothers or sisters (1,68,100). Such a large difference between a person and his or her first-degree relatives suggests the contribution of a recessive gene having a large effect. This hypothesis can be tested by complex segregation analysis.

Table 10.6 summarizes results of complex segregation analyses of the BMI, weight adjusted for stature, or percentage body fat in various studies. All studies are unanimous in reporting evidence for a multifactorial component and a major effect, the latter accounting for 20% to 45% of the variance. However, in two reports, the major effect was non-Mendelian. In a third paper, the major effect became Mendelian only when age- and sex-related variations in the major effect were considered. In five of the seven studies, the estimated gene frequency of the putative recessive gene ranged from about 0.2 to 0.3.

Genetic-Environment Interactions

Genotype-environment interaction (GxE) arises when the response of a phenotype (e.g., fat mass) to environmental changes (e.g., dietary restriction) depends on the genotype of the individual. Although it is well known that there are interindividual differences in responses to various dietary interventions, whether in terms of serum cholesterol changes in high-fat diets (3) or of weight gains following chronic overfeeding (111),

Table 10.6
Overview of Results of Segregation Analysis for the BMI or Body Fat Phenotypes

	Multifactorial transmission (%)	Major effect	Major gene	Gene frequency
Province et al. (97)	41	Yes, 20%	Yes	0.25
Price et al. (96)	34	Yes	Yes	0.21
Moll et al. (76)	42	Yes, 35%	Yes	0.25
Tiret et al. (122)[a]	39	Yes	No	Non-Mendelian
Rice et al. (102)	42	Yes, 20%	No	Non-Mendelian
Borecki et al. (5)	Yes	Yes	Age- and sex-related	0.22
Rice et al. (103)[b]	25	Yes, 45%	Yes	0.30

[a]Weight adjusted for height.
[b]Percentage body fat.

Reproduced from Bouchard (10), with permission from CRC Press.

very few attempts have been made to test whether these differences are dependent on genotype. Most genetic epidemiology studies of human obesity have assumed the absence of genotype-environment interaction simply because of the difficulty in handling such interaction effects in quantitative genetic models. Methods from both genetic epidemiology (unmeasured genotype approach) and from molecular epidemiology (measured genotype approach) can now be used to detect GxE effects in humans. The methods have been reviewed in chapter 4.

Evidence from both the top-down and bottom-up approaches indicates the presence of GxE effects in obesity-related phenotypes. Using appropriate statistical modeling, three studies reported major gene effects for measures of height-adjusted weight, but only after accounting for age and/or gender effects in the model (5,102,122). This suggests that the effect of the putative gene on body mass is dependent on the sex and age of the individual, which is a special case of genotype-environment interaction effect.

The role of the genotype in determining the response to changes in energy balance was studied in an unmeasured genotype approach by submitting both members of MZ twin pairs either to positive energy balance induced by overfeeding (23,91,92) or negative energy balance induced by exercise training (25,93). The objective of these studies was to determine whether the sensitivity of individuals to gain fat when exposed to positive energy balance, or to lose fat when exposed to negative energy balance, was modulated by the genotype. The results of these studies, which are reviewed subsequently, revealed significant interaction effects between genotype and energy balance for body weight, body fat, and fat distribution phenotypes. Thus, genetic factors are apparently important in determining how an individual will respond to alterations in energy balance.

The positive energy balance experiments

It is well recognized that some individuals are prone to excessive accumulation of fat, for which losing weight represents a continuous battle, and others who are seemingly well protected against such a problem. The following two questions were tested: Are there differences in the tendency of individuals to gain fat when chronically exposed to positive energy balance? Are such differences, if they exist, dependent on genotype? Affirmative answers to these questions would demonstrate a significant interaction effect between genotype and energy balance. Results from two experiments suggest that such an effect may exist for body weight, body fat, and fat distribution.

A short-term experiment: For 22 consecutive days, six pairs of male MZ twins were exposed to a surplus energy intake of 1,000 kcal per day

(24,92). Individual differences in gains of body weight, fat mass, subcutaneous fat, and site of fat deposition were observed, but these differences were not randomly distributed. Significant intrapair resemblance was observed for changes in most body composition and fat distribution variables, despite the fact that the treatment was of short duration and that the changes induced by the treatment were not large. The intrapair resemblance in response to overfeeding, as assessed by intraclass coefficients computed with the individual changes, reached 0.88 for total fat mass and 0.76 for fat-free mass. Subjects gained body weight and body fat, but there was a nonsignificant 7% increase in resting metabolic rate (91).

A long-term experiment: Twelve pairs of male MZ twins consumed a surplus 1,000 kcal per day, six days a week, for a period of 100 days (23). Significant increases in body weight and fat mass were observed after the period of overfeeding. The data showed considerable individual differences in the adaptation to excess calories. The observed variation, however, was not randomly distributed, as indicated by significant within-pair resemblance in response. For instance, there was at least three times more variance in responses between pairs than within pairs for gains in body weight, fat mass, and fat-free mass (figure 10.7, *left panel*). These data, and those of the short-term overfeeding experiment, demonstrate that some individuals are more at risk than others to

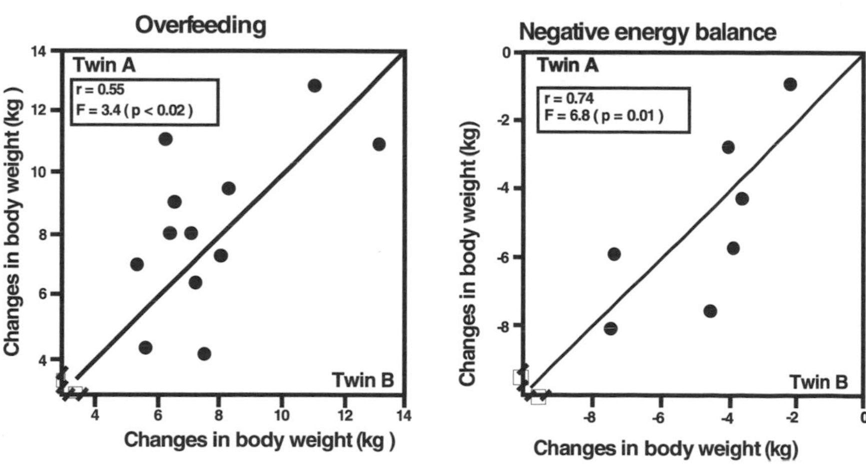

Figure 10.7 Intrapair resemblance in response of identical twins to long-term changes in energy balance. *Left panel:* Twelve pairs of identical twins were submitted to an 84,000-kcal energy intake surplus over 100 days. *Right panel:* Seven pairs were subjected to a negative energy balance protocol caused by exercise. The energy deficit was 58,000 kcal over 93 days. Reproduced from Bouchard et al. (23,25) with permission.

gain fat when energy intake surplus is maintained at the same level for everyone and when all subjects are confined to a sedentary lifestyle. The response to the standardized caloric surplus within pairs of identical twins suggests that the amount of fat stored is probably influenced by genotype.

For changes in upper-body fat and in abdominal visceral fat as measured by computerized tomography (CT) scan, the long-term overfeeding study also revealed six times more variance between pairs than within pairs. Both measures were adjusted for the gain in total fat mass. These observations indicate that some individuals are storing fat predominantly in selected fat depots, primarily as a result of undetermined genetic characteristics. It also suggests that variations in regional fat distribution are more closely related to the genotype of the individuals than are variations in body mass and in overall body composition.

At the beginning of the long-term overfeeding treatment, almost all of the daily caloric surplus was recovered as body energy gain, but the proportion decreased to 60% at the end of the 100-day protocol (37). The weight-gain pattern followed an exponential curve, with a half-duration of about 86 days. It was estimated that the weight gained in the experiment reached about 55% of the maximal weight gain anticipated if the overfeeding protocol had continued indefinitely (37). The mean body mass gained by the 24 subjects of the 100-day overfeeding experiment was 8.1 kg, of which 5.4 kg were fat mass and 2.7 kg were fat-free mass increases. Assuming that the energy content of body fat is about 9,300 kcal per kilogram and that of fat-free tissue is 1,020 kcal per kilogram, then about 63% of the excess energy intake was recovered on the average as body mass changes. This proportion is of the same order as that reported by other investigators (82,99), that is, between 60% and 75% of total excess energy intake. There were, however, individual differences among the 24 subjects in the amount of fat and fat-free tissues gained.

Absolute resting metabolic rate increased by about 10% with overfeeding. However, the increase was only marginal when it was expressed per unit of fat-free mass (36,124). The intrapair resemblance for changes in resting metabolic rate brought about by overfeeding was significant, but became nonsignificant when changes in body mass or body composition were taken into account. The thermic response to food, as assessed by indirect calorimetry for a period of four hours following ingestion of a 1,000-kcal meal of mixed composition, did not increase with overfeeding when resting metabolic rate was subtracted from postprandial energy expenditure (124). In contrast, postprandial energy expenditure and the total energy cost of weight maintenance increased significantly, but the increments were mostly due to the gain in body mass.

The negative energy balance experiments

A short-term experiment: A short-term negative energy balance experiment was also undertaken with six pairs of MZ twins. The energy deficit was achieved by exercise performed twice a day, for 22 consecutive days, for about 50 minutes per session, on the cycle ergometer. The exercise prescription was precisely controlled for each subject during each exercise session and was designed to induce an extra energy expenditure of 1,000 kcal per day over resting metabolic rate while baseline energy intake was maintained throughout (93). The changes in body composition were generally small and were not related to the twin lines. The only exception was for changes in fat-free mass, which were more similar within pairs compared to the between-pairs variance.

A long-term experiment: Seven pairs of young adult male MZ twins completed a negative energy balance protocol during which they exercised on cycle ergometers twice a day, 9 out of 10 days, over a period of 93 days while being kept on a constant daily energy and nutrient intake. The mean total energy deficit caused by exercise above the estimated energy cost of body weight maintenance reached 58,000 kcal. Baseline energy intake was estimated over a period of 17 days preceding the negative energy balance protocol. Mean body weight loss was 5.0 kg and was entirely accounted for by the loss of fat mass. Fat-free mass was unchanged. Mean body energy loss reached 46,000 kcal, which represented about 78% of the estimated energy deficit. Decreases in metabolic rates and in energy expenditure of activity not associated with the cycle ergometer protocol must have occurred in order to explain the difference between the estimated energy deficit and body energy losses. Subcutaneous fat loss was slightly more pronounced on the trunk than on the limbs as estimated from skinfolds, circumferences, and computed tomography. The reduction in abdominal visceral fat area was quite striking, from 81 cm^2 to 52 cm^2. At the same submaximal power output level, subjects oxidized more lipids than carbohydrates after the program as indicated by changes in the respiratory exchange ratio.

Intrapair resemblance was observed for the changes in body weight (figure 10.7, *right panel*), fat mass, percentage body fat, body energy content, sum of 10 skinfolds, abdominal visceral fat, and respiratory exchange ratio during submaximal work. Even though there were large individual differences in response to the negative energy balance and exercise protocol, subjects with the same genotype were more alike in responses than subjects with different genotypes, particularly for changes in body fat, body energy, and abdominal visceral fat. High lipid oxidizers and low lipid oxidizers during submaximal exercise were also seen, despite the fact that all subjects had experienced the same exercise and nutritional conditions for about three months.

Thus, changes in body mass, body fat, and body energy content are characterized by more heterogeneity between twin pairs than within pairs. These results are remarkably similar to those reported earlier for body mass, body fat, and body energy gains with 12 pairs of twins subjected to a 100-day overfeeding protocol.

Contribution of Specific Genes

Support for a role of genes in human obesity or variation in body fat content has been obtained from the six lines of evidence identified in table 10.7. All of them will be addressed in some detail. The evidence accumulated from transgenic rodent models and gene knockout experiments will, however, not be fully reviewed. Although the latter studies provide support for the potential role of a gene in body fat content fluctuation when it is overexpressed in some or all tissues, or when the gene is inactivated or results in a nonfunctional peptide, the true contribution of these genes to human obesity remains to be clarified. An overview of the contribution of specific genes and molecular markers to obesity was recently published by us; this section is a summary of the review (16).

Mendelian disorders

Genetic syndromes characterized by the presence of obesity could be useful in the identification of loci contributing to obesity. Among 24 such disorders (16), 9 have been mapped to the following eight chromosomes: 3, 4, 8, 11, 15, 16, 20, and X. The other 15 Mendelian syndromes have not yet been mapped to a chromosomal area of the

Table 10.7
Lines of Evidence for Human Obesity Genes

- Mendelian disorders (MIM)
- Single-gene rodent models
- QTL from crossbreeding experiments
- Transgenic and knockout models
- Association studies
- Linkage studies

genome. With the Bardet-Biedl syndrome alone having been linked to four different chromosomal regions, a total of 12 loci are linked to these Mendelian disorders (table 10.8). Although the disorders account for a very small fraction of the obese population, they illustrate that single-gene defects could lead to the development of obesity in humans.

In order to determine whether the identified loci contribute to obesity in otherwise clinically normal people, Reed et al. (101) examined 44 families that had obese individuals. They tested linkage relationships between BMI and 17 genetic markers spanning the chromosomal regions of the Prader-Willi, the Bardet-Biedl (BBS1, 2, and 3), the Cohen, the Borjeson, and the Wilson-Turner syndromes. Analyses based on a

Table 10.8
Obesity-Related Mendelian Disorders With a Known Chromosome Map Location

MIM number	Disorder	Locus
Autosomal dominant		
100800	Achondroplasia (ACH)	4p16.3
122000	Posterior polymorphous corneal dystrophy (PPCD)	20q11
176270	Prader-Willi syndrome (PWS)	15q11.2-12
Autosomal recessive		
	Bardet-Biedl syndrome (BBS)	
209901	BBS1	11q13
209900	BBS2	16q21
600151	BBS3	3p13-12
600374	BBS4	15q22.3-23
216550	Cohen syndrome	8q22-23
X-linked		
301900	Borjeson-Forssman-Lehmann syndrome (BFLS)	Xq26-q27
303110	Choroideremia with deafness and obesity	Xq21
309585	Wilson-Turner syndrome (WTS)	Xq21.1-22
312870	Simpson-Golabi-Behmel syndrome (SGBS)	Xq25-27

Adapted from OMIM (Online Mendelian Inheritance in Man) computerized database.
From Bouchard and Pérusse (16).

total of 207 pairs of siblings revealed no evidence of linkage between any of the markers and obesity in the families. The results suggest that the genetic loci contributing to obesity in these families are not the same as those involved in the Mendelian disorders previously reported. However, before excluding these chromosomal regions as potential carriers of loci predisposing to obesity in clinically normal subjects, other studies with a larger number of genetic markers need to be performed. Further, these negative results do not exclude the possibility that individuals carrying one copy of a gene responsible for an autosomal recessive disorder are at a greater risk of becoming obese than individuals with no copy of the defective gene. For example, it has recently been shown that the prevalence of obesity was significantly higher among heterozygous carriers (parents of affected individuals) of a gene responsible for the Bardet-Biedl syndrome compared to age- and sex-matched normal noncarriers, suggesting that this gene may predispose to obesity in the general population (34).

Single-gene rodent models

Because of the high degree of homology among the genomes of mammalian species, it is possible to use rodent models of obesity to identify genes potentially involved in the etiology of human obesity. Several different single-gene mutations have been shown to cause obesity in rodents: diabetes (db), fat (fat), obese (ob), tubby (tub), adipose (Ad) and Yellow (A^y) in the mouse, and fatty (fa) in the rat (45). Although all of these mutations result in obesity in the animals, the time of onset and severity of obesity vary among the various mutations. Furthermore, the obesity phenotype is frequently associated with metabolic abnormalities, including insulin resistance, hyperinsulinemia, hyperglycemia, noninsulin dependent diabetes mellitus, hypertension, and hyperlipidemias (56).

Table 10.9 lists the six single-gene rodent models of obesity and the gene products recently identified in a series of sophisticated studies. The mouse agouti gene was the first to be cloned. In the normal mouse, the agouti gene encodes a paracrine signaling protein (ASP) that causes hair follicle melanocytes to synthesize pheomelanin, a yellow pigment. In the yellow mouse mutations the agouti gene is expressed in nearly every tissue, with obesity being one of the consequences (28,135).

The fat/fat mouse has a single ser202pro mutation that results in the lack of carboxypeptidase E (CPE) activity. The mutated CPE homozygotes exhibit a defect in the processing of proinsulin into insulin. The obesity of the fat mouse seems to be secondary to hyperproinsulinemia (79).

Table 10.9
Single-Gene Rodent Models of Obesity, Gene Products, and Possible Synteny Between Rodent Obesity Genes and Human Chromosome Regions

Locus	Gene product	Trans-mission	Mouse chromo-some	Human chromo-some
Mouse				
Diabetes (db)	Leptin receptor (LEPR)	Recessive	4	1p35-31
Fat (fat)	Carboxypeptidase E (CPE)	Recessive	8	4q21
Obese (ob)	Leptin (LEP)	Recessive	6	7q31.3
Tubby (tub)	Phosphodiesterase (?)	Recessive	7	11p15.1
Yellow (A^Y)	Agouti signaling protein (ASP)	Dominant	2	20q11.2
Rat				
Fatty (fa)	Leptin receptor (OB-R)	Recessive	5	1p35-31

A higher level of enthusiasm has been raised by the cloning of the ob gene (140) and the characterization of its product, the hormone leptin (52). Leptin decreases food intake in a variety of conditions and animal models (29,52,87). One study has also indicated that administration of leptin increases metabolic rate and augments locomotor activity in initially obese mice (52). These results have raised great expectations for the understanding and treatment of obesity. The recent news that a high-affinity leptin receptor gene ob-r was cloned and that it mapped within the interval of the mouse db locus has provided new impetus to this concept (121). Indeed, since most obese humans have normal or elevated levels of leptin, the view was that obesity may have been caused by a "resistance" to leptin.

The six mutations listed in table 10.9 were found to have homologous regions in the human genome. None of these regions, however, has yet been related to obesity in humans. Recently, 16 markers located on chromosome 20 in a region homologous to the Yellow mutation in the mouse (20q12-13.11) were tested for linkage with the BMI and with percentage body fat (as estimated by bioelectrical impedance) in a maximum of 210 sib pairs from 45 obese families (137). No evidence of linkage was found, suggesting that this gene does not contribute to obesity in humans. However, other genes in the same chromosomal areas appear to influence body fat content in humans as shown recently (66).

QTL from crossbreeding experiments

A method originally developed for plant genetics is now commonly used with rodents to identify loci that influence quantitative phenotypes such as obesity. It is the quantitative trait locus mapping method, or the QTL, which was described in chapter 4.

To date, eight QTLs have been identified by this method (table 10.10). Fisler et al. (44) obtained a backcross between the strains Mus spretus and C57BL/6J, which they called the BSB mouse. BSB exhibits a wide range of carcass lipid, from 1% to 50% and more. On the basis of the QTL approach with a large number of markers, Warden et al. (129,130) identified four "multigenic obesity" (Mob) loci on four different mouse chromosomes. One locus (Mob-1) on BSB chromosome 7 determined the lipid content of the carcass (lod score of 4.2). A second locus (Mob-2) on chromosome 6 affected only subcutaneous fat pads (lod = 4.8 with the femoral fat pad). A third locus (Mob-3) on chromosome 12 was linked to percentage lipid in the carcass (lod = 4.8), while Mob-4 (chromosome 15) was linked primarily to mesenteric fat (lod = 3.4). Syntenic regions with these four mouse QTLs are on human chromosomes 10q21-26, 11p14-ter, and 16p13-11 for Mob-1, 7q22-36 for Mob-2, 14q13-32 for Mob-3, and 5q11-13 for Mob-4 (table 10.10).

A mouse polygenic model of differential susceptibility to dietary fat has been developed by crossing a dietary-lipid-sensitive strain (AKR/J) with a resistant strain (SWR/J). After 12 weeks of feeding on a moderately high fat diet, the AKR/J strain had an approximately sixfold higher carcass fat content than the SWR/J strain (131). F2 animals and backcross data were used and, to date, three dietary obesity (Do) QTLs have been identified (132,133). Do-1 (chromosome 4), Do-2 (chromosome 9), and Do-3 (chromosome 15) are linked to level of adiposity, and Do-2 is also linked to mesenteric fat. Although regions of homology in the human genome have not been described, it appears likely that syntenic areas can be found on human chromosomes 1p36-32 and 9pter-q32 for Do-1, 3p21 for Do-2, and 5p14-12 for Do-3.

Finally, one QTL on mouse chromosome 2 was identified in a cross between mouse NZB/B1NJ and SM/J strains (43). The human equivalent of this locus appears to be on human chromosome 20p11.2-q13.2. Other loci have been reported recently for the various QTL experiments previously defined, but they are as yet unpublished and are not reported herein.

Table 10.10
Quantitative Trait Loci (QTL) Linked to Body Fat Phenotypes

Mouse cross	Locus	Lod score	Effect on adiposity	Mouse chromosome	Human location	Reference
AKR/J × SWR/J	Do-1	4.5	NA	4	1p36-p32	West et al. (132)
	Do-2	4.8	7% adiposity 47% mesenteric fat	9	3p21	West et al. (132)
	Do-3	3.9	4% adiposity	15	5p14-12	West et al. (132)
C57BL//6J × Mus spretus	Mob-1	4.2	7% of percent fat	7	10q21-26 11p14-ter 16p13-11	Warden et al. (130)
	Mob-2	4.8	7% femoral fat	6	7q22-36	Warden et al. (130)
	Mob-3	4.8	7% of percent fat	12	14q13-32	Warden et al. (130)
	Mob-4	3.4	6% mesenteric fat	15	5q11-13	Warden et al. (130)
NZB/B1NJ × SM/J	D2Mit22 D2Mit28		36% of percent fat	2	20q11.2-q13.2	Fisler et al. (43)

Do = dietary obese; Mob = multigenic obesity; NA = not available.

Reprinted, by permission, from C. Bouchard, A. Tremblay, J.P. Després et al., 1988, "Sensitivity to overfeeding:The Quebec experiment with identical twins," *Prog. Food Nutrition Science* 12:45-72.

Transgenic models

Results from transgenic mouse models also reveal that dysfunction in the expression of certain genes can cause obesity. For example, a transgenic mouse with impaired corticosteroid receptor function was created by a partial knockout of the type II glucocorticoid receptor with an antisense RNA transgene (88). The transgenic animals had increased fat deposition, with a body mass twice as high as that of control animals by six months of age. This was observed despite the fact that the transgenic animals ate about 15% less than the normal mice. It has also been reported that a transgenic mouse that expresses Glut-4 constitutively in the adipose tissue becomes quite fat, with only a moderate elevation in body mass (110). In the latter case, nutrient partitioning appears to be altered in favor of fat deposition.

Association studies

Association and linkage studies are important tools for the delineation of the genetic basis of overweight and obesity. The concept of association refers to a situation in which the correlation of a genetic polymorphism with a phenotype is investigated (see chapter 4).

It is important to recognize that the strength of an association is critical in the appraisal of its relevance as a marker for the susceptibility to obesity. Although there are no commonly agreed upon standards, an association can be qualified as strong when the p value is $< .001$, or when the mutation accounts for at least 10% of the phenotypic variance adjusted for the proper concomitants. In contrast, a weak association is one characterized by a p value $< .05$, or when the locus is associated with less than 5% of the phenotypic variance. In all cases, replication studies are highly desirable. Note, however, that these criteria are somewhat arbitrary.

The evidence for the presence of a significant association between a candidate gene and the BMI or body fat phenotypes is summarized in table 10.11. Only five markers have an association with a p value $< .01$: These are APOB, APOD, TNF-α (a closely linked marker of the gene), DRD2, and LDLR. Among these genes, only APOB was confirmed in an independent study, albeit at a lower p value (.05). In the case of the D2 dopamine receptor, an independent study found no relationship with the BMI (81). On the other hand, the weak associations with HSD3B1 or UCP were observed only with changes in fatness over time in adults in the Quebec Family Study and not with body fat content at a given time (86,127). As for fat distribution phenotypes, there are significant associations with HSD3B1 and specific skinfold sites among unrelated adults (127).

Table 10.11
Evidence for the Presence of an Association With the BMI or Body Fat Phenotypes

		EVIDENCE			
Gene	**Location**	***N* cases**	**Phenotype**	***p* value**	**Reference**
HSD3B1	1p13.1	132	12-year changes in sum of 6 skinfolds	.04	Vohl et al. (127)
ATP1A2	1q21-23	122	Percent fat	.05	Dériaz et al. (35)
APOB	2p24-23	132	BMI	.005	Rajput-Williams et al. (98)
		181	BMI	.05	Saha et al. (107)
ACP1	2p25	75	BMI in children	.02	Lucarini et al. (69)
APOD	3q26-ter	114	BMI	.006	Vijayaraghavan et al. (126)
UCP	4q28-31	123	High fat gainers over 12 years	.05	Oppert et al. (86)
TNF-α	6p21.3	304	BMI but not percent fat	.01	Norman et al. (83)
LPL	8p22	236	BMI	.05	Jemaa et al. (55)
DRD2	11q23.1	392	Relative weight	.002	Comings et al. (33)
LDLR	19p13	84	BMI in hypertensives	.004	Zee et al. (139)

Gene abbreviations and their chromosomal location are from the Human Genome Data Base. HSD3B1 = 3-beta hydroxysteroid dehydrogenase; ATP1A2 = sodium potassium adenosine triphosphatase alpha-2 subunit; APOB = apoliproprotien B; ACP1 = acid phosphatase; APOD = apolipoprotein D; UCP = uncoupling protein; TNF-α = tumor necrosis factor alpha; LPL = lipoprotein lipase; DRD2 = dopamine D_2 receptor; LDLR = low-density lipoprotien receptor.

Reprinted, by permission, from C. Bouchard, A. Tremblay, J.P. Després et al., 1988, "Sensitivity to overfeeding: The Quebec experiment with identical twins." *Prog. Food Nutrition Science* 12:45-72.

The table does not include the significant associations reported previously with the ABO (9q34) blood group and the human leukocyte antigen (HLA) systems. A review of the earlier studies (58) suggests that the results are rather ambiguous: Some found association of ABO with body weight, while others did not. The few studies that used the BMI or skinfold thicknesses as phenotypes reported negative results. In an attempt to further clarify the issue, the association between ABO blood type and body weight was recently investigated in four culturally distinct population samples; no evidence to support an association between ABO blood types and body weight was found (57). A few studies have also looked at association between obesity and class-I HLA markers. Although frequencies of the HLA B18, Bw35, and Cw4 antigens were significantly higher in obese subjects compared to controls in some studies (40,48), no association between HLA antigens and percentage body fat or subcutaneous fat was found in a larger population (21).

Linkage studies

Linkage refers to the cosegregation of a marker and a trait locus together in families. Linkage analysis can be performed with candidate gene markers or with a variety of other polymorphic markers such as microsatellites. Evidence for linkage becomes more apparent as the marker loci get closer to the true trait locus that cosegregates with the phenotype.

Since most of the results summarized in table 10.12 were derived from the single locus sib-pair linkage method, it is useful to consider the strength of evidence for such data. Linkage can be considered as strong if the p value $\leq .001$ and preferably with a replication. A weak linkage ($p < .05$) may be relevant if the finding is replicated by several laboratories. It is obvious in table 10.12 that only a few of the linkages reported have a reasonable degree of robustness. ACP1 (2p25), TNF-α (closely linked marker) (6p21.3), KEL (7q33) and ADA (20q12-13.11) are the only markers exhibiting such a linkage relationship. The others are supported by weaker evidence at this time. One marker on 1q31-32 was linked with a lod score of 3.6 at a recombination fraction of 0.05 in a three-generation pedigree in which the prevalence of obesity was high (78). Since these findings are sensitive to the assumptions made about mode of transmission, penetrance, and other characteristics, more information is needed to warrant a conclusion on the presence of a linkage with a marker on 1q.

Association and linkage studies that have generated negative results are numerous and have been summarized elsewhere (16). Most of the negative results were reported only once, and replication studies are needed to ensure that the findings are not false negative. The only clear

Table 10.12
Evidence for the Presence of Linkage With the BMI or Body Fat Phenotypes Based on the Sib-Pair Method

Gene or marker	Location	EVIDENCE			
		N pairs	Phenotype	*p* value	Reference
D1S202	1q31-32	3-generation pedigree	BMI	Lod = 3.6 at $\theta = .05$	Murray et al. (78)
ACP1	2p25	>300	BMI	.004	Bailey-Wilson et al. (2)
BF	6p21	>168	Triceps, subscapular, and suprailiac skinfold	$.01 < p < .03$	Wilson et al. (136)
GLO1	6p21	>168	Suprailiac skinfold and relative weight	$.004 < p < .05$	Wilson et al. (136)
TNF-α	6p21.3	304	% body fat	.002	Norman et al. (83)
KEL	7q33	402	BMI and the sum of 6 skinfolds	<.0001	Borecki et al. (6)
ESD	13q14.1-14.2	194	% body fat and sum of skinfolds	<.04	Borecki et al. (6)
ADA	20q13.11	428	BMI and sum of 6 skinfolds	$.02 < p < .001$	Borecki et al. (6)
P1	22q11	>168	Relative weight	.03	Wilson et al. (136)

See previous tables for abbreviations of loci already referred to.

Gene abbreviations and their chromosomal location are from the Human Genome Data Base. BF = properdin factor B; GLOI = glyoxylase I; KEL = blood group Kell; ESD = esterase D; ADA = adenosine deaminase; P1 = blood group P.

Reprinted, by permission, from C. Bouchard, A. Tremblay, J.P. Després et al., 1988, "Sensitivity to overfeeding: The Quebec experiment with identical twins." *Prog. Food Nutrition Science* 12:45-72.

case appears to be that of the ADRB3 gene, for which the same mutation was shown to be independent of the BMI in three different studies and independent of percentage body fat in one study (31,128,134). It should be noted that most association studies are based on small sample sizes.

This overview reveals that several genes can cause obesity or are linked with body fat content in animals. Table 10.13 lists those genes, loci, or markers for which the evidence of an association or a linkage with obesity, the BMI, or body fat content is more robust. The list includes the loci from single-gene rodent models, QTLs from cross-breeding experiments, Mendelian disorders exhibiting obesity as one of the clinical features, and genes supported by robust evidence from association and linkage studies in human populations. Four chromosomal arms (2p, 6p, 7q, and 20q) are of particular interest, since strong evidence for linkage with a marker in each of these regions has been reported. To date, however, no study has replicated the linkage for any of the four chromosomal areas. Other regions of considerable interest include 1p, 3p, 11p, 15q, and perhaps Xq. The compendium of markers and genes related to obesity or body fat is likely to grow significantly in the coming years.

Genetics and Regional Fat Distribution

Regional fat distribution is an important determinant of the relationship between obesity and health, and is an independent risk factor for various morbid conditions, such as cardiovascular diseases or noninsulin dependent diabetes mellitus. The genetics of fat topography phenotypes is considered subsequently.

Truncal-abdominal subcutaneous fat

Upper-body obesity is more prevalent in males than in females, and it increases in frequency with age in males and after menopause in females. It is moderately correlated with total body fat and appears to be more prevalent in individuals habitually exposed to stress. It is also associated in females with the levels of plasma androgens and cortisol. In addition, the activity of lipoprotein lipase in abdominal adipose tissue is elevated with higher levels of truncal-abdominal fat (15).

Evidence for familial resemblance in body fat distribution has been reported (41). Based on skinfold measurements obtained in 173 MZ and 178 DZ pairs of male twins, Selby et al. (109) concluded that there was a significant genetic influence on central deposition of body fat. Using data from the Canada Fitness Survey and the strategy of path analysis, the transmissible effect across generations reached about 40% for trunk

Table 10.13
A Summary of the Rodent (Human Homologous Regions) and Human Loci Potentially Associated or Linked With Obesity or Body Fat

Human chromosome	Locus	Human chromosome	Locus
1p36-32	mouse Do-1	11p15.1	mouse Tubby
1p35-31	mouse Db	11p14-ter	mouse Mob-1
1p35-31	rat Fa	11q13	BBS1
1q31-32	D1S202	13q14.1-14.2	ESD
2p24-23	APOB	14q13-32	mouse Mob-3
2p25	ACP1*	15q11.2-12	PWS
3p21	mouse Do-2	15q22.3-23	BBS4
3p13-12	BBS3	16p13-11	mouse Mob-1
4p16.3	ACH	16q21	BBS2
4q21	mouse Fat	20	NZB QTL
5p14-12	mouse Do-3	20q11	PPCD
5q11-13	mouse Mob-4	20q11.2	mouse Yellow
6p21	BF	20q13.1	ADA*
6p21	GLO1	22q11	P1
6p21.3	TNF-α*	Xq21	Choroideremia
7q31	mouse Ob	Xq21.1-22	WTS
7q22-36	mouse Mob-2	Xq25-27	SGBS
7q33	KEL*	Xq26-27	BFLS
8q22-23	Cohen syndrome		
10q21-26	mouse Mob-1		

See previous tables for abbreviations.
*Shows the strongest evidence of linkage in human studies.
Reprinted, by permission, from C. Bouchard, A. Tremblay, J.P. Després et al., 1988, "Sensitivity to overfeeding: The Quebec experiment with identical twins." *Prog. Food Nutrition Science* 12:45-72.

skinfolds (sum of subscapular and suprailiac skinfolds), limb skinfolds (sum of biceps, triceps, and medial calf skinfolds), and the ratio of trunk to limb skinfolds; it reached 28% for the waist-to-hip circumference ratio (90).

The biological and cultural components of transmission in regional fat distribution were further assessed with data from the Quebec Family Study (20). Two indicators of regional fat distribution were considered: the ratio of trunk to limb skinfolds and the ratio of subcutaneous fat to fat mass, obtained by dividing the sum of the six skinfolds by fat mass estimated from hydrostatic body density measurements. Genetic effects of 25% to 30% were obtained. When the influence of total body fat was taken into account, the profile of subcutaneous fat deposition was characterized by higher heritability estimates, which reached about 40% to 50% of the residual variance (12,13). These results imply that for a given level of fatness, some individuals store more fat on the trunk or abdominal area than others.

Results from two other studies suggest an influence of major genes for regional fat distribution phenotypes. In one study, Hasstedt et al. (53) reported a major gene effect explaining 42% of the variance in a relative fat pattern index defined as the ratio of the subscapular skinfold to the sum of the subscapular and suprailiac skinfold thicknesses. Recent results from the Quebec Family Study suggest major gene effects for the ratio of trunk to extremity skinfolds, adjusted for total fat mass, which account for about 35% of the phenotypic variance (7).

Abdominal visceral fat

Abdominal visceral fat (AVF) is recognized as the most atherogenic and diabetogenic fat depot of the human body. The first to propose that computerized tomography (CT) scanning could be used to assess AVF level was Borkan and colleagues (8). Tokunaga et al. (123) extended the method and proposed to use CT to assess overall fat distribution. However, it remained for H. Kvist and L. Sjöström to define a practical CT methodology specifically designed to approximate AVF from a single scan (63,113). More recently, magnetic resonance imaging (MRI) was demonstrated as a valid and reliable technique to assess AVF in humans (108,125).

AVF is consistently a slightly better correlate of glucose intolerance, insulin resistance, high plasma triglyceride, and low HDL cholesterol levels than body weight, the BMI, or general indicators of body fat (4,14,38,39,59,75). Subjects with elevated AVF levels also appear to have diminished skeletal muscle capacity to utilize free fatty acids (FFA) during fasting conditions (32).

AVF levels increase with age in both men and women and even more so in the obese (42). Men have more visceral fat on the average than women, but women gain substantially more AVF after menopause (60,67,71). Intervention studies have demonstrated that changes in total fat mass are not necessarily accompanied by parallel fluctuations in AVF. Thus, the relationship between gains in weight or fat mass and in AVF with chronic overfeeding was only moderate (23). The same observation was made for losses of body weight or fat mass and decreases in AVF under chronic negative energy balance (25). Moreover, Björntorp (4) has proposed that visceral obesity was a reflection of a so-called "civilization syndrome" in which factors such as physical inactivity, stress, smoking, and alcohol consumption contribute to the accumulation of AVF.

Several studies have indicated that variation in AVF in the population is characterized by a significant genetic component. Thus, familial aggregation accounts for almost 60% of the variance in AVF without or with adjustment for total fat mass (89). A major gene hypothesis for abdominal visceral fat was also examined in the Quebec Family Study (22). A putative recessive locus accounted for 51% of the variance, with 21% due to a multifactorial component. However, after adjustment for fat mass, support for a major gene was reduced. Although the major effect was significant, the Mendelian and no-transmission (i.e., environmental) models were not resolved. A bivariate analysis conducted between fat mass and AVF (104) supported a genetic pleiotropy hypothesis. The bivariate (cross-trait) familiality ranged from 29% to 50%. Moreover, since the univariate familialities for each trait were even higher (55% to 77% for fat mass and 55% to 65% for abdominal visceral fat), each trait was assumed to be influenced by additional familial factors that were specific to each. Given the significant spousal cross-trait correlations, at least some of the bivariate familial effect may be environmental in origin. In studies of MZ twins, significant intrapair resemblance was also observed for both increases in AVF with overfeeding and reductions in AVF with negative energy balance (23,25).

The studies with identical twins revealed that AVF was amenable to changes with gains or losses in body weight or body fat. The phenomenon has been reproduced in several recent studies using diet-induced or exercise-induced weight loss. For instance, small (30) and larger weight losses (47,64,65) brought about by caloric restriction are accompanied by a significant reduction in AVF, a reduction that may be slightly greater than the fat loss registered in subcutaneous fat. On the other hand, when fat loss is induced by a negative energy balance protocol that includes regular exercise, the loss of AVF tends to be even more pronounced (25,106). Interestingly, it was noted in one study that the final level of AVF was not significantly altered following a full cycle of weight loss (about 13 kg on average) followed by weight regain (about 12 kg) in 32 obese subjects of both sexes (125).

Contribution of specific genes

In the Quebec Family Study, three markers are weakly linked with the ratio of trunk to extremity skinfolds adjusted for total body fat content. The MN (*GYPA*) (4q28-31) and the KEL (7q33) were both found to be linked ($p = .03$ and .04) with the ratio based on a minimum of 160 pairs of sibs (6). Another linkage was reported between the alpha 2α adrenergic receptor gene (10q24-26) and the same phenotype (85).

QTLs have also been uncovered for fat distribution phenotypes. Thus, Mob-2, Mob-4, and Do-2 have been shown to be linked with specific fat depots, including mesenteric fat for Do-2 (129,130,132,133). Mob-2 accounted for 7% of femoral fat, while Mob-4 was associated with about 6% of mesenteric fat. The strongest linkage was between Do-2 and mesenteric fat, the QTL accounting for 47% of the variance in this fat depot. The syntenic regions of these QTLs are thought to be located on human 7q22-q36, 5p15-q11, and 3p21.

Is Obesity Inherited?

The increasing prevalence of obesity is viewed as a major public health problem. Recent evidence from the third phase of the National Health and Nutrition Examination Survey of the American population indicates a dramatic increase in the prevalence of obesity in adults from earlier surveys in the 1960s to the most recent survey in the early 1990s (62). The magnitude of concern in the problem of obesity and related dietary behaviors was highlighted in the cover story of the January 16, 1995, issue of *Time* magazine: "Girth of a Nation—Here's some news that's hard to swallow: Despite the health craze, Americans are fatter than ever." About one-third of adults in the United States and Canada are classified as overweight (BMI > 27) or obese (BMI > 30), and the prevalence of such cases has doubled since the beginning of this century. There are considerable differences in the prevalence of obesity, however, among nations of the developed and developing world (26).

Excessive body fat content or obesity is a complex multifactorial trait that develops under the interactive influences of many affectors from the social, behavioral, physiological, metabolic, cellular, and molecular domains. Segregation of genes for obesity is not easily detected in familial or pedigree studies; whatever the influence of genotype on the etiology of obesity, it is generally attenuated or exacerbated by nongenetic factors.

Efforts to understand the genetic basis of such traits can be successful only if they are based on an appropriate conceptual framework, adequate phenotype and intermediate phenotype measurements, proper samples of unrelated persons and nuclear families or extended pedigrees, and typing of large panels of candidate genes and other molecular markers. In this context, the distinction between "necessary" genes and "susceptibility" genes is particularly relevant. In the case of obesity, there are several examples of necessary loci resulting in excess body mass or body fat for stature, that is, carriers of the deficient alleles have the disease (e.g., patients with the Prader-Willi syndrome). However, these represent only a very small fraction of the obese population.

A susceptibility gene is defined as one that increases susceptibility or risk for a disease but that is not necessary for the expression of the disease. An allele at a susceptibility gene may make it more likely that the carrier will become affected, but the presence of that allele is not sufficient by itself to explain the occurrence of the disease. It merely lowers the threshold for a person to develop the disease. This scenario seems to apply particularly well to the case of obesity.

It is likely that body fat content is also modulated over the lifetime of a person by a variety of gene-environment interaction effects. These effects result from the fact that sensitivity to environmental exposures or lifestyle differences varies from person to person because of genetic individuality. These factors include dietary fat, energy intake, level of habitual physical activity, smoking, and alcohol intake, among many others. In addition, gene-gene interaction effects must be considered. However, little research bearing directly on this topic has been reported thus far.

Association and linkage genetic studies are more frequently reported. Based on current understanding of the pathophysiology of human obesity, it is expected that the candidate gene approach will yield useful association and linkage results. Moreover, with the advent of a comprehensive human genetic linkage map, linkage studies with a large number of markers covering most of the chromosomes of the human genome are likely to be helpful in identification of putative obesity genes or chromosomal regions. Recent progress in animal genetics, transfection systems, transgenic animal models, recombinant DNA technologies applied to positional cloning, and methods to identify loci contributing to quantitative traits have given a new impetus to this field. The stage is now set for further advances in understanding the genetic and molecular basis of human obesities.

From the research currently available, a good number of genes seem to have the capacity to cause obesity and/or to increase the likelihood of becoming obese when they are altered or are dysfunctional in mammals. Even though it is obvious that investigation of the molecular markers of obesity has barely begun, several genes, loci, or chromosomal regions that appear to play a role in determining obesity phenotypes have been identified. They are located on a large number of chromosomes. Many additional genes will likely be identified in the future so that the panel of human obesity genes, based on association, linkage, or animal models, will grow and become quite large. This is a reflection of how most cases of human obesity may come about. In other words, the susceptibility genotypes may result from allelic variations in a large number of genes.

Figure 10.8 attempts to depict the situation described here. The multigenic nature is illustrated by the mixture of "susceptibility" and "resistant" alleles that define the genotype. We do not know the number of genes involved yet, but four are proposed here for the purpose of the illustration. In addition, the influence of nongenetic factors is represented as being superimposed on given genetic characteristics, characteristics which seem to impose limits on one's ability to shift the body weight or body fat phenotype upward or downward. The reality may be even more complex than suggested by the figure, as a result of a larger number of susceptibility genes and likely gene-gene and gene-environment interaction effects. We believe that the rationale described here and the conditions illustrated in the figure also apply to visceral obesity.

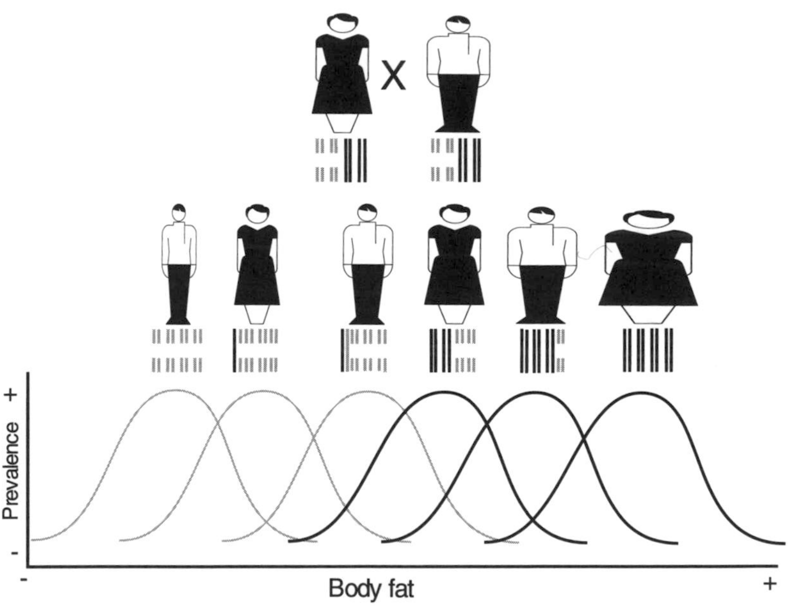

Figure 10.8 A schematic illustration of the contribution of a multigenic and a nongenetic component. ‖ represents susceptibility alleles while ⁞⁞ stands for resistance alleles. The multigenic component is represented by only four genes. The phenotypic value of each genotype can be shifted toward a higher or a lower body fat content under the influence of a variety of environmental and lifestyle factors that affect energy balance. From Bouchard (25a).

The search for these genes is a major undertaking. The task is even more complex when the important issues of gene-environment, gene-energy intake, gene-nutrient intake, gene-exercise, and, particularly, gene-gene interactions are added. The resources of human genetics, experimental genetics, and molecular biology are needed to elucidate the genetic basis of the predisposition to obesity.

Summary

Body fat content and fat distribution are complex phenotypes influenced by genetic and environment or lifestyle factors. Body composition and fat topography vary with age and exhibit striking gender differences. Body fat increases with age, on the average, and amount of abdominal visceral fat augments with age in men and particularly after menopause in women. Genetic epidemiology studies reveal that heritability of the BMI or body fat reaches about 25% to 40% of age- and gender-adjusted phenotypic variance. The level of heritability seems to be higher for fat distribution phenotypes. Identical twins are significantly similar for changes in body composition and fat topography associated with prolonged exposure to positive or negative energy balance conditions.

The search for the genes associated with the susceptibility to obesity has begun. Evidence from rodent studies (single-gene obesity models, crossbreeding experiments with inbred strains, transgenic studies, and knockout studies) and from human research (Mendelian syndromes, association studies with candidate genes, linkage studies) support the notion that obesity is influenced by many genes and that the disorder is causally heterogeneous. Despite the fact that only a few studies have been reported about the contribution of specific genes to upper-body fat or abdominal visceral fat levels, a similar pattern is emerging. It is fair to say at this time that research supports the conclusion that there are individuals who are more prone than others to become obese. Conversely, the increase in the prevalence of obesity during this century also reveals that nongenetic influences are extremely important in the determination of long-term energy balance.

References

1. Adams, T.D.; Hunt, S.C.; Mason, L.A.; Ramirez, M.E.; Fischer, A.G.; Williams, R.R. Familial aggregation of morbid obesity. Obes. Res. 1:261-70; 1993.

2. Bailey-Wilson, J.E.; Wilson, A.F.; Bamba, V. Linkage analysis in a large pedigree ascertained due to essential familial hypercholesterolemia. Genet. Epidemiol. 10:665-69; 1993.
3. Beynen, A.C.; Katan, M.B.; Van Zutphen, L.F. Hypo- and hyperresponders: Individual differences in the response of serum cholesterol concentration to changes in diet. Adv. Lipid. Res. 22:115-71; 1987.
4. Björntorp, P. Visceral obesity: A "civilization syndrome." Obes. Res. 1:206-22; 1993.
5. Borecki, I.B.; Bonney, G.E.; Rice, T.; Bouchard, C.; Rao, D.C. Influence of genotype-dependent effects of covariates on the outcome of segregation analysis of the body mass index. Am. J. Hum. Genet. 53:676-87; 1993.
6. Borecki, I.B.; Rice, T.; Pérusse, L.; Bouchard, C.; Rao, D.C. An exploratory investigation of genetic linkage with obesity phenotypes: The Quebec Family Study. Obes. Res. 2:213-19; 1994.
7. Borecki, I.B.; Rice, T.; Pérusse, L.; Bouchard, C.; Rao, D.C. Major gene influence on the propensity to store fat in trunk versus extremity depots: Evidence from the Québec Family Study. Obes. Res. 3:1-8; 1995.
8. Borkan, G.A.; Gerzof, S.G.; Robbins, A.H.; Hults, D.E.; Silbert, C.K.; Silbert, J.E. Assessment of abdominal fat content by computer tomography. Am. J. Clin. Nutr. 36:172-77; 1982.
9. Bouchard, C. Genetics of human obesities: Introductory notes. In: Bouchard, C., ed. The genetics of obesity. Boca Raton, FL: CRC Press; 1994:1-15.
10. Bouchard, C. Genetics of obesity: Overview and research directions. In: Bouchard, C., ed. The genetics of obesity. Boca Raton: CRC Press; 1994:223-33.
11. Bouchard, C. Human obesities: Chaos or determinism. In: Ailhaud, G.; Guy-Grand, B.; Lafontan, M.; Ricquier, D., eds. Obesity in Europe 91. Proceedings of the 3rd European Congress on Obesity. London: John Libbey; 1992:7-14.
12. Bouchard, C. Inheritance of human fat distribution. In: Bouchard, C.; Johnson, F.E., eds. Fat distribution during growth and later health outcomes. New York: Alan R. Liss; 1988:103-25.
13. Bouchard, C. Variation in human body fat: The contribution of the genotype. In: Bray, G.A.; Ricquier, D.; Spiegelman, B.M., eds. Obesity: Towards a molecular approach. New York: Alan R. Liss; 1990:17-28.
14. Bouchard, C.; Després, J.P.; Mauriège, P. Genetic and nongenetic determinants of regional fat distribution. Endocr. Rev. 14:72-93; 1993.
15. Bouchard, C.; Després, J.P.; Mauriège, P.; Marcotte, M.; Chagnon, M.; Dionne, F.T.; Bélanger, A. The genes in the constellation of determinants of regional fat distribution. Int. J. Obes. 15:9-18; 1991.
16. Bouchard, C.; Pérusse, L. Current status of the human obesity gene map. Obes. Res. 4:81-90; 1996.
17. Bouchard, C.; Pérusse, L. Genetics of obesity. Ann. Rev. Nutr. 13:337-54; 1993.

18. Bouchard, C.; Pérusse, L. Genetics of obesity: Family studies. In: Bouchard, C., ed. The genetics of obesity. Boca Raton, FL: CRC Press; 1994:72-92.

19. Bouchard, C.; Pérusse, L. Heredity and body fat. Ann. Rev. Nutr. 8:259-77; 1988.

20. Bouchard, C.; Pérusse, L.; Leblanc, C.; Tremblay, A.; Thériault, G. Inheritance of the amount and distribution of human body fat. Int. J. Obes. 12:205-15; 1988.

21. Bouchard, C.; Pérusse, L.; Rivest, J.; Roy, R.; Morissette, J.; Allard, C.; Thériault, G.; Leblanc, C. HLA system body fat and fat distribution in children and adults. Int. J. Obes. 9:411-22; 1985.

22. Bouchard, C.; Rice, T.; Lemieux, S.; Després, J.P.; Pérusse, L.; Rao, D.C. Major gene for abdominal visceral fat area in the Québec Family Study. Int. J. Obes. 20:420-27; 1996.

23. Bouchard, C.; Tremblay, A.; Després, J.P.; Nadeau, A.; Lupien, P.J.; Thériault, G.; Dussault, J.; Moorjani, S.; Pineault, S.; Fournier, G. The response to long-term overfeeding in identical twins. N. Engl. J. Med. 322:1477-82; 1990.

24. Bouchard, C.; Tremblay, A.; Després, J.P.; Poehlman, E.T.; Thériault, G.; Nadeau, A.; Lupien, P.; Moorjani, S. Sensitivity to overfeeding: The Quebec experiment with identical twins. Prog. Food Nutr. Sci. 12:45-72; 1988.

25. Bouchard, C.; Tremblay, A.; Després, J.P.; Thériault, G.; Nadeau, A.; Lupien, P.J.; Moorjani, S.; Prud'homme, D.; Fournier, G. The response to exercise with constant energy intake in identical twins. Obes. Res. 2:400-10; 1994.

25a. Bouchard, C. Advances in the genetics of obesity: impact on therapeutic perspectives. In: Caterson, I., ed. Treating eating patterns nutrition and overweight. No.17. Australia: Servier Laboratories (AUST) Pty. Ltd.; 1996: 1-18.

26. Bray, G.A. Etiology and prevalence of obesity. In: Bouchard, C., ed. The genetics of obesity. Boca Raton, FL: CRC Press; 1994:17-33.

27. Bray, G.A. The inheritance of corpulence. In: Cioffi, L.A., ed. The body weight regulators system: Normal and disturbed mechanism. New York: Raven Press; 1981:185-95.

28. Bultman, S.J.; Michaud, E.J.; Woychik, R.P. Molecular characterization of the mouse agouti locus. Cell. 71:1195-1204; 1992.

29. Campfield, L.A.; Smith, F.J.; Guisez, Y.; Devos, R.; Burn, P. Recombinant mouse OB protein: Evidence for a peripheral signal linking adiposity and central neural networks. Science. 269:546-49; 1995.

30. Chowdhury, B.; Kvist, H.; Andersson, B.; Björntorp, P.; Sjöström, L. CT-determined changes in adipose tissue distribution during a small weight reduction in obese males. Int. J. Obes. 17:685-91; 1993.

31. Clément, K.; Vaisse, C.; Manning, B.S.J.; Basdevant, A.; Guy-Grand, B.; Ruiz, J.; Silver, K.D.; Shuldiner, A.R.; Froguel, P.; Strosberg, D. Genetic

variation in the β3-adrenergic receptor and an increased capacity to gain weight in patients with morbid obesity. N. Engl. J. Med. 333:352-54; 1995.

32. Colberg, S.R.; Simoneau, J.A.; Thaete, F.L.; Kelley, D.E. Skeletal muscle utilization of free fatty acids in women with visceral obesity. J. Clin. Invest. 95:1846-53; 1995.
33. Comings, D.E.; Flanagan, S.D.; Dietz, G.; Muhleman, D.; Knell, E.; Gysin, R. The dopamine D2 receptor (DRD2) as a major gene in obesity and height. Biochem. Med. Metab. Biol. 50:176-85; 1993.
34. Croft, J.B.; Morrell, D.; Chase, C.L.; Swift, M. Obesity in heterozygous carriers of the gene for the Bardet-Biedl syndrome. Am. J. Med. Genet. 55:12-15; 1995.
35. Dériaz, O.; Dionne, F.; Pérusse, L.; Tremblay, A.; Vohl, M.C.; Côté, G.; Bouchard, C. DNA variation in the genes of the Na,K-adenosine triphosphatase and its relation with resting metabolic rate, respiratory quotient, and body fat. J. Clin. Invest. 93:838-43; 1994.
36. Dériaz, O.; Fournier, G.; Tremblay, A.; Després, J.P.; Bouchard, C. Lean-body-mass composition and resting energy expenditure before and after long-term overfeeding. Am. J. Clin. Nutr. 56:840-47; 1992.
37. Dériaz, O.; Tremblay, A.; Bouchard, C. Non linear weight gain with long term overfeeding in man. Obes. Res. 1:179-85; 1993.
38. Després, J.P.; Moorjani, S.; Lupien, P.J.; Tremblay, A.; Nadeau, A.; Bouchard, C. Genetic aspects of susceptibility to obesity and related dyslipidemias. Mol. Cell. Biochem. 113:151-69; 1992.
39. Després, J.P.; Moorjani, S.; Lupien, P.J.; Tremblay, A.; Nadeau, A.; Bouchard, C. Regional distribution of body fat, plasma lipoproteins, and cardiovascular disease. Arteriosclerosis. 10:497-511; 1990.
40. Digy, J.P.; Raffoux, C.; Pointel, J.P.; Perrier, P.; Drouin, P.; Mejean, L.; Streiff, F.; Debry, G. HLA and familial obesity: Evidence for a genetic origin. In: Hirsch, J.; Van Itallie, T.B., eds. Recent advances in obesity research: IV. London: John Libbey; 1983:171-75.
41. Donahue, R.P.; Prineas, R.J.; Gomez, O.; Hong, C.P. Familial resemblance of body fat distribution: The Minneapolis children's blood pressure study. Int. J. Obes. 16:161-67; 1992.
42. Enzi, G.; Gasparo, M.; Biondetti, P.R.; Fior, D.; Semisa, M.; Zurlo, F. Subcutaneous and visceral fat distribution according to sex, age, and overweight, evaluated by computed tomography. Am. J. Clin. Nutr. 44:739-46; 1986.
43. Fisler, J.S.; Purcell-Huynh, D.A.; Cuevas, M.; Lusis, A.J. The agouti gene may promote obesity in a polygenic mouse model. Int. J. Obes. [Abstract]. 18:104; 1994.
44. Fisler, J.S.; Warden, C.H.; Pace, M.J.; Lusis, A.J. BSB: A new mouse model of multigenic obesity. Obes. Res. 1:271-80; 1993.
45. Friedman, J.M.; Leibel, R.L.; Bahary, N. Molecular mapping of obesity genes. Mamm. Genome. 1:130-44; 1991.

46. Friis-Hansen, B. Hydrometry of growth and aging. In: Brozek, J., ed. Human body composition: Approaches and applications. Oxford: Pergamon Press; 1965:191-209.

47. Fujioka, S.; Matsuzawa, Y.; Tokunaga, D.; Kawamoto, T.; Kabatake, T.; Keno, Y.; Kotani, K.; Yoshida, S. Improvement of glucose and lipid metabolism associated with selective reduction of intra-abdominal visceral fat in premenopausal women with visceral fat obesity. Int. J. Obes. 15:853-59; 1991.

48. Fumeron, F.; Apfelbaum, M. Association between HLA-B18 and the familial obesity syndrome. N. Engl. J. Med. 305:645; 1981.

49. Garn, S.M.; Leonard, W.R.; Hawthorne, V.M. Three limitations of the body mass index. Am. J. Clin. Nutr. 44:996-97; 1986.

50. Garn, S.M.; Sullivan, T.V.; Hawthorne, V.M. Fatness and obesity of the parents of obese individuals. Am. J. Clin. Nutr. 50:1308-13; 1989.

51. Grilo, C.M.; Pogue-Geile, M.F. The nature of environmental influences on weight and obesity: A behavior genetic analysis. Psychol. Bull. 110:520-37; 1991.

52. Halaas, J.L.; Gajiwala, K.S.; Maffei, M.; Cohen, S.L.; Chait, B.T.; Rabinowitz, D.; Lallone, R.L.; Burley, S.K.; Friedman, J.M. Weight reducing effects of the plasma protein encoded by the obese gene. Science. 269:543-46; 1995.

53. Hasstedt, S.J.; Ramirez, M.E.; Kuida, H.; Williams, R.R. Recessive inheritance of a relative fat pattern. Am. J. Hum. Genet. 45:917-25; 1989.

54. Heller, R.; Garrison, R.J.; Havlik, R.J.; Feinleib, M.; Padgett, S. Family resemblances in height and relative weight in the Framingham Heart Study. Int. J. Obes. 8:399-405; 1984.

55. Jemaa, R.; Tuzet, S.; Portos, C.; Betoulle, D.; Apfelbaum, M.; Fumeron, F. Lipoprotein lipase gene polymorphisms: Associations with hypertriglyceridemia and body mass index in obese people. Int. J. Obes. 19:270-74; 1995.

56. Johnson, P.R.; Gregoire, F. Animal models of genetic obesity: Peripheral tissue changes. In: Bouchard, C., ed. The genetics of obesity. Boca Raton, FL: CRC Press; 1994:161-79.

57. Kelso, A.J.; Maggi, W.; Belas, K.L. Body weight and ABO blood types: Are AB females heavier? Am. J. Hum. Biol. 6:385-87; 1994.

58. Kelso, A.J.; Siffert, T.; Maggi, W. Association of ABO phenotypes and body weight in a sample of Brazilian infants. Am. J. Hum. Biol. 4:607-11; 1992.

59. Kissebah, A.H.; Krakower, G.R. Regional adiposity and morbidity. Physiol. Rev. 74:761-811; 1994.

60. Kotani, K.; Tokunaga, K.; Fujioka, S.; Kobatake, T.; Keno, Y.; Yoshida, S.; Shimomura, I.; Tarui, S.; Matsuzawa, Y. Sexual dimorphism of age-related changes in whole-body fat distribution in the obese. Int. J. Obes. 18:207-12; 1994.

61. Krzywicki, H.J.; Chinn, K.S.K. Human body density and fat of an adult male population as measured by water displacement. Am. J. Clin. Nutr. 20:305-10; 1967.

62. Kuczmarski, R.J.; Flegal, K.M.; Campbell, S.M.; Johnson, C.L. Increasing prevalence of overweight among US adults. The national health and nutrition examination surveys, 1960 to 1991. JAMA. 272:205-11; 1994.

63. Kvist, H.; Sjöström, L.; Tylen, U. Adipose tissue volume determinations in women by computed tomography: Technical considerations. Int. J. Obes. 10:53-67; 1986.

64. Leenen, R.; van der Kooy, D.; Deurenberg, P.; Seidell, J.C.; Weststrate, J.A.; Schouten, F.J.; Hautvast, J.G. Visceral fat accumulation in obese subjects: Relation to energy expenditure and response to weight loss. Amer. J. Physiol. 263:E913-19; 1992.

65. Leenen, R.; van der Kooy, K.; Droop, A.; Seidell, J.C.; Deurenberg, P.; Weststrate, J.A.; Hautvast, J.G.A.J. Visceral fat loss measured by magnetic resonance imaging in relation to changes in serum lipid levels of obese men and women. Arterioscler. Thromb. 13:487-94; 1993.

66. Lembertas, A.V.; Pérusse, L.; Chagnon, Y.C.; Fisler, J.S.; Warden, C.H.; Ting, K.; Purcell-Huynh, D.A.; Dionne, F.T.; Gagnon, J.; Nadeau, A.; Lusis, A.J.; Bouchard, C. Genes contributing to body fat on mouse chromosome 2 and the homologous region of human chromosome 20. Submitted.

67. Lemieux, S.; Prud'homme, D.; Bouchard, C.; Tremblay, A.; Després, J.P. Sex differences in the relation of visceral adipose tissue accumulation to total body fatness. Am. J. Clin. Nutr. 58:463-67; 1993.

68. Lissner, L.; Sjöström, L.; Bengtsson, C.; Bouchard, C.; Larsson, B. The natural history of obesity in an obese population and associations with metabolic aberrations. Int. J. Obes. 18:441-47; 1994.

69. Lucarini, N.; Finocchi, G.; Gloria-Bottini, F.; Macioce, M.; Borgiani, P.; Amante, A.; Bottini, E. A possible genetic component of obesity in childhood: Observations on acid phosphatase polymorphism. Experientia. 46:90-91; 1990.

70. MacDonald, A.; Stunkard, A.J. Body mass indexes of British separated twins. N. Engl. J. Med. 322:1530; 1990.

71. Malina, R.M. Regional body composition: Age, sex, and ethnic variation. In: Roche, A.F.; Heymsfield, S.B.; Lohman, T.G., eds. Human body composition. Champaign, IL: Human Kinetics; 1996:217-55.

72. Malina, R.M.; Bouchard, C. Growth, maturation and physical activity. Champaign, IL: Human Kinetics; 1991.

73. Malina, R.M.; Bouchard, C. Subcutaneous fat distribution during growth. In: Bouchard, C.; Johnston, F.E., eds. Fat distribution during growth and later health outcomes. New York: Wiley-Liss; 1988:63-84.

74. Malina, R.M.; Bouchard, C.; Beunen, G. Human growth: Selected aspects of current research on well-nourished children. Ann. Rev. Anthropol. 17:187-219; 1988.

75. Matsuzawa, Y.; Fujioka, S.; Tokunaga, K.; Tarui, S. Classification of obesity with respect to morbidity. Proc. Soc. Exp. Biol. Med. 200:197-201; 1992.

76. Moll, P.P.; Burns, T.L.; Lauer, R.M. The genetic and environmental sources of body mass index variability: The Muscatine ponderosity family study. Am. J. Hum. Genet. 49:1243-55; 1991.

77. Mueller, W.H. The genetics of human fatness. Yearbook. Phys. Anthropol. 26:215-30; 1983.

78. Murray, J.D.; Bulman, D.E.; Ebers, G.C.; Lathrop, G.M.; Rice, G.P.A. Linkage of morbid obesity with polymorphic microsatellite markers on chromosome 1q31 in a three-generation Canadian kinbred. Am. J. Hum. Genet. 55:A197; 1994.

79. Naggert, J.K.; Fricker, L.D.; Varlamov, O.; Nishina, P.M.; Rouille, Y.; Steiner, D.F.; Carroll, R.J.; Paigen, B.J.; Leiter, E.H. Hyperproinsulinaemia in obese fat/fat mice associated with a carboxypeptidase E mutation which reduces enzyme activity. Nat. Genet. 10:135-42; 1995.

80. Najjar, M.F.; Rowland, M. Anthropometric reference data and prevalence of overweight. Vital Health Stat. Ser. 11, No. 238; 1987.

81. Noble, E.P.; Noble, R.E.; Ritchie, T.; Syndulko, K.; Bohlman, M.C.; Noble, L.A.; Zhang, Y.; Sparkes, R.S.; Grandy, D.K. D2 dopamine receptor gene and obesity. Int. J. Eat. Disord. 15:205-17; 1994.

82. Norgan, N.G.; Durnin, J.V.G.A. The effect of 6 weeks of overfeeding on the body weight, body composition, and energy metabolism of young men. Am. J. Clin. Nutr. 33:978-88; 1980.

83. Norman, R.A.; Bogardus, C.; Ravussin, E. Linkage between obesity and a marker near the tumor necrosis factor-α locus in Pima Indians. J. Clin. Invest. 96:158-62; 1995.

84. Norris, A.H.; Lundy, T.; Shock, N. Trends in selected indices of body composition in men between the ages of 30 and 80 years. Ann. N.Y. Acad. Sci. 110:623-39; 1963.

85. Oppert, J.M.; Tourville, J.; Chagnon, M.; Mauriège, P.; Dionne, F.T.; Pérusse, L.; Bouchard, C. DNA polymorphisms in α2- and β2-adrenoceptor genes and regional fat distribution in humans: Association and linkage studies. Obes. Res. 3:249-55; 1995.

86. Oppert, J.M.; Vohl, M.C.; Chagnon, M.; Dionne, F.T.; Cassard-Doulcier, A.M.; Ricquier, D.; Pérusse, L.; Bouchard, C. DNA polymorphism in the uncoupling protein (UCP) gene and human body fat. Int. J. Obes. 18:526-31; 1994.

87. Pelleymounter, M.A.; Cullen, M.J.; Baker, M.S.; Hecht, R.; Winters, D.; Boone, T.; Collins, F. Effects of the obese gene product on body regulation in ob/ob mice. Science. 269:540-43; 1995.

88. Pépin, M.C.; Pothier, F.; Barden, N. Impaired type II glucocorticoid receptor function in mice bearing antisense RNA transgene. Nature. 235:725-28; 1992.

89. Pérusse, L.; Després, J.P.; Lemieux, S.; Rice, T.; Rao, D.C.; Bouchard, C. Familial aggregation of abdominal visceral fat level: Results from the Quebec Family Study. Metabolism. 45:378-82; 1996.
90. Pérusse, L.; Leblanc, C.; Bouchard, C. Inter-generation transmission of physical fitness in the Canadian population. Can. J. Sport. Sci. 13:8-14; 1988.
91. Poehlman, E.T.; Després, J.P.; Marcotte, M.; Tremblay, A.; Thériault, G.; Bouchard, C. Genotype dependency of adaptation in adipose tissue metabolism after short-term overfeeding. Am. J. Physiol. 250:E480-85; 1986.
92. Poehlman, E.T.; Tremblay, A.; Després, J.P.; Fontaine, E.; Pérusse, L.; Thériault, G.; Bouchard, C. Genotype-controlled changes in body composition and fat morphology following overfeeding in twins. Am. J. Clin. Nutr. 43:723-31; 1986.
93. Poehlman, E.; Tremblay, A.; Marcotte, M.; Pérusse, L.; Thériault, G.; Bouchard, C. Heredity and changes in body composition and adipose tissue metabolism after short-term exercise-training. Eur. J. Appl. Physiol. 56:398-402; 1987.
94. Price, R.A.; Cadoret, R.J.; Stunkard, A.J.; Troughton, E. Genetic contributions to human fatness: An adoption study. Am. J. Psych. 144:1003-8; 1987.
95. Price, R.A.; Gottesman, I.I. Body fat in identical twins reared apart: Roles for genes and environment. Behav. Genet. 21:1-7; 1991.
96. Price, R.A.; Ness, R.; Laskarzewski, P. Common major gene inheritance of extreme overweight. Hum. Biol. 62:747-65; 1990.
97. Province, M.A.; Amqvist, P.; Keller, J.; Higgins, M.; Rao, D.C. Strong evidence for a major gene for obesity in the large, unselected, total Community Health Study of Tecumseh. Am. J. Hum. Genet. 47:A143; 1990.
98. Rajput-Williams, J.; Wallis, S.C.; Yarnell, J.; Bell, G.I.; Knott, T.J.; Sweetnam, P.; Cox, N.; Miller, N.E. Variation of apolipoprotein-B gene is associated with obesity, high blood cholesterol levels, and increased risk of coronary heart disease. Lancet. 2:1442-46; 1988.
99. Ravussin, E.; Schutz, Y.; Acheson, K.J.; Dusmet, M.; Dusmet, M.; Bourquin, L.; Jequier, E. Short-term, mixed-diet overfeeding in man: No evidence for "luxuskonsumption." Am. J. Physiol. 249:E470-77; 1985.
100. Reed, D.R.; Bradley, E.C.; Price, R.A. Obesity in families of extremely obese women. Obes. Res. 1:167-72; 1993.
101. Reed, D.R.; Ding, Y.; Xu, W.; Cather, C.; Price, R.A. Human obesity does not segregate with the chromosomal regions of Prader-Willi, Bardet-Biedl, Cohen, Borjeson or Wilson-Turner syndromes. Int. J. Obes. 19:599-603; 1995.
102. Rice, T.; Borecki, I.B.; Bouchard, C.; Rao, D.C. Segregation analysis of body mass index in an unselected French-Canadian sample: The Québec Family Study. Obes. Res. 2:288-94; 1993.
103. Rice, T.; Borecki, I.B.; Bouchard, C.; Rao, D.C. Segregation analysis of fat mass and other body composition measures derived from underwater weighing. Am. J. Hum. Genet. 52:967-73; 1993.

104. Rice, T.; Pérusse, L.; Bouchard, C.; Rao, D.C. Familial clustering of abdominal visceral fat and total fat mass: The Québec Family Study. Obes. Res. 4:253-61; 1996.

105. Roche, A.F.; Heymsfield, S.B.; Lohman, T.G. Human body composition. Champaign, IL: Human Kinetics; 1996.

106. Ross, R.; Rissanen, J. Mobilization of visceral and subcutaneous adipose tissue in response to energy restriction and exercise. Am. J. Clin. Nutr. 60:695-703; 1994.

107. Saha, N.; Tay, J.S.H.; Heng, C.K.; Humphries, S.E. DNA polymorphisms of the apolipoprotein B gene are associated with obesity and serum lipids in healthy Indians in Singapore. Clin. Genet. 44:113-20; 1993.

108. Seidell, J.C.; Bakker, C.J.G.; van der Kooy, K. Imaging techniques for measuring adipose-tissue distribution—A comparison between computed tomography and 1.5-T magnetic resonance. Am. J. Clin. Nutr. 51:953-57; 1990.

109. Selby, J.V.; Newman, B.; Quesenberry, C.P., Jr.; Fabsitz, R.R.; King, M.C.; Meaney, J.M. Evidence of genetic influence on central body fat in middle-aged twins. Hum. Biol. 61:179-93; 1989.

110. Shepherd, P.R.; Gnudi, L.; Tozzo, E.; Yand, H.; Kahn, B.B. Enhanced glucose disposal and obesity in transgenic mice overexpressing Glut-4 selectively in fat. Diabetes. 42:239; 1993.

111. Sims, E.A.H.; Goldman, R.F.; Gluck, C.M.; Hortin, C.M.; Kelleher, P.C.; Rowe, D.W. Experimental obesity in man. Trans. Assoc. Am. Physicians. 81:153-69; 1968.

112. Sjöström, L. Measurement of fat distribution. In: Bouchard, C.; Johnston, F.E., eds. Fat distribution during growth and later health outcomes. New York: Wiley-Liss; 1988:43-61.

113. Sjöström, L.; Kvist, H.; Cederblad, A.; Tylen, U. Determination of the total adipose tissue volume in women by computed tomography: Comparisons with 40K and tritium techniques. Am. J. Physiol. 250:E736-45; 1986.

114. Sorensen, T.I.A.; Holst, C.; Stunkard, A.J. Childhood body mass index—Genetic and familial environmental influences assessed in a longitudinal adoption study. Int. J. Obes. 16:705-14; 1992.

115. Sorensen, T.I.A.; Holst, C.; Stunkard, A.J.; Theil, L. Correlations of body mass index of adult adoptees and their biological relatives. Int. J. Obes. 16:227-36; 1992.

116. Sorensen, T.I.A.; Price, R.A.; Stunkard, A.J.; Schulsinger, F. Genetics of obesity in adult adoptees and their biological siblings. Br. J. Med. 298:87-90; 1989.

117. Stunkard, A.J.; Foch, T.T.; Hrubec, Z. A twin study of human obesity. JAMA. 256:51-54; 1986.

118. Stunkard, A.J.; Harris, J.R.; Pedersen, N.L.; McClearn, G.E. The body-mass index of twins who have been reared apart. N. Engl. J. Med. 322:1483-87; 1990.

119. Stunkard, A.J.; Sorensen, T.I.A.; Hannis, C.; Teasdale, T.W.; Chakraborty, R.; Schull, W.J.; Schulsinger, F. An adoption study of human obesity. N. Engl. J. Med. 314:193-98; 1986.

120. Tambs, K.; Moum, T.; Eaves, L.; Neale, M.; Midthjell, K.; Lund-Larsen, P.G.; Naess, S.; Holmen, J. Genetic and environmental contributions to the variance of the body mass index in a Norwegian sample of first- and second-degree relatives. Am. J. Hum. Biol. 3:257-67; 1991.

121. Tartaglia, L.; Dembski, M.; Weng, X.; et al. Identification and expression cloning of a leptin receptor OB-R. Cell. 83:1263-71; 1995.

122. Tiret, L.; André, J.L.; Ducimetière, P.; Herbeth, B.; Rakotovao, R.; Gueguen, R.; Spycherelle, Y.; Cambien, F. Segregation analysis of height-adjusted weight with generation- and age-dependent effects: The Nancy Family Study. Genet. Epidemiol. 9:389-403; 1992.

123. Tokunaga, K.; Matsuzawa, Y.; Ishikawa, K.; Tarui, S. A novel technique for the determination of body fat by computed tomography. Int. J. Obes. 7:437-45; 1983.

124. Tremblay, A.; Després, J.P.; Thériault, G.; Fournier, G.; Bouchard, C. Overfeeding and energy expenditure in humans. Am. J. Clin. Nutr. 56:857-62; 1992.

125. van der Kooy, K.; Leenen, R.; Seidell, J.C.; Deurenberg, P.; Hautvast, J.G.A.J. Effect of a weight cycle on visceral fat accumulation. Am. J. Clin. Nutr. 58:853-57; 1993.

126. Vijayaraghavan, S.; Hitman, G.A.; Kopelman, P.G. Apolipoprotein D polymorphism: A genetic marker of obesity and hyperinsulinemia. J. Clin. Endocrinol. Metab. 79:568-70; 1994.

127. Vohl, M.C.; Dionne, F.T.; Pérusse, L.; Dériaz, O.; Chagnon, M.; Bouchard, C. Relation between BglII polymorphism in 3β-hydroxysteroid dehydrogenase gene and adipose-tissue distribution in humans. Obes. Res. 2:444-49; 1994.

128. Walston, J.; Silver, K.; Bogardus, C.; Knowler, W.C.; Celi, F.S.; Austin, S.; Manning, B.; Strosberg, A.D.; Stern, M.P.; Raben, N.; Sorkin, J.D.; Roth, J.; Shuldiner, A.R. Time of onset of non-insulin-dependent diabetes mellitus and genetic variation in the β3-adrenergic receptor gene. N. Engl. J. Med. 333:343-47; 1995.

129. Warden, C.H.; Fisler, J.S.; Pace, M.J.; Svenson, K.L.; Lusis, A.J. Coincidence of genetic loci for plasma cholesterol levels and obesity in a multifactorial mouse model. J. Clin. Invest. 92:773-79; 1993.

130. Warden, C.H.; Fisler, J.S.; Shoemaker, S.M.; Wen, P.Z.; Svenson, K.L.; Pace, M.J.; Lusis, A.J. Identification of four chromosomal loci determining obesity in a multifactorial mouse model. J. Clin. Invest. 95:1545-52; 1995.

131. West, D.B.; Boozer, C.N.; Moody, D.L.; Atkinson, R.L. Dietary obesity in nine inbred mouse strains. Am. J. Physiol. 262:R1025-32; 1992.

132. West, D.B.; Goudey-Lefevre, J.; York, B.; Truett, G.E. Dietary obesity linked to genetic loci on chromosomes 9 and 15 in a polygenic mouse model. J. Clin. Invest. 94:1410-16; 1994.

133. West, D.B.; Waguespack, J.; York, B.; Goudey-Lefevre, J.; Price, R.A. Genetics of dietary obesity in AKR/J x SWR/J mice: Segregation of the trait and identification of a linked locus on chromosome 4. Mamm. Genome. 5:546-52; 1994.

134. Widen, E.; Lehto, M.; Kannine, T.; Walston, J.; Shuldiner, A.R.; Groop, L.C. Association of a polymorphism in the β3-adrenergic-receptor gene with features of the insulin resistance syndrome in Finns. N. Engl. J. Med. 333:348-51; 1995.

135. Wilkison, W.; Hansmann, I.; Woychik, R.P. Molecular structure and chromosomal mapping of the human homolog of the agouti gene. Proc. Natl. Acad. Sci. 91:9760-64; 1994.

136. Wilson, A.F.; Elston, R.C.; Tran, L.D.; Siervogel, R.M. Use of the robust sib-pair method to screen for single-locus, multiple-locus, and pleiotropic effects: Application to traits related to hypertension. Am. J. Hum. Genet. 48:862-72; 1991.

137. Xu, W.; Reed, D.R.; Ding, Y.; Price, R.A. Absence of linkage between human obesity and the mouse agouti homologous region (20q11.2) or other markers spanning chromosome 20q. Obes. Res. 3:557-60; 1995.

138. Young, C.M.; Blondin, J.; Tensuan, R.; Fryer, J.H. Body composition studies of "older" women thirty to seventy years of age. Ann. N.Y. Acad. Sci. 110:589-607; 1963.

139. Zee, R.Y.L.; Griffiths, L.R.; Morris, B.J. Marked association of a RFLP for the low density lipoprotein receptor gene with obesity in essential hypertensives. Biochem. Biophys. Res. Commun. 189:965-71; 1992.

140. Zhang, Y.; Proenca, R.; Maffei, M.; Barone, M.; Leopold, L.; Friedman, J.M. Positional cloning of the mouse obese gene and its human homologue. Nature. 372:425-32; 1994.

CHAPTER 11

Genetics and Skeletal Muscle Phenotypes

Skeletal muscle is the largest tissue mass in the body. It is also the main energy-consuming and work-producing tissue, providing the propulsive force to perform physical activities. There are more than 500 skeletal muscles in the body, and each consists of many smaller units—muscle fibers—which are in turn composed of myofibrils. This chapter considers changes in muscle tissue through the life cycle and then addresses genetic aspects of mass, size, fiber composition, glycolytic and oxidative properties, and finally responsiveness to training.

Gender and Age Difference in Muscle

Muscle mass

Estimates of the size of the skeletal muscle compartment of the body's composition are most often derived from creatinine excretion or potassium concentration. Age- and sex-associated variation in estimated muscle mass derived from urinary creatinine excretion is shown in figure 11.1.

The use of creatinine excretion to estimate muscle mass has limitations; for example, it is influenced by diet and physical activity. Nevertheless, it provides a reasonable estimate of changes in muscle mass through most of the life span. Sex differences are small prior to the adolescent spurt, but males gain considerably more muscle mass than females during adolescence, and the sex difference persists throughout the life span. Estimated muscle mass continues to increase through the third decade of life and then is rather stable into the 40s and 50s, after which estimated muscle mass tends to decline (31,32). Note, however, that the age-associated decline in muscle mass is dependent in part upon

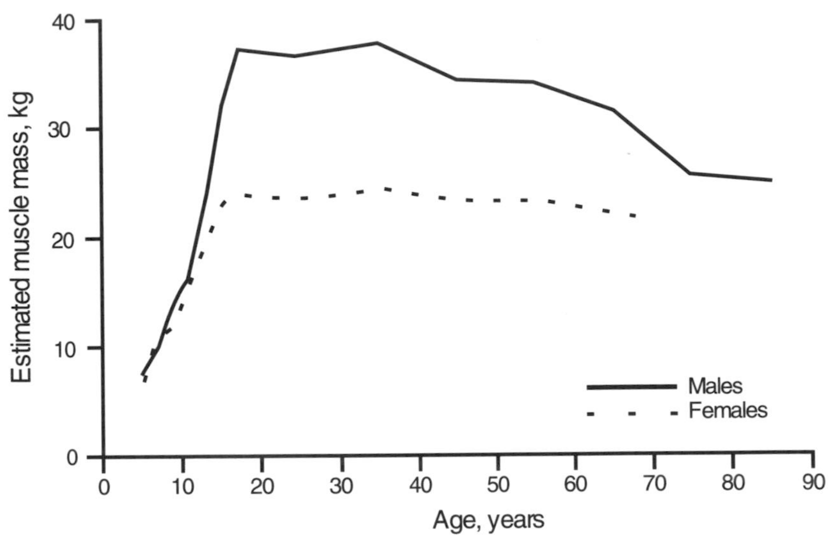

Figure 11.1 Changes in estimated muscle mass with age. Data are derived from 24-hour urinary creatinine excretion. See Malina (32) for the primary references used to calculate the estimates.

lifestyle and specifically upon habits of regular physical activity. More-active older adults tend to have a larger muscle mass and a slower estimated rate of decline with advancing age.

Midarm muscle circumference corrected for the thickness of the triceps skinfold is often used to estimate muscular development. Age- and sex-associated variation in estimated midarm muscle area derived from a nationally representative sample of the U.S. population are shown in figure 11.2. The trends are quite similar to those for estimated muscle mass derived from creatinine excretion. Note, however, that anthropometric estimates of limb musculature generally overestimate muscle areas compared with measures derived from computerized tomography scanning procedures (33).

Of the three tissues that contribute to most of the variation in body composition, skeletal muscle and bone tissue are major components of FFM. Both tissues show patterns of age- and sex-associated variation during childhood, adolescence, adulthood, and aging that are similar to those of FFM. The sex difference is considerable in both tissues. The decline in bone mineral with advancing age in females is, however, greater than in males (33).

Body regions contribute differentially to total muscle mass during growth, especially an increasing contribution of the musculature of the lower extremities to the total weight of the musculature (42). Other

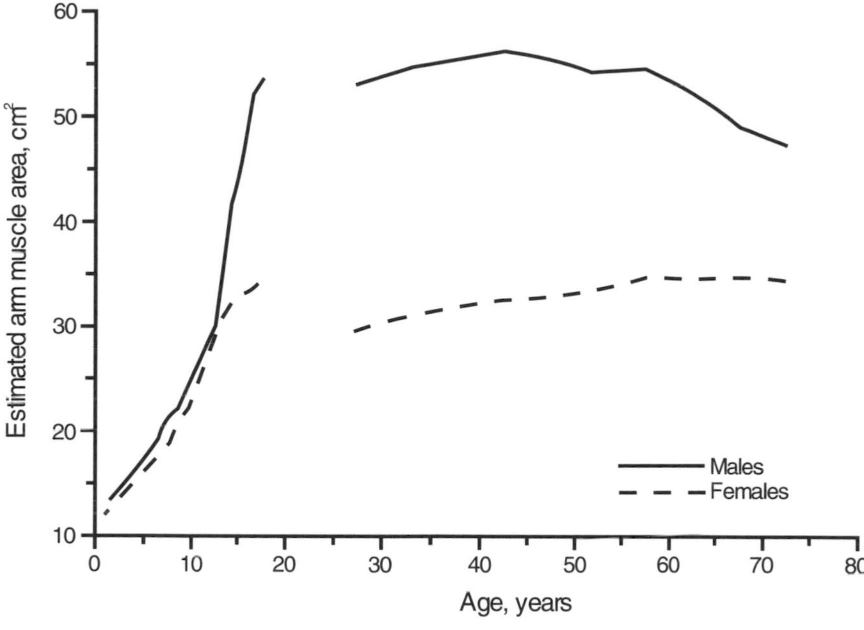

Figure 11.2 Changes in estimated midarm muscle areas with age in a representative sample of the U.S. population. Estimated muscle areas are derived from relaxed arm circumference corrected for the thickness of the triceps skinfold. Drawn from medians reported by Frisancho (13).

studies of regional variation in the distribution of muscle tissue are based on radiography, computerized tomography, magnetic resonance imaging, ultrasound, and dual-photon absorptiometry. Males, for example, on average have proportionally more upper-extremity muscle mass than females in childhood and especially during adolescence. The sex difference persists through adulthood (33). There are also racial differences. American black women have about 10% to 15% more appendicular skeletal muscle than white women matched for age, stature, and weight, and the ethnic difference is more marked in the upper than in the lower extremity (15,38). American white women tend to have greater muscle thicknesses (based on ultrasound) than Japanese women of the same age, and the differences persist when adjusted for FFM and BMI (25,26).

Muscle fiber types

In mature human muscle, there is a variety of skeletal muscle fiber types, depending on the histochemical, contractile, or metabolic criteria used to classify fibers. There are thus small and large fibers, fibers with various speeds of contraction and relaxation, fibers with a high or low myoglobin content, red and white fibers, fibers with high or low glycolytic potential, and so on (41).

It is common to classify human muscle fibers into two or three categories based on histochemical characteristics related to the myosin ATPase reaction. Type I, or slow-twitch (ST), fibers are generally characterized by high mitochondrial oxidative enzyme activity and low phosphorylase and myosin ATPase reactions. Type II, or fast-twitch (FT), fibers are characterized by high phosphorylase and ATPase reactions, a lower number of mitochondria, and less oxidative potential. Type II fibers are also subdivided into type IIA (FTa) and type IIB (FTb). The former are often viewed as both oxidative and glycolytic, while the latter are viewed as having primarily glycolytic potential. However, and particularly in humans, there is an almost complete overlap of the subpopulations of type II fibers when they are characterized in terms of oxidative capacity. Hence, it cannot be concluded that type IIA fibers are solely oxidative or that type IIB are not oxidative in humans.

The number of type I and II fibers increases during the first postnatal month, and the increase is probably an extension of prenatal muscle differentiation. By about one year of age, there is little difference in the proportional fiber distribution in the muscle tissue of children and adults (10), and the proportion of type I and II fibers does not change significantly with age (30).

The percentage of type I fibers varies among specific muscles. Among children from birth to 8 years of age, there is no sex difference in the proportion of type I fibers in the quadriceps, deltoid, rectus abdominis, and diaphragm. Type I fibers are more numerous in the quadriceps, deltoid, and rectus abdominis, while the opposite is characteristic of the diaphragm (61). In a muscle of mixed composition, such as the vastus lateralis, the proportion of type I fibers may increase with age more in females than in males, since young adult females have about 5% to 8% more type I fibers than young adult males (16,51). It is of interest that from muscle biopsies of the vastus lateralis taken in subjects at 16 and then at 27 years of age, the percentage of type I fibers tended to increase in 28 females (51 ± 9% to 55 ± 12%, NS), but decreased in 55 males (55 ± 12% to 48 ± 13%, $p < .001$) (16). A comparison of several studies of the vastus lateralis muscle in samples of young (about 20-year-old) and older (about 70- and 80-year-old) adults indicated considerable overlap in the proportion of type I fibers in the younger (41%-58%) and older (48%-75%) groups (30). Note, however, that about 15% of the total variance in the proportion of type I fibers in the vastus lateralis can be explained by muscle tissue sampling in biopsy procedures and by technical variance (46) and that the vast majority of research on the fiber distribution, size, and metabolic characteristics is based solely upon the vastus lateralis muscle.

Overall, data from a variety of sources suggest that the fiber type distribution is quite heterogeneous among individuals and probably among populations. It is estimated that about 25% of North American men and women of European ancestry (white) have either less than 35% or more than 65% of type I fibers in the vastus lateralis muscle (46). On the other hand, the vastus lateralis muscle of sedentary black African subjects (western and central African ancestry) has a slightly greater percentage of type II fibers (1). The extent to which the proportion of muscle fibers in the vastus lateralis is under the control of genetic factors is discussed later in the chapter.

Muscle fiber size

Muscle fibers increase rapidly in diameter with age and body size postnatally, but the increase varies with the muscle studied. There are no sex differences in fiber diameter during childhood, and adult diameters are apparently attained during adolescence. Fiber areas are, on average, larger in males than in females during adolescence and young adulthood. Cross-sectional data indicate that fiber areas increase into the mid-20s in males (33), but one longitudinal study from 16 to 27 years indicates no significant changes in mean fiber areas in males and females (16). Mean muscle fiber areas decrease with age beginning at about 30 years of age, and the age-associated decline occurs primarily in type II fibers (30).

Contractile properties

The contractile properties of a muscle are characterized in terms of the speed of contraction and relaxation and the maximal force that can be generated. Speed of contraction and relaxation is related to motor units that regulate a group of muscle fibers. Contractile properties of individual motor units are often defined in terms of speed of contraction (time to peak tension) and relaxation time (one half of the relaxation time). In humans, ST and FT muscle fibers have an approximate mean time to peak tension of 90 ms and 45 ms, respectively (17,41). Currently available data, though limited, suggest that the contractile properties of skeletal muscle become mature early in postnatal life. Electrophysiological data suggest that the number of motor units decreases with age, and this decrease in part underlies the observed age-associated reduction in contractile strength (12).

The maximal force that can be generated by an adult muscle is largely a function of muscle size. In children, growth per se and neuromuscular maturation are important factors affecting force generation. Maximal force is generally independent of the proportion of type I and type II fibers, although this issue is still a matter of debate. Some evidence suggests that the maximal voluntary force of mature human skeletal muscle reaches about 40 N/cm^2 of muscle area.

Metabolic properties

The predominant characteristics of type I and type II fibers in mature humans that have implications for metabolism of muscle tissue are summarized in table 11.1. Little information is currently available on changes that occur in most of these characteristics during growth and maturation (34). Some of the metabolic properties show sex differences. For example, levels of glycolytic enzyme markers in the vastus lateralis are higher in young adult males than in females; in contrast, activities of aerobic-oxidative enzyme markers are similar in males and females (47). The lower glycolytic potential in females, in conjunction with the higher proportion of type I fibers and smaller fiber areas described earlier, may underlie some of the metabolic differences routinely observed between males and females during short-term or long-term exercise (9,14,57,58).

There is also evidence for population differences in some metabolic properties of skeletal muscle. One study indicates higher enzyme activities of the anaerobic energy processes in adult sedentary black subjects of western and central African ancestry compared to sedentary white subjects. However, enzyme markers of aerobic-oxidative metabolism are comparable in both ethnic groups (1). The higher glycolytic potential and greater percentage of type II fibers in sedentary black subjects is a phenotype not observed in all ethnic groups of African origin. Despite the observation that the best sprinters in the world are black individuals whose ancestry can, in general, be traced to western and central Africa, a cause-and-effect relationship cannot be established between a greater percentage of type II fibers and ability in sprinting events.

Genetics of Skeletal Muscle Tissue Phenotypes

Muscle mass and size

There is little information on the genotypic contribution to individual differences in muscle mass. Muscle tissue, however, is the major component of FFM, and familial data indicate a significant genetic contribution to FFM.

Nine different types of relatives by descent or by adoption had their body composition estimated from underwater weighing in the Quebec

Table 11.1
Summary of the Main Metabolic Characteristics of the Two Major Human Skeletal Muscle Fiber Types

Characteristics	Type I (ST)	Type II (FT)
Speed of contraction	Slow	Fast
Relaxation time	Long	Short
Myosin ATPase	Low	High
Lipid content	High	Low
Glycogen content	Low	High
ATP content	Same	Same
Creatine phosphate content	Same	Same
Mitochondrial content	High	Low
Capillary density	High	Low
Creatine kinase activity	Low	High
Phosphorylase activity (glycogenolysis)	Low	High
Phosphofructokinase (glycolysis)	Low	High
Krebs cycle enzymes	High	Low
Anaerobic capacity	Low	High
Aerobic capacity	High	Low

Differences obtained when muscle fibers of the same muscle in a given individual are compared.

Family Study (7). The resulting measure of body density was converted to estimates of FFM. Correlations for the various pairs of relatives suggest a larger contribution of biological inheritance to FFM than to fat mass and percent body fat. Path analysis (BETA) of familial resemblance indicated a total transmissible variance of 40% to 50% for FFM (7). The genetic component of the transmissible variance was about 30% for FFM. Subsequent commingling analysis of these data has shown that FFM was characterized by a single distribution in parents but not in the offspring (3). More recently, segregation analysis of the data did not provide evidence for a major locus effect on FFM (39). In contrast, about 60% of the variance in FFM was accounted for by a non-Mendelian major effect, which may reflect environmentally based commingling or which may be, in part, a function of gene-environment interactions or correlations (39). It may be inferred, therefore, that some undetermined genetic characteristics contribute to individuality in estimated muscle mass.

This suggestion is consistent with limited familial data for radiographic and anthropometric indicators of limb musculature. Radiographic measurements of calf muscle diameter in biological siblings between 0.5 and 5 years of age reach 0.56 in males and 0.63 in females (19). Parent-child correlations for estimated muscle diameters of the arm and calf (arm and calf circumferences corrected for the thickness of the triceps and medial calf skinfolds, respectively) also suggest significant familial resemblance. Midparent–natural child correlations for both estimated muscle diameters approximate 0.30 ($p = .001$), while foster midparent–adopted child correlations reach only 0.10 (4). These results suggest that muscle size is influenced by genetic characteristics and should come as no surprise, given the extensive animal husbandry literature on selective breeding for meat content and composition.

Muscle fiber distribution

Limited data on the extent to which human skeletal muscle fiber types are under the control of genetic factors give variable results. The early study of Komi et al. (27) on small samples of MZ ($n = 15$ pairs) and DZ ($n = 16$ pairs) twins yielded surprisingly high heritability coefficients for the proportion of type I fibers in the vastus lateralis: 0.995 in males and 0.928 in females. Such high estimates would suggest that fiber type distribution is almost entirely genotype dependent. This is not consistent with observations of variation in tissue sampling and technical errors associated with biopsy procedures. Repeated measurements within the vastus lateralis muscle in adults indicate that sampling variability and technical error together account for about 15% of the variance in the proportion of type I muscle fibers (46,50).

Data for a larger series of twins and siblings indicate markedly lower estimates of heritability. Intraclass correlations for the proportion of different fibers in the vastus lateralis of 32 pairs of brothers, 26 pairs of DZ twins, and 35 pairs of male and female MZ twins are summarized in table 11.2. Estimated broad heritabilities ranged from 66% (twice the biological sibship correlation) to 55% (directly from the MZ twin sibship correlation) to 6% (twice the difference between MZ and DZ correlations) (6,8). Although brothers and DZ twins share about one-half of their genes by descent, comparison of the respective within-pair correlations suggests that increased environmental similarity—that is, DZ twins experience more similar environmental circumstances than brothers separated in age—translates into increased phenotypic resemblance for the proportion of type I fibers. It is interesting that the respective within-pair correlations for the proportion of type I fibers in MZ ($r = 0.55$) and DZ ($r = 0.52$) twins are quite similar to the correlation between the proportion of type I fibers in the right and left legs of the same individual ($r = 0.67$) (49). Corresponding correlations indicate no genetic effect in type IIA fibers, but more resemblance within MZ twin pairs for type IIB fibers but not within DZ twin and sibling pairs (table 11.2).

Table 11.2
Intraclass Coefficients for the Proportions of Fiber Types in the Vastus Lateralis Muscle of Twins and Brothers

Fiber type	MZ ($n = 35$)	DZ ($n = 26$)	Brothers ($n = 32$)
I	0.55**	0.52**	0.33*
IIA	0.18	0.39*	0.00
IIB	0.56**	0.26	0.26

*p < .05; ** p < .01

Adapted from Bouchard et al. (8).

Estimated heritabilities of type I fibers from the studies of Komi et al. (27) and Bouchard et al. (8) are thus widely divergent, ranging from a low of 6% to a high of almost 100%. Allowing for sampling and technical variation, it appears that factors other than the genotype are involved in the modulation of the proportions of type I and type II fibers in the vastus lateralis muscle. Nevertheless, examination of the variability in the percentage of type I fibers within specific pairs of twins suggests that there are genetic factors that predispose some individuals to have a high or low prevalence of type I or type II fibers. The mean difference in percentage of type I fibers between a member of an MZ pair and his or her twin control reached 9.5 ± 6.9%. The difference in the percentage of type I fibers between members of an MZ pair was less than 6% in 16 pairs, between 6% and 12% in 11 pairs, between 12% and 18% in 5 pairs, and between 18% and 23% in the remaining 8 pairs (a total of 40 MZ pairs, including the 35 pairs in table 11.2 and 5 additional pairs). The largest difference between members of an MZ twin pair was 23%, which was observed in three pairs and was of the same magnitude as the largest differences when samples were taken from the right and left vastus lateralis of the same individual (46).

The preceding results are generally consistent with animal experiments. In an inbred strain of mice, genetic factors accounted for about 75% of the variation in the proportion of type I fibers of the soleus muscle, but the 95% confidence intervals ranged from 55% to 89% (37). In a selective breeding study for a high percentage of type I muscle fibers in the gastrocnemius muscle in four generations of rats, only about 17% of the variation in the proportion of fibers was determined by the genes (36). However, the variance in the proportion of type I fibers in the gastrocnemius muscle was about one-half of that commonly observed for the vastus lateralis muscle in humans, which is a situation that may have reduced the estimated heritability. Placing the results of human studies in the context of those from animal experiments leads to the suggestion that a genetic component accounts for about 40% to 50% of the variation in the proportion of type I muscle fibers in humans (46). A summary of the genetic, environmental, and methodological sources of variation in the proportion of type I fibers in human skeletal muscle is illustrated in figure 11.3.

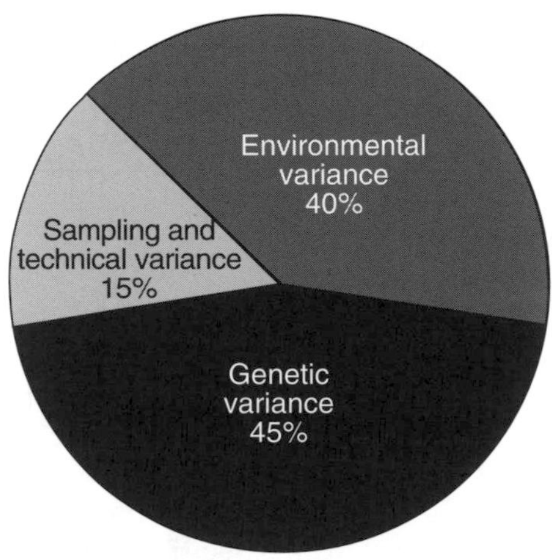

Figure 11.3 Estimates of the sampling and technical, environmental, and genetic variances for the proportion of type I fibers in human skeletal muscle (52). Reprinted, by permission, from J.A. Simoneau and D. Pette, 1988, "Species-specific effects of chronic nerve stimulation upon tibialis anterior muscle of mouse, rat, guinea pig and rabbit," *Pflügers Arch* 412:86-92.

An interesting question relates to the sources of environmental variation in the proportion of muscle fibers in MZ twins. The prenatal environment of MZ twins may be variable. For example, more than 20% of MZ twins share one chorion; as a result, one member of the twin pair may experience a reduction in the cytoplasmic mass of parenchymal organs and in hemoglobin and serum protein (60). It is possible that the chorionic status of MZ twins, or some form of gametic or embryonic imprinting, may account for some of the intrapair variation in the muscle fiber proportions observed in MZ twins. The potential role of training is discussed later in the chapter.

Metabolic properties

Advances in molecular biology and medical genetics provide important insights into the genetic basis of many diseases, including a variety of gene-associated defects in skeletal muscle tissue. Genetically based muscle diseases include the dystrophies, myotonias, and myopathies, among others, and inherited disorders of glycogen and lipid metabolism in skeletal muscle (44). The spectrum of gene-associated diseases involving muscle tissue imply that genes exert significant influences upon contractile proteins and metabolic properties of skeletal muscle. Specific implications of skeletal muscle diseases for exercise performance are discussed in detail by Lewis and Haller (29).

Structural genes for contractile proteins of skeletal muscle are numerous. Variation in the expression of these genes would seem to underlie the coexistence of mixed forms of contractile proteins in a given fiber and perhaps influence responses to altered functional demands, such as those associated with use and disuse. Myosin heavy- and light-chain isoforms exist in human skeletal muscle. Several genes are involved and are transcribed differentially depending on the age of the individual (43,54). Both myosin heavy- and light-chain isoforms are involved in determining the maximal velocity of shortening of skeletal muscles (43). Further, exercise, detraining, or inactivity results in sequential alterations in myosin expression within fast fibers of skeletal muscle, and perhaps similar alterations occur in slow fibers under prolonged functional demand or disuse (54).

There is considerable variation among individuals in the enzymatic activity profile of skeletal muscle. There are high and low activity levels of enzyme markers of the catabolism of different substrates in skeletal muscle of healthy sedentary and moderately active individuals of both sexes (47). Many factors probably contribute to this variation among individuals, and it is likely that the genotype plays a role in the amount of protein for several important enzymes in skeletal muscle.

Data on the extent of genetic determination of enzyme markers in skeletal muscle are limited. Earlier studies based on small samples of MZ and DZ twins (23,27) suggested no gene-associated variation in several enzymes. However, data for a larger sample of MZ ($n = 35$) and DZ ($n = 26$) twins of both sexes and pairs of biological brothers ($n = 32$) indicate that variation in several key enzymes of skeletal muscle is inherited to a significant extent (table 11.3). There is significant within-pair resemblance in MZ twins for all skeletal muscle enzyme activities ($r = 0.30$ to 0.68), but the within-pair correlations for DZ twins and brothers suggest that variation in several enzyme activities are related to nongenetic factors and environmental conditions. Nevertheless, after adjusting for variation in enzyme activities associated with age and sex, it appears that genetic factors are responsible for about 25% to 50% of the total phenotypic variation in the activities of the regulatory enzymes of the glycolytic (PFK) and citric acid cycle (OGDH) pathways and of the variation in the ratio of glycolytic to oxidative activity (PFK/OGDH ratio) (8). Such genetic effects could not be accounted for by charge

Table 11.3
Intraclass Coefficients for Enzyme Activities in the Vastus Lateralis Muscle of Twins and Brothers

Enzyme	MZ (n = 35)	DZ (n = 26)	Brothers (n = 32)
Creatine kinase (CK)	0.61**	0.53**	—
Hexokinase (HK)	0.41**	0.75**	–0.22
Phosphofructokinase (PFK)	0.55**	0.35*	0.27
Lactate dehydrogenase (LDH)	0.68**	0.03	0.50**
Malate dehydrogenase (MDH)	0.58**	–0.14	0.15
Oxoglutarate dehydrogenase (OGDH)	0.53**	0.19	0.09
3-hydroxyacyl CoA dehydrogenase (HADH)	0.43**	–0.33	0.48*
PFK/OGDH ratio	0.30*	0.06	0.34*

*$p < .05$; ** $p < .01$

Adapted from Bouchard et al. (8).

variation in the enzyme molecules (5,35); they likely depend upon individual differences in transcription rate or translation.

Genotype-Training Interaction Effects in Skeletal Muscle

The role of the genotype in the response of skeletal muscle tissue to standardized training programs needs elucidation in humans. Experimental evidence from animals suggests that genetic influences on the phenotypic expression of most of the skeletal muscle proteins involved in energy metabolism are associated with differences in the regulatory processes of gene expression (46). For example, maximal increases or decreases in specific enzyme activities of skeletal muscle in the rat and rabbit in response to chronic electrical stimulation are closely related to induced changes in gene transcription (2,21,63). In a study of rabbits, enhanced translation of available mRNA was responsible for about one-eighth of the 340% electrical stimulation–induced increase in the activity of citrate synthase in skeletal muscle (45). The changes in citrate synthase activity brought about through alterations at the level of translation were not as large as those induced by alterations at the level of transcription. Nevertheless, the experimental results suggest that the processes of transcription and translation can contribute to the phenotypic responses to stimulation. It may be speculated, therefore, that simultaneous increases in transcription of genes and translation of mRNA-encoding muscle proteins of energy metabolism are desirable conditions in the adaptation to training and that they may be characteristics of individuals who respond strongly to training (high responders).

In the experimental study of rat skeletal muscle, the stimulation-induced changes in cytochrome-c oxidase result from a parallel rise in mitochondrial and nuclear encoded subunit mRNAs (21). Under the same experimental conditions, the nuclear encoded subunit of cytochrome-c oxidase increased to a similar extent, about 1.8-fold, in skeletal muscle of the rat and rabbit (21,62). However, in the rabbit, the mitochondrial gene transcript increased out of proportion with the nuclear encoded transcript (62). This result coincides with the observation that aerobic-oxidative metabolism of rabbit skeletal muscle can adapt to a larger extent (relatively and absolutely) than that of rat skeletal muscle in response to chronic increase in functional demands (20,52). The contribution of such differences in the control of transcription and translation and in the mitochondrial and nuclear genome to human variation in performance in general, and to skeletal muscle metabolism in particular, remains to be delineated under sedentary conditions and in response to regular exercise.

Three experimental studies of the responses of young adult MZ twins to different training programs provide some insights into potential genotypic contributions to the response of skeletal muscle to training. All three included male and female MZ twins, and there were no sex differences in gains associated with the respective training programs. The results from a study of intermittent training are summarized in table 11.4. The training program involved both continuous and interval work on a cycle ergometer for 15 weeks by 12 pairs of MZ twins of both sexes. Although there were significant changes in the proportions of type I and type IIB fibers, the intraclass correlations showed no significant within-pair resemblance for fiber type proportions. The responses of all enzymes except PFK and HADH were at least three times more similar within MZ pairs. The evidence suggests that about 50% to 60% of the training response in HK, LDH, MDH, OGDH, and the PFK/OGDH ratio in response to the intermittent training program were associated with the genotype; about 80% of the training response in CK appeared to be determined by the genotype (49).

Results of a 10-week isokinetic strength training program in four pairs of young adult MZ twins (one pair of males, three pairs of females) are summarized in table 11.5. Of the five skeletal muscle enzymes considered, only the response of OGDH activity to strength training appeared to be significantly genotype dependent. The responses of CK

Table 11.4
Intraclass Coefficients for the Resemblance of Responses of Skeletal Muscle Enzymes to High-Intensity Intermittent Training Within MZ Twin Pairs (n = 12 pairs)

Enzyme	Intraclass correlation
Creatine kinase (CK)	0.82**
Hexokinase (HK)	0.59*
Phosphofructokinase (PFK)	0.38
Lactate dehydrogenase (LDH)	0.64**
Malate dehydrogenase (MDH)	0.50*
3-hydroxyacyl CoA dehydrogenase (HADH)	0.15
Oxoglutarate dehydrogenase (OGDH)	0.50*
PFK/OGDH ratio	0.64**

*$p < .05$; ** $p < .01$

Adapted from Simoneau et al. (49).

Table 11.5
Intraclass Coefficients for the Resemblance of Responses of Skeletal Muscle Enzymes to Isokinetic Strength Training Within MZ Twin Pairs (n = 4 pairs)

Enzyme	Intraclass correlation
Creatine kinase (CK)	0.37
Hexokinase (HK)	–0.16
Malate dehydrogenase (MDH)	0.26
3-hydroxyacyl CoA dehydrogenase (HADH)	0.45
Oxoglutarate dehydrogenase (OGDH)	0.76*

*$p < .05$

Adapted from Thibault et al. (59).

and HADH to training appeared to be dependent upon genotype to a lesser extent, as reflected in the lower genotype-training interaction F-ratios (59). The results emphasize the need for further study of the genetic basis of the response of skeletal muscle to strength training.

The third study followed six pairs of young adult MZ twins (three males and three females) through 15 weeks of ergocycle endurance training (18). Genotype-training interactions were evaluated for the proportion of muscle fibers and several enzymes after 7 weeks and after 15 weeks of training. Relative changes in the proportions of type I and type IIA and IIB fibers were not significant. More important, there was no significant genotype-training interaction (i.e., intrapair resemblance in response to training) for the distribution of fiber types. Genotype-training interactions in the responses of skeletal muscle tissue enzymes are summarized in table 11.6. Changes observed during the first half of training were not related to genotype, with the exception of that in PFK activity. However, during the second part of the training program, changes in the activities of PFK, MDH, HADH, and OGDH were partly determined by the genotype, as evident in the significant intraclass correlations. Thus, within-pair similarity in the response of enzyme activities apparently varies with time during a training program. This in turn suggests that adaptation to endurance training early in the program may be under less stringent genetic control; however, as training continues and perhaps nears maximal trainability, the response to training could be more genotype dependent.

Two of the studies just discussed (intermittent and endurance training) indicate minor changes in the muscle fiber population associated

Table 11.6
Intraclass Coefficients for the Resemblance of Responses of Skeletal Muscle Enzymes to Endurance Training Within MZ Pairs ($n = 6$ pairs)

Enzyme	INTRACLASS CORRELATION	
	After 7 weeks	From 7-15 weeks
Creatine kinase (CK)	0.58	–0.25
Hexokinase (HK)	–0.06	0.59
Phosphofructokinase (PFK)	0.69*	0.63*
Lactate dehydrogenase (LDH)	–0.15	–0.69
Malate dehydrogenase (MDH)	0.31	0.81**
3-hydroxyacyl CoA dehydrogenase (HADH)	0.00	0.89**
Oxoglutarate dehydrogenase (OGDH)	0.50	0.72[a]
PFK/OGDH ratio	0.29	0.58

[a] $n = 4$ pairs
*$p < .05$; ** $p < .01$
Adapted from Hamel et al. (18).

with the relatively short-term training programs. These observations are consistent with the literature. Several exercise-training studies have failed to induce any alteration of the percentage of type I fibers (41), while other experimental data indicate that the proportion of type I fibers in the human vastus lateralis may increase in response to high-intensity intermittent training (48), to intense endurance training (24,53), and to strength training (40,55,56). Further, the proportion of muscle fiber types is also altered with long-term inactivity (28), limb immobilization in trained individuals (22), and prolonged drug treatment (11). The reported changes are generally of modest magnitude. Nevertheless, they warrant the conclusion that the proportion of type I fibers can be modified in response to training and other environmental variables (46).

The issue of interest is whether the changes in proportion of fibers as a result of training or other environmental conditions are genotype dependent. Results of the training studies with MZ twins indicate an absence of significant intrapair resemblance in the changes. This would suggest that changes in fiber type composition are independent of genotype. However, caution is warranted in concluding that fiber type transformation is not genotype dependent. The variable results among studies could be related to the methodological sources of variation in estimating the proportion of type I fibers (see figure 11.3). Further research is needed, particularly with training regimens lasting years rather than months.

On the other hand, the three studies indicate that the activities of muscle enzymes are modified by training and that there is considerable interindividual variability in response to training. The genotype dependency of specific enzymes to training (i.e., genotype-training interaction), however, is variable and is apparently dependent in part on the type of training stimulus (endurance, high intensity, strength) and perhaps the duration of the training protocol among other factors.

Summary

The identification of the genetic and molecular basis of the skeletal muscle characteristics and properties, particularly in terms of their response to regular exercise, has just begun. The notion that DNA sequence variation at specific sites influences the expression of the genes that are responsible for the contractile and metabolic properties of human skeletal muscle has not been explored to any extent. It is, however, an important research issue, but one that may require complex experimental designs and elaborate protocols. From a variety of lines of evidence, it has been estimated that the genetic heritability of the proportion of type I fibers in the vastus lateralis muscle reaches about 45% of the phenotype variance.

It appears that sequence variation in the coding exons of the genes expressed in human skeletal muscle plays only a marginal role. This is suggested by the lack of polymorphism in the enzymes of key metabolic pathways as revealed by isoelectrofocusing studies of extracted skeletal muscle proteins. However, coding sequence variation and other forms of DNA polymorphisms can bring about dramatic changes in skeletal muscle function as exemplified by inherited disorders that are manifest in skeletal muscle. Finally, an important issue is that of the role of DNA sequence variation in the response of skeletal muscle properties to the increased demands of regular exercise. These studies should provide us with a wealth of information about the molecular mechanisms involved in skeletal muscle adaptation.

References

1. Ama, P.F.M.; Simoneau, J.A.; Boulay, M.R.; Serresse, O.; Thériault, G.; Bouchard, C. Skeletal muscle characteristics in sedentary black and Caucasian males. J. Appl. Physiol. 61:1758-61; 1986.
2. Annex, B.H.; Kraus, W.E.; Dohm, G.L.; Williams, R.S. Mitochondrial biogenesis in striated muscles: Rapid induction of citrate synthase mRNA by nerve stimulation. Amer. J. Physiol. 260:266-70; 1991.
3. Borecki, L.B.; Rice, T.; Bouchard, C.; Rao, D.C. Commingling analysis of generalized body mass and composition measures: The Québec Family Study. Int. J. Obes. 15:763-73; 1991.
4. Bouchard, C. Genetic aspects of anthropometric dimensions relevant to assessment of nutritional status. In: Himes, J., ed. Anthropometric assessment of nutritional status. New York: Alan R. Liss; 1991:213-31.
5. Bouchard, C.; Chagnon, M.; Thibault, M.-C.; Boulay, M.R.; Marcotte, M.; Simoneau, J.A. Absence of charge variants in human skeletal muscle enzymes of the glycolytic pathway. Hum. Genet. 78:100; 1988.
6. Bouchard, C.; Dionne, F.T.; Simoneau, J.A.; Boulay, M.R. Genetics of aerobic and anaerobic performances. Exerc. Sport Sci. Rev. 20:27-58; 1992.
7. Bouchard, C.; Pérusse, L.; Leblanc, C.; Tremblay, A.; Thériault, G. Inheritance of the amount and distribution of human body fat. Int. J. Obes. 12:205-15; 1988.
8. Bouchard, C.; Simoneau, J.A.; Lortie, G.; Boulay, M.R.; Marcotte, M.; Thibault, M.-C. Genetic effects in human skeletal muscle fiber type distribution and enzyme activities. Can. J. Physiol. Pharmacol. 64:1245-51; 1986.

9. Chen, J.D.; Bouchard, C. Sex differences in metabolic adaptation to exercise. In: Saltin, B., ed. Biochemistry of exercise VI, international series on sport sciences. 16:227-38; 1986.
10. Colling-Saltin, A.S. Some quantitative biochemical evaluations of developing skeletal muscles in the human foetus. J. Neurol. Sci. 39:187-98; 1978.
11. Danneskiold-Samsoe, B.; Grimby, G. The influence of prednisone on the muscle morphology and muscle enzymes in patients with rheumatoid arthritis. Clin. Sci. 71:693-701; 1986.
12. Doherty, T.J.; Vandervoort, A.A.; Taylor, A.W.; Brown, W.F. Effects of motor unit losses on strength in older men and women. J. Appl. Physiol. 74:868-74; 1993.
13. Frisancho, A.R. Anthropometric standards for the assessment of growth and nutritional status. Ann Arbor, MI: University of Michigan Press; 1990.
14. Froberg, K.; Pedersen, P.K. Sex differences in endurance capacity and metabolic response to prolonged, heavy exercise. Eur. J. Appl. Physiol. 52:446-50; 1984.
15. Gasperino, J.A.; Wang, J.; Pierson, R.N.; Heymsfield, S.B. Age-related changes in musculoskeletal mass between black and white women. Metabolism. 44:30-34; 1995.
16. Glenmark, B.; Hedberg, G.; Jansson, E. Changes in muscle fibre type from adolescence to adulthood in women and men. Acta Physiol. Scand. 146:251-59; 1992.
17. Gollnick, P.D. Relationship of strength and endurance with skeletal muscle structure and metabolic potential. Int. J. Sports Med. 3:26-32; 1982.
18. Hamel, P.; Simoneau, J.-A.; Lortie, G.; Boulay, M.R.; Bouchard, C. Heredity and muscle adaptation to endurance training. Med. Sci. Sports Exerc. 18:690-96; 1986.
19. Hewitt, D. Sib resemblance in bone, muscle and fat measurements of the human calf. Ann. Hum. Genet. 22:213-21; 1957.
20. Hood, D.A.; Pette, D. Chronic long-term electrostimulation creates a unique metabolic profile in rabbit fast-twitch muscle. FEBS Lett. 247:471-74; 1989.
21. Hood, D.A.; Zak, R.; Pette, D. Chronic stimulation of rat skeletal muscle induces coordinate increases in mitochondrial and nuclear mRNAs of cytochrome c oxidase subunits. Eur. J. Biochem. 179:275-80; 1989.
22. Howald, H. Training-induced morphological and functional changes in skeletal muscle. Int. J. Sports Med. 3:1-12; 1982.
23. Howald, H. Ultrastructure and biochemical function of skeletal muscle in twins. Ann. Hum. Biol. 3:455-62; 1976.
24. Howald, H.; Hoppeler, H.; Claasen, H.; Mathieu, O.; Straub, R. Influence of endurance training on the ultrastructural composition of the different muscle fiber types in humans. Pflügers Arch. 403:369-76; 1985.

25. Ishida, Y.; Kanehisa, H.; Fukunaga, T.; Pollock, M.L. A comparison of fat and muscle thickness in Japanese and American women. Ann. Physiol. Anthropol. 11:29-35; 1992.
26. Ishida, Y.; Kanehisa, H.; Kondo, M.; Fukunaga, T.; Carroll, J.F.; Pollock, M.L.; Graves, J.E.; Leggett, S.H. Body fat and muscle thickness in Japanese and Caucasian females. Am. J. Hum. Biol. 6:711-18; 1994.
27. Komi, P.V.; Viitasalo, J.H.T.; Havu, M.; Throstensson, A.; Sjodin, B.; Karisson, J. Skeletal muscle fibres and muscle enzyme activities in monozygotous and dizygous twins of both sexes. Acta Physiol. Scand. 100:385-92; 1977.
28. Larsson, L.; Ansved, T. Effects of long-term physical training and detraining on enzyme histochemical and functional skeletal muscle characteristics in man. Muscle Nerve. 8:714-22; 1985.
29. Lewis, S.F.; Haller, R.G. Skeletal muscle disorders and associated factors that limit exercise performance. Exerc. Sport Sci. Rev. 17:67-113; 1989.
30. Lexell, J. Human aging, muscle mass, and fiber type composition. J. Gerontol. Series A, 50A:11-16; 1995.
31. Malina, R.M. Growth of muscle tissue and muscle mass. In: Falkner, F.; Tanner, J.M., eds. Human growth. Volume 2. Postnatal growth, neurobiology. New York: Plenum Press; 1986:77-99.
32. Malina, R.M. Quantification of fat, muscle and bone in man. Clin. Orthop. 65:9-38; 1969.
33. Malina, R.M. Regional body composition: Age, sex, and ethnic variation. In: Roche, A.F.; Heymsfield, S.B.; Lohman, T.G., eds. Human body composition. Champaign, IL: Human Kinetics; 1996:217-55.
34. Malina, R.M.; Bouchard, C. Growth, maturation and physical activity. Champaign, IL: Human Kinetics; 1991.
35. Marcotte, M.; Chagnon, M.; Côté, C.; Thibault, M.-C.; Boulay, M.R.; Bouchard, C. Lack of genetic polymorphism in human skeletal muscles enzymes of the tricarboxylic acid cycle. Hum. Genet. 77:200; 1987.
36. Nakamura, T.; Masui, S.; Wada, M.; Katoh, H.; Mikami, H.; Katsuta, S. Heredity of muscle fibre composition estimated from a selection experiment in rats. Eur. J. Appl. Physiol. 66:85-89; 1993.
37. Nimmo, M.A.; Wilson, R.H.; Snow, D.H. The inheritance of skeletal muscle fibre composition in mice. Comp. Biochem. Physiol. 81A:109-15; 1985.
38. Ortiz, O.; Russell, M.; Daley, T.L.; Baumgartner, R.N.; Waki, M.; Lichtman, S.; Wang, J.; Pierson, R.N.; Heymsfield, S.B. Differences in skeletal muscle and bone mineral mass between black and white females and their relevance to estimates of body composition. Am. J. Clin. Nutr. 55:8-13; 1992.
39. Rice, T.; Borecki, I.B.; Bouchard, C.; Rao, D.C. Segregation analysis of fat mass and other body composition measures derived from underwater weighing. Am. J. Hum. Genet. 52:967-73; 1993.

40. Sale, D.G.; MacDougall, J.D.; Jacobs, I.; Garner, S. Interaction between concurrent strength and endurance training. J. Appl. Physiol. 68:260-70; 1990.

41. Saltin, B.; Gollnick, P.D. Significance for metabolism and performance. In: Peachy, L.D.; Adrian, R.H.; Geiger, S.R., eds. Skeletal muscle. Handbook of physiology, section 10. Bethesda, MD: American Physiological Society; 1983:555-631.

42. Scammon, R.E. A summary of the anatomy of the infant and child. In: Abt, I.A., ed. Pediatrics. Philadelphia: Saunders; 1923:257-444.

43. Schiaffino, S.; Reggiani, C. Myosin isoforms in mammalian skeletal muscle. J. Appl. Physiol. 77:493-501; 1994.

44. Scriver, C.R.; Beaudet, A.L.; Sly, W.S.; Valle, D. The metabolic basis of inherited disease. Volumes I and II. 6th ed. New York: McGraw-Hill; 1989.

45. Seedorf, U.; Leberer, E.; Kirschbaum, B.J.; Pette, D. Neural control of gene expression in skeletal muscle: Effect of chronic stimulation on lactate dehydrogenase isoenzymes and citrate synthase. Biochem. J. 239:115-20; 1986.

46. Simoneau, J.A.; Bouchard, C. Genetic determinism of fiber type proportion in human skeletal muscle. FASEB J. 9:1091-95; 1995.

47. Simoneau, J.A.; Bouchard, C. Human variation in skeletal muscle fiber type proportion and enzyme activities. Am. J. Physiol. Endocr. Metab. 257:E567-72; 1989.

48. Simoneau, J.A.; Lortie, G.; Boulay, M.R.; Marcotte, M.; Thibault, M.-C.; Bouchard, C. Human skeletal muscle fiber type alteration with high-intensity intermittent training. Eur. J. Appl. Physiol. 54:250-53; 1985.

49. Simoneau, J.A.; Lortie, G.; Boulay, M.R.; Marcotte, M.; Thibault, M.-C.; Bouchard, C. Inheritance of human skeletal muscle and anaerobic capacity adaptation to high-intensity intermittent training. Int. J. Sports Med. 7:167-71; 1986.

50. Simoneau, J.A.; Lortie, G.; Boulay, M.R.; Thibault, M.-C.; Bouchard, C. Repeatability of fibre type and enzyme activity measurements in human skeletal muscle. Clin. Physiol. 6:347-56; 1986.

51. Simoneau, J.A.; Lortie, G.; Boulay, M.R.; Thibault, M.-C.; Thériault, G.; Bouchard, C. Skeletal muscle histochemical and biochemical characteristics in sedentary male and female subjects. Can. J. Physiol. Pharmacol. 63:30-35; 1985.

52. Simoneau, J.A.; Pette, D. Species-specific effects of chronic nerve stimulation upon tibialis anterior muscle of mouse, rat, guinea pig and rabbit. Pflügers Arch. 412:86-92; 1988.

53. Sjöström, M.; Friden, J.; Ekblom, B. Endurance, what is it? Muscle morphology after an extremely long distance run. Acta Physiol. Scand. 130:513-20; 1987.

54. Staron, R.S.; Johnson, P. Myosin polymorphism and differential expression in adult human skeletal muscle. Comp. Biochem. Physiol. 106:463-75; 1993.

55. Staron, R.S.; Leonardi, M.J.; Karaponlo, D.S.; Malicky, E.S.; Falkel, J.E.; Hagerman, F.C.; Hikida, R.S. Strength and skeletal muscle adaptations in heavy-resistance-trained women after detraining and retraining. J. Appl. Physiol. 70:631-40; 1991.

56. Staron, R.S.; Malicky, E.S.; Leonardi, M.J.; Falkel, J.E.; Hagerman, F.C.; Dudley, G.A. Muscle hypertrophy and fast fiber conversions in heavy resistance-trained women. Eur. J. Appl. Physiol. Occup. Physiol. 60:71-79; 1990.

57. Tarnopolsky, L.J.; MacDougall, J.D.; Atkinson, S.A.; Tarnopolsky, M.A.; Sutton, J.R. Gender differences in substrate for endurance exercise. J. Appl. Physiol. 68:302-8; 1990.

58. Tarnopolsky, M.A.; Atkinson, S.A.; Phillips, S.M.; MacDougall, J.D. Carbohydrate loading and metabolism during exercise in men and women. J. Appl. Physiol. 78:1360-68; 1995.

59. Thibault, M.-C.; Simoneau, J.A.; Côté, C.; Boulay, M.R.; Lagassé, P.; Marcotte, M.; Bouchard, C. Inheritance of human muscle enzyme adaptation to isokinetic strength training. Hum. Hered. 36:341-47; 1986.

60. Vogel, F.; Motulsky, A.G. Human genetics. Problems and approaches. New York: Springer Verlag; 1986.

61. Vogler, C.; Bove, K.E. Morphology of skeletal muscle in children. Arch. Pathol. Lab. Med. 109:238-42; 1985.

62. Williams, R.S.; Garcia-Moll, M.; Mellor, J.; Salmons, S.; Harlan, W. Adaptation of skeletal muscle to increased contractile activity. J. Biol. Chem. 262:2764-67; 1987.

63. Williams, R.S.; Salmons, S.; Newsholme, E.A.; Kaufmann, R.E.; Mellor, J. Regulation of nuclear and mitochondrial gene expression by contractile activity. J. Biol. Chem. 261:376-80; 1986.

CHAPTER 12

Genetics of Cardiorespiratory Fitness Phenotypes

Cardiorespiratory fitness has traditionally been viewed by many as the most important component of performance- and health-related fitness. It is commonly defined in terms of several more specific phenotypes, the most important of which were summarized earlier in table 5.1. This chapter focuses on phenotypes related to cardiac structure and function and to pulmonary function. The genetic contribution to submaximal exercise capacity and maximal aerobic power are discussed in chapter 15 in the context of other performance phenotypes.

Life Span and Gender Variability

Cardiac dimensions, blood pressure, and pulmonary function changes over the years are discussed in the following sections.

Cardiac structure and function

Cardiac dimensions increase postnatally until maturity and vary largely as a function of body size. Heart volume is about 40 cm^3 at birth, doubles by 6 months of age, quadruples by age 2, and reaches approximately 600 to 800 cm^3 in the young adult. These changes are closely related to changes in body weight. The ratio of heart volume to body weight tends to be constant at about 10 cm^3/kg during childhood and adolescence. This is shown in table 12.1, which is based on radiograph measurements of heart diameters in 237 boys.

Heart rate, electrocardiographic (ECG) features, stroke volume, cardiac output, and blood pressures are indicators of cardiac function. Changes in these functions during growth have been described in detail by Malina and Bouchard (55). Briefly, between birth and adulthood, resting heart rate decreases by about 50%, amplitudes of the various

Table 12.1
Heart Dimensions of Boys 8 to 18 Years of Age

Age	Body weight (kg)	Heart length (cm)	Heart width (cm)	Heart depth (cm)	Heart volume (cm^3)	Heart volume/ body weight ratio (cm^3/kg)
8	28.4	11.2	9.1	7.8	282	10.0
9	30.8	11.8	9.3	8.1	312	10.3
10	32.3	11.8	9.6	8.2	328	10.1
11	35.6	12.2	9.8	8.6	362	10.3
12	38.6	12.4	10.0	9.0	395	10.3
13	44.8	13.2	10.4	9.1	444	10.1
14	49.0	13.7	10.9	9.5	503	10.3
15	56.1	14.1	11.5	9.6	551	9.8
16	63.0	14.8	11.9	9.7	603	9.6
17	66.7	14.8	12.2	10.2	646	9.7
18	66.8	15.3	12.3	10.1	671	10.1

Heart dimensions are based on measurements from two chest X-rays.
Adapted from Bouchard et al. (13).

deflections of the ECG increase, and cardiac output increases about 10-fold. With advancing age, cardiac function declines. Although resting heart rate remains unchanged, the stroke volume decreases, leading to a concomitant reduction in cardiac output. From a value of about 5 L/min in young adulthood, cardiac output declines to about 3 L/min by 80 years of age (15). To compensate for the reduced cardiac output, an increase in peripheral resistance occurs with age. The elevated resistance, combined with increasing rigidity of blood vessels (due to reduced elasticity of arteries), may contribute to the rise of blood pressures commonly observed in aging individuals.

Changes in systolic and diastolic blood pressures from birth to old age are shown in figure 12.1. During growth, systolic and diastolic blood pressures increase in males and females. On the average, systolic blood pressure increases from about 80 mmHg to 110 mmHg during this period of life, while diastolic blood pressure increases from about 50 mmHg to 70 mmHg. Differences between males and females become apparent during adolescence with higher average values of both blood pressures in males compared with females. Systolic blood pressure continues to increase throughout the remainder of life, while diastolic

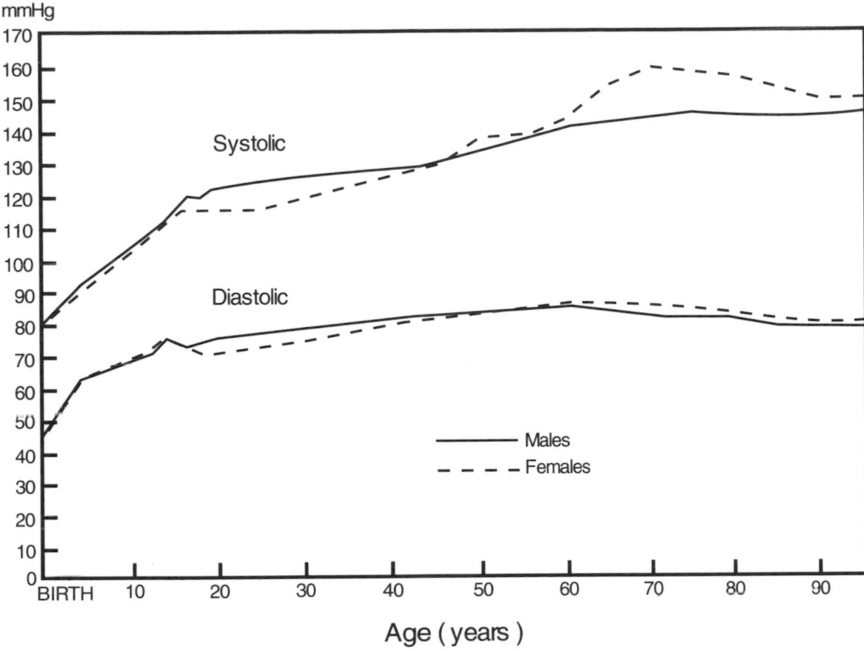

Figure 12.1 Average blood pressures in healthy subjects from birth to old ages. Reprinted, by permission, from Brest and Moyer, 1961, *Hypertension: Recent advances* (Baltimore: Williams and Wilkins).

blood pressure tends to reach its peak in the early 60s and declines very slightly afterward. Sex differences are maintained until the early 50s and are reversed in the 60s with higher blood pressures (especially systolic blood pressure) in females than in males.

Pulmonary function

The respiratory system, by its ability to extract oxygen from inspired air, ensures an adequate oxygen supply to all tissues of the body. The amount of oxygen taken up by the blood from the lungs and transported to the tissues during exercise decreases with age. The decline of oxygen uptake observed with aging could be partly attributed, in some individuals, to the deterioration of pulmonary functions.

Pulmonary functions are assessed by measurements of several volumes and capacities. During growth, respiratory volumes and capacities increase primarily as a function of stature. Changes in respiratory volumes and capacities during adulthood are depicted in figure 12.2. Overall, they tend to deteriorate with age. Despite no changes in the total amount of air that the lungs can hold (i.e., total lung capacity), there are significant reductions in the various volumes of air moved in

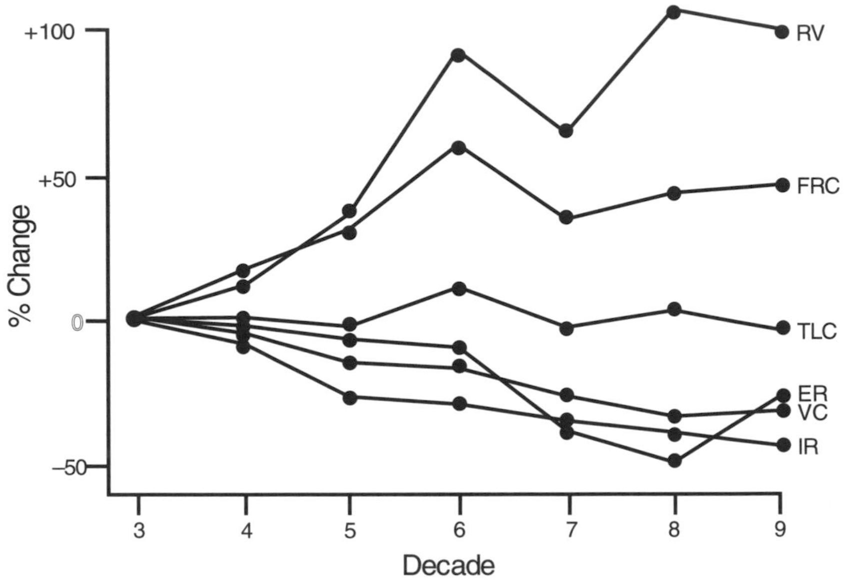

Figure 12.2 Percentage changes in lung volumes at various ages compared to values observed in the third decade of life. RV = residual volume; FRC = functional residual capacity; TLC = total lung capacity; ER = expiratory reserve; VC = vital capacity; IR = inspiratory reserve. Reprinted, by permission, from *Surgery of the aged and debilitated patient*, 1968, edited by J.D. Powers (Philadelphia: W.B. Saunders).

and out of the lungs (vital capacity, expiratory reserve, inspiratory reserve) as well as increases in the volumes of air remaining in the lungs after normal (functional reserve capacity) and maximal (residual volume) expirations. The decline of pulmonary function is generally attributed to a decrease in the elastic component of the lung tissue and to a diminution of the strength of the respiratory muscles.

Genetics of Cardiac Structures and Functions

Genetic epidemiology studies

Cardiac structures and functions are critical limiting factors of aerobic performance. It is well documented that endurance athletes have larger hearts, larger stroke volume and cardiac output during maximal exercise, bradycardia at rest, and reduced heart rate at the same submaximal work load.

The genetics of heart structures and functions, however, has not been extensively studied. There is evidence that genetic factors play a significant role in the vascular wall thickness of the coronary arteries (69) and in the branching pattern of the coronary and pulmonary arteries (38,64). The role of heredity in cardiac function at rest, assessed from ECG measurements, has been considered in several twin studies (35,58,92). A significant genetic effect was observed for the duration of the P-R interval, but not for the QRS and Q-T intervals (35). There is also evidence from family studies that the duration of the P-R interval aggregates in families and may be determined by a single gene (59).

Echographic measurements of cardiac dimensions performed in members of nuclear families suggest familial resemblance in several ventricular dimensions (1,93). Familial correlations among various kinds of relatives by descent and adoption were assessed for echocardiographically derived heart dimensions (93). For most cardiac dimensions, correlations were significant in both biological relatives and relatives by adoption, suggesting that genetic and environmental factors contribute to the phenotypic variance.

The inheritance of left ventricular structure and function assessed by M-mode echocardiography and Doppler echocardiography at the aorta and mitral valve has been studied in 32 MZ pairs and 21 DZ pairs of healthy male twins (9). A path analysis model in which phenotypic variance could be partitioned into genetic (h^2), shared environmental (c^2), and nonshared environmental (e^2) components was used (table 12.2). All of the data were adjusted for significant effects of age and weight. Except for left ventricular internal diameter, for which no significant genetic effect was observed, all other heart structures were significantly influenced by genetic factors, with heritability levels ranging from 29% to 68%. In addition to the cardiac structures, several functions (heart rate, peak aortic flow velocity, stroke distance, and minute distance) also showed a significant genetic effect, which ranged from about 50% to 60%. Environmental factors, represented by the c^2 and e^2 components, were also important determinants of interindividual differences in cardiac structure and function. These components accounted for a substantial portion of the phenotypic variance: 30% to 70%. Although cardiac structures and functions are significantly affected by the genotype, it should be noted that they can be modified by environmental factors.

Since cardiac dimensions are largely dependent on body size, a significant genetic effect observed for cardiac dimensions may, in part, be attributable to the inheritance of body size. Therefore, after adjustment for height or body mass, the genetic effect for cardiac dimensions is generally reduced (27,93). The strong relationship between body size

Table 12.2
Contribution (%) of Genetic and Environmentals Factors to Several Cardiac Structures and Functions

Variable	h^2	c^2	e^2
Left ventricular internal diameter (mm)	0	40	60
Posterior wall thickness (mm)	68	0	32
Septal wall thickness (mm)	29	0	71
Left atrial diameter (mm)	34	0	66
Aortic diameter (mm)	68	0	32
Heart rate (beats/min)	61	0	39
Aortic peak flow (cm/s)	48	0	52
Stroke distance (cm)	55	0	45
Minute distance (m/min)	47	9	44

Data were adjusted for age and weight. h^2 = genetic component; c^2 = shared environment component; e^2 = nonshared environmental component.

Adapted from Bielen et al. (9).

and cardiac dimensions raises the question of how much of the covariation between these two variables is explained by shared genetic factors. This question was addressed by Verhaaren et al. (94) in a bivariate genetic analysis of left ventricular mass and body weight in 147 MZ and 107 DZ pubertal twin pairs of both sexes. Heritabilities of left ventricular mass reached 60% in males and 73% in females. After adjustment of left ventricular mass for body weight and sexual maturity status, the genetic effect was reduced but remained significant, with heritabilities of 39% and 59% in males and females, respectively. Bivariate genetic analyses showed that the correlation between left ventricular mass and body weight was almost entirely of genetic origin, as 90% was attributed to common genes (94).

The preceding studies suggest that genetic factors are important in determining cardiac structures and functions under resting conditions. The role of the genotype in the response of cardiac dimensions to acute exercise or exercise training is less well documented. Bielen et al. (8) studied the inheritance of cardiac changes during a submaximal supine cycle exercise at a heart rate of 110 beats per minute in 21 pairs of MZ twins and 12 pairs of DZ twins. In resting conditions and after adjustment for body weight, only the average cardiac wall thickness was significantly influenced by genetic factors ($h^2 = 57\%$). On the other hand, the increases of left ventricular internal diameter and fractional shortening in response to exercise showed a genetic effect of 24% and 47%, respectively. The results were interpreted as an indication that adaptation of cardiac function during submaximal exercise may be partly determined by the genotype (8).

In a study on genotype-training interaction effects in cardiac dimensions, Landry et al. (49) submitted 10 pairs of sedentary MZ twins to a 20-week program of endurance training on a cycle ergometer. Changes in cardiac dimensions were assessed by echocardiography and were compared within and between pairs. Despite the absence of significant changes in heart dimensions following exercise training, the within-pair resemblance was greater after training than before, suggesting that the variability of the response of cardiac dimensions to endurance training is potentially influenced by genes.

Molecular studies

The molecular basis of cardiac structures and functions has been investigated primarily in pathological conditions, such as in familial hypertrophic cardiomyopathy (i.e., conditions characterized by increased left ventricular mass). Several molecular genetic studies have shown that inherited cardiomyopathies may be linked to gene loci involved in cardiac energy metabolism (mitochondrial oxidative phosphorylation and fatty acid beta-oxidation) and contractile and structural cardiac proteins (45). Several missense mutations of the cardiac beta-myosin heavy chain (MHC) located on chromosome 14q12 have been reported in families with hypertrophic cardiomyopathy (2,76,88). The deletion/insertion (D/I) polymorphism in intron 16 of the angiotensin-converting enzyme (ACE) gene, the ACE/ID polymorphism, has also been linked with hypertrophic (57) and dilated (77) cardiomyopathy. In a recent study, Schunkert et al. (83) compared the genotype distribution of the ACE/ID polymorphism among 290 men and women with evidence of left ventricular hypertrophy and 290 controls matched for age, sex, and blood pressure status. An increased frequency of the DD genotype among cases compared to controls was observed. The association of the DD genotype was stronger in men (odds ratio = 2.6) than women (odds ratio = 1.2); within each gender, it was stronger in normotensives than hypertensives. It was concluded that the DD genotype of the ACE was associated with an increased risk of left ventricular hypertrophy in middle-aged men (83).

Very few association or linkage studies have been reported for cardiac function in the absence of cardiomyopathy. In one study, the influence of the ACE/ID polymorphism on the variation of left ventricular mass and function was investigated in 86 healthy subjects free of clinical heart disease (47). Left ventricular mass adjusted for body size and the various indices of left ventricular function were similar across the ACE/ID genotypes. The authors concluded that, in the absence of heart disease, the ACE/ID polymorphism has no impact on cardiac size and function. In the Quebec Family Study, linkage relationships between cardiac dimensions and polymorphisms of 14 red blood cell antigens and enzymes located on eight different chromosomes were investigated (72). Echocardiographic measurements of 730 individuals from 189 randomly ascertained nuclear families were used in a sib-pair linkage analysis. Significant evidence of linkage was found between the phosphoglucomutase-1 locus on chromosome 1p22.1 and left ventricular mass, left ventricular internal diameter, and left ventricular volume adjusted for age and body surface area (72). The results suggest that a gene located in this region of the genome could be important in determining several cardiac structures and functions.

Genetics of Pulmonary Function

The genetics of pulmonary function has been primarily investigated in the context of abnormal function, such as in emphysema, chronic obstructive pulmonary disease (COPD), asthma, and bronchitis. Such investigations suggest the presence of familial aggregation in pulmonary diseases (37,39,50,91,101). Pulmonary function in healthy individuals also aggregates in families. In the Quebec Family Study, familial resemblance was observed for maximum pulmonary flow, 1- and 2-s forced expiratory volumes (FEV-1 and FEV-2), and forced vital capacity (FVC), after adjustment for age and stature, in 1,258 healthy subjects from 304 nuclear families (71). Spouse, parent-child, and sibling correlations were all significant, suggesting that both genetic factors and familial environment accounted for a significant portion of normal variation in pulmonary functions.

Twin studies generally show higher intrapair correlations for pulmonary measures in MZ twins than in DZ twins, suggesting a significant genetic effect. The results of these studies, often based on a small number of twins, provide a wide range of heritability estimates (5,29,40,44,51,56). For example, heritabilities ranging from 0% (29) to 77% (40) have been reported for FEV-1. In one study based on a large number of DZ ($n = 158$) and MZ ($n = 256$) adult twins of both sexes, the intraclass correlations for DZ twins (0.16-0.39) were approximately one-half the magnitude of those in MZ twins (0.52-0.76) for various pulmonary measurements adjusted for age, sex, height, and smoking status (78). These results suggest that genetic factors are more important than environmental factors in determining the familial aggregation of pulmonary function.

Genetic and nongenetic determinants of pulmonary functions were also evaluated with the more powerful method of path analysis (24-26,52). Using the XTAU path analysis model, which incorporates sex differences in transmissibility of the phenotype from parents to offspring, Devor and Crawford (26) estimated that 20% to 30% of the variance in normal pulmonary function could be accounted for by genetic and cultural factors transmitted across generations with no sex differences in the transmissibility estimates. Furthermore, the relative effect of shared sibling environment accounted for the majority (about 80%) of the resemblance observed between siblings, which is consistent with the high heritability estimates derived from twin data. A commingling analysis of the same data set also revealed that the ventilatory capacities in this population fitted a single normal distribution, which suggests the absence of major gene effects for pulmonary function (26).

Based on another path analysis model, which allows the partitioning of transmissible factors into genetic and common familial environmental factors, Lewitter et al. (52) evaluated pulmonary function in 404 nuclear families living in the Boston area. Genetic heritability of FEV-1 reached 47% and was the same for parents and their children. Common familial environment explained 1% of the variance in children and 28% in parents. In a population-based sample of New Mexico Hispanics, using path analysis Coultas et al. (25) reported heritabilities of 43% and 42% for FVC and FEV-1, respectively. The estimates of heritability were similar in parents and children and were unaltered by smoking status.

Two segregation analyses of pulmonary functions have been reported. In one study based on data from COPD families, Rybicki et al. (81) showed that variation in FEV-1 could be entirely accounted for by a single gene, while no evidence of a genetic effect was observed in families of nonpulmonary patients. Segregation analysis of two indices of lung function, FEV-1 and maximal midexpiratory flow rate, was also carried out in 799 healthy individuals from 214 nuclear families (20). After adjustment of both phenotypes for age, sex, height, and smoking status, evidence of a major effect with a polygenic component was observed only for FEV-1. However, the Mendelian and the nontransmission models provided an equal fit to the data, suggesting that the genetic effect found for these pulmonary phenotypes were polygenic (20).

Results from segregation analyses suggest that specific gene loci could be involved in determining lung function in subjects with pulmonary disease. Because of the relationship between the serum enzyme alpha-1-antitrypsin and obstructive lung disease (22), this locus (P_i) is a candidate gene of pulmonary function. In a study of 203 twin pairs, the

P_i polymorphism was associated with lung function (forced expiratory volume corrected for the effects of age and height) and accounted for about 9% of the variance in female twins (30). No effect of this polymorphism was, however, observed in males.

Adaptive responses of pulmonary functions to various environments have also been considered. In one study, for example, Arkinstall et al. (4) measured the ventilatory response to inhaled CO_2 in MZ and DZ twin pairs. At three levels of end tidal PCO_2, a significant genetic effect was observed for tidal volume response but not for breathing frequency or ventilatory responses. Significant correlations between hypercapnic ventilatory responses of championship swimmers and their siblings have also been reported (82). In another study, Scoggin et al. (84) investigated the ventilatory responses to hypoxia and hypercapnia in five outstanding male long-distance runners, 16 of their parents and siblings who were classified as nonathletes, and 34 sedentary controls. Runners and members of their families exhibited similar ventilatory responses to hypoxia, while control subjects were significantly different from both groups. Ventilatory responses to hypercapnia followed the same pattern but did not reach statistical significance. In another study from the same group, Collins et al. (23) reported more similar ventilatory responses to progressive hypoxia in MZ twins than in DZ twins. Thus, a role for genetic factors in determining ventilatory adaptations to hypoxia is suggested.

Studies of pulmonary functions at different altitudes suggest that environmental factors are also important determinants of these functions. By studying ventilation and lung volumes of individuals living at high altitude and sea level, Lahiri et al. (48) showed that the diminished ventilatory response to hypoxia and the large lung volumes of adult high-altitude native residents were more environmentally than genetically determined. In another study of pulmonary function in relatives residing at different altitudes in South America, Mueller et al. (65) observed that correlations between family members decreased when one or both relatives migrated to a different altitude. These results are consistent with those of Anderson et al. (3) who reported that children born to parents living at sea level, but living at altitudes of 1,500 to 2,000 m, had pulmonary functions comparable to those of children born and raised at altitude. Thus, the hypoxia associated with residence at various altitudes influences pulmonary functions.

Genetic influences on lung function and genetic determinants of the susceptibility of the lungs to smoking represent two different genetic effects. Although smoking and passive smoking could affect lung function, few studies have investigated the role of the genotype in the lungs' susceptibility to smoking. Results of twin studies also suggest that genetic factors are important determinants of the susceptibility of the lungs to smoking (34,96).

Genetics of Blood Pressure

Genetic epidemiology studies

Compared to other cardiorespiratory fitness phenotypes, the genetics of blood pressure has been extensively studied in both animals and humans (60,66,75,86,87,95,98).

Both systolic and diastolic blood pressures aggregate in families (table 12.3). Correlations for various kinds of relatives tend to increase with increasing level of genetic relatedness, with highest correlations in MZ twins. Correlations in genetically unrelated individuals approximate zero.

Family, twin, and adoption studies in combination with various multivariate analytical strategies have been used to quantify the contribution of heredity to blood pressures (95). Heritability estimates vary with study design and methods of analysis. Estimates from twin studies are generally higher (50% to 70%) than those from family or adoption studies (30% to 50%). In the Minnesota study of MZ twins reared apart, a correlation of 0.64 was reported for systolic blood pressure (14).

Most studies are consistent with a multifactorial model of inheritance, with both genetic and cultural transmissible effects and with

Table 12.3
Familial Correlations for Blood Pressure

Relative	Systolic	Diastolic
MZ twins	0.55-0.72	0.58
DZ twins	0.25	0.27
Siblings	0.17-0.33	0.18-0.29
Parent-child	0.13-0.32	0.17-0.37
Uncle/aunt-nephew/niece	0.05	0.06
Spouses	0.09-0.15	0.07-0.14
Parent-adoptees	0.09	0.10
Unrelated siblings	0-0.15	0-0.29

From Mongeau (60); Williams et al. (98); Ward (95).

nontransmissible environmental factors contributing to the phenotypic variance. The genetic effect detected in these studies is polygenic (i.e., attributable to the contribution of several genes, each having a small effect on the phenotype).

The hypothesis that blood pressures could be influenced by a major gene effect was evaluated in several studies using segregation analysis (17,46,63,68,73,79,85). Evidence of a mixture of distributions in blood pressure variation was found in the Quebec Family Study. The results were compatible with the presence of a major gene effect; however, Mendelian segregation of this putative major gene was not supported by the results of the segregation analysis (79).

Only three studies have reported evidence for Mendelian segregation of a major gene effect affecting blood pressure variability. In the first study (17), an autosomal recessive gene was associated with lower blood pressure levels, suggesting that individuals who inherit this gene could be protected against the development of hypertension. In the second study, systolic blood pressure was influenced not only by a polygenic effect accounting for 36% of the phenotypic variance, but also by allelic variation at a single gene with gender- and age-specific effects (73). The age- and gender-specific effects of this putative gene suggest that individuals carrying the gene are characterized by a steeper increase of systolic blood pressure with age and that age-related changes in systolic blood pressure attributable to this gene are different in males and females. Based on the best fit segregation model, the contribution of this major gene to systolic blood pressure variation increased with age. It accounted for about 1% and 6% of the variance at 5 years of age and about 61% and 55% of the variance at 50 years of age in males and females, respectively (73). In the third study, age-related changes in diastolic blood pressure, measured over a 7-year follow-up period in 965 adult subjects from 73 pedigrees, were influenced by an autosomal recessive locus with a frequency of 0.23 (85). In subjects inferred to be homozygotes for the allele associated with large blood pressure increases, diastolic blood pressure increased 32% compared to only 1% in homozygotes for the nonsusceptible allele.

The results from the last two studies (73,85) suggest that variation in a single gene may be important in determining susceptibility to hypertension. In order to test whether genetic variation in the sodium-potassium adenosine triphosphatase beta gene (ATP1B1) could be responsible for this interaction effect, association and linkage analyses with 11-year changes in blood pressures were performed on 74 families from the Quebec Family Study (70). Significant differences among genotypes were observed for changes in systolic blood pressure, and linkage analysis based on the sib-pair method revealed a strong linkage ($p = .0005$) with this marker locus. No evidence of association or linkage was observed for any of the cross-sectional values of blood pressures. These results suggest that sodium-potassium ATPase could be one of the genes responsible for the major gene effect reported in the segregation analyses (73,85).

Genetic studies of blood pressures are complicated by the complex and multifactorial nature of the phenotypes. Such phenotypes are influenced by several intermediate quantitative phenotypes, which are themselves determined by genetic and environmental factors. Intermediate phenotypes for blood pressure include the transport of electrolytes, the renin-angiotensin-aldosterone system, the sympathetic nervous system, and other metabolic phenotypes. These intermediate phenotypes aggregate in families and are influenced by polygenes as well as major genes (16,33,98,99).

Although no evidence of a genotype-training interaction effect has been reported thus far for blood pressure, genetic factors have been implicated in determining blood pressure responses to a variety of other stimuli from the physical, mental, behavioral, and biological domains. Evidence from twin and family studies suggest that blood pressure "reactivity" is genetically conditioned (74).

Blood pressure responses to dietary salt represents the most documented example of genotype-environment interaction for blood pressure-related phenotypes. There is now substantial evidence that salt sensitivity and salt resistance in humans is genetically determined (31,33,54). The haptoglobin gene (21) and the gene coding for the precursor of the atrial natriuretic peptide (43) are apparently associated with this interaction effect.

Molecular studies

Several random genetic markers and candidate genes have been tested for association and linkage with blood pressure and hypertension. A summary of the studies is presented in table 12.4.

Antigens of the major histocompatibility complex (HLA) have been the most widely investigated. Results from several studies are both positive and negative (89,99). Some antigens (B12, B15, and B18) were occasionally found more frequently in hypertensive patients than in controls, but in most cases the associations lost significance after correction for the number of antigens tested. Other associations have been reported with blood groups, particularly with the MN system (36) and the haptoglobin (*HP*) locus (33,97), while linkage was reported with the 6-phosphogluconate dehydrogenase (*PGD*) gene (100).

Table 12.4
Summary of Association and Linkage Studies of Hypertension and Blood Pressure

Chromosome location	Marker locus	Phenotype	Association	Linkage	Reference
1p36	PGD	DBP		Yes	Wilson et al. (100)
1p36	ANF	HPT	No		Zee et al. (102)
1q23-25	ATP1B1	changes in SBP	Yes	Yes	Pérusse et al. (70)
1q23-25	AT3	HPT	No		Zee et al. (102)
1q32	REN	HPT	No		Soubrier et al. (89)
		HPT	No		Soubrier et al. (90)
		HPT	No		Morris and Griffiths (61)
		HPT		No	Naftilan et al. (67)
		HPT	No		Zee et al. (102)
1q42-43	AGT	HPT	No		Zee et al. (102)
		HPT	Yes	Yes	Jeunemaître et al. (42)
		HPT	No	Yes	Caulfield et al. (18)
		HPT	Yes	No	Caulfield et al. (19)
3q21-25	AGTR1	HPT	Yes		Bonnardeaux et al. (11)
		HPT	No		Rolfs et al. (80)
4q28-31	MN	HPT	Yes		Heise et al. (36)
4q31.3-32	NPYY1R	HPT	No	Yes	Zee et al. (102)

6p21.3	HLA	SBP, DBP	Yes		Gerbase-DeLima et al. (28)
6p24-23	EDN1	HPT	No	No	Berge and Berg (7)
7q36	NOS	HPT	No		Bonnardeaux et al. (12)
10q24-26	ADRA2R	HPT	No		Zee et al. (102)
	ADRB1R	HPT	Yes		Lockette et al. (53)
		HPT	No		Zee et al. (102)
11p15	INS	HPT	No		Zee et al. (102)
16q22.1	HP	HPT	Yes		Weinberger et al. (97)
		HPT	Yes		Grim and Robinson (33)
17q23	ACE	SBP	Yes		Grim and Robinson (33)
		SBP, DBP	No	No	Berge and Berg (6)
		HPT	No		Morris and Zee (62)
		HPT			Jeunemaître et al. (41)
17q22-24	GH	HPT	No		Zee et al. (102)
19p13.3-13.4	LDLR	HPT	No		Zee et al. (102)
19q13.3	INSR	HPT	No		Zee et al. (102)
19q13.2-13.4	KLKR	HPT	No		Zee et al. (102)

SBP = systolic blood pressure; DBP = diastolic blood pressure; HPT = hypertension.
See text for other abbreviations.

More interesting results were obtained with candidate genes, particularly genes related to the renin-angiotensin system. The renin-angiotensin system plays an important role in the regulation of blood pressure, and genes encoding the proteins involved in this system are good candidates for molecular studies. Angiotensinogen (AGT), a protein synthesized by the liver, interacts with the kidney enzyme renin (REN) to produce angiotensin I, which is then cleaved by the angiotensin-converting enzyme (ACE) into angiotensin II. Angiotensin II, via a cell surface receptor called the angiotensin II type 1 receptor (AGTR1), promotes sodium retention and vasoconstriction and is thus important in blood pressure regulation. All of the genes implicated in this system have been tested for association or linkage with blood pressure or hypertension. In two different studies involving populations from France and the United States (42) and from the United Kingdom (18), the AGT gene was linked with hypertension. Evidence of association was found in only one of these studies (42). The AGT gene also contributed to differences of blood pressures between white and black children (10) and was found to be associated with hypertension in African Caribbeans (18).

Some studies, however, have not demonstrated association or linkage. For example, studies undertaken with the REN gene polymorphism did not show any evidence of association or linkage with hypertension (61,67,89,90). Although there are discrepancies among findings with the ACE/ID polymorphism from different populations, most studies failed to provide evidence of association or linkage of this polymorphism with blood pressures (6,41,62,102). Mixed results were also reported for polymorphisms of AGTR1 in association and linkage with hypertension (table 12.4). One study based on a relatively small number of subjects (20 cases and 67 controls) found no association between the AGTR1 polymorphism and essential hypertension (80). In a second case-control study of 298 normotensives and 206 hypertensives, significant association was reported for three of five AGTR1 polymorphisms, while no linkage was found in 267 sib-pairs from 138 pedigrees (11).

Allele frequencies of 13 polymorphisms at 12 cardiovascular candidate gene loci were compared between patients with severe familial hypertension and normotensive controls (102). The candidate genes included atrial natriuretic factor (ANF), antithrombin III (AT3), renin (REN), angiotensinogen (AGT), neuropeptide-Y Y1 receptor (NPYY1R), insulin (INS), alpha-2 adrenoreceptor (ADRA2R), beta-1 adrenoreceptor (ADRB1R), growth hormone (GH), low-density lipoprotein receptor (LDLR), insulin receptor (INSR), and renal kallikrein (KLKR). No significant differences in allele frequencies were found between normotensives and hypertensives, suggesting that none of these candidate genes is associated with hypertension. More recently, Lockette et al. (53) reported a significant association between a genetic polymorphism of the alpha-2 adrenergic receptor and essential hypertension in African Americans.

Endothelin-1 (one of the three isoforms of endothelin) is a potent vasoconstrictor peptide produced by vascular endothelial cells that could, therefore, play a role in the regulation of blood pressure. However, a restriction fragment length polymorphism at the endothelin-1 (EDN1) locus was tested for association with blood pressures with negative results (6). Nitric oxide released by blood vessel walls mediates the endothelium-derived relaxing factor, which plays an important role in the regulation of blood flow and blood pressure. It has been suggested that hypertensive subjects have a blunted vasodilatation secondary to a decrease in nitric oxide production. Nitric oxide synthase (NOS) is the enzyme responsible for the synthesis of nitric oxide in the endothelium and represents a good candidate gene for hypertension. However, using a highly polymorphic marker of NOS, Bonnardeaux et al. (12) found no evidence of association and linkage with essential hypertension.

Summary

Genetic factors play a significant role in explaining interindividual differences in cardiorespiratory fitness phenotypes. Heart size and cardiac functions are influenced by genetic factors, and heritabilities of these phenotypes assessed by echocardiography vary between 30% and 70%. Cardiac changes in response to acute exercise and to exercise training are also influenced by the genotype. Molecular studies of inherited cardiomyopathies have led to the identification of a few genetic loci contributing to cardiac structures and functions. The cardiac beta-myosin heavy chain and the angiotensin-converting-enzyme genes are among the genes that may be linked to these phenotypes. Pulmonary functions appear to be moderately influenced by genetic factors. They also adapt to changes in environmental conditions, such as hypoxia and altitude. Results from twin and family studies suggest that these adaptations may be partly determined by the genotype.

The contribution of genetic factors in blood pressure variability and hypertension is well established. Heritabilities around 30% are generally reported for blood pressure, and this effect appears to be mainly attributable to polygenes. Two studies have shown that the increase of blood pressure with age could be determined by a major locus.

Association and linkage studies using either blood pressure levels or hypertension as phenotypes have been conducted with genes implicated in the renin-angiotensin system. Despite positive findings in some studies, the majority of candidate genes identified to date were not associated nor linked with blood pressure or hypertension.

References

1. Adams, T.D.; Yanowitz, F.G.; Fisher, G.; Ridges, J.D.; Nelson, A.G.; Hagan, A.D.; Williams, R.R.; Hunt, S.C. Genetics and cardiac size. In: Malina, R.M.; Bouchard, C., eds. Sport and human genetics. Champaign, IL: Human Kinetics; 1986:131-45.
2. Anan, R.; Greve, G.; Thierfelder, L.; Watkins, H.; McKenna, W.J.; Solomon, S.; Vecchio, C.; Shono, H.; Nakao, S.; Tanaka, H.; Mares, A.; Towbin, J.A.; Spirito, P.; Roberts, R.; Seidman, J.G.; Seidman, C.E. Prognostic implications of Novel β cardiac myosin heavy chain gene mutations that cause familial hypertrophic cardiomyopathy. J. Clin. Invest. 93:280-85; 1994.
3. Anderson, H.R.; Anderson, J.R.; King, H.O.M.; Cotes, J.E. Variations in the lung size of children in Papua New Guinea: Genetic and environmental factors. Ann. Hum. Biol. 5:209-18; 1978.
4. Arkinstall, W.W.; Nirmel, K.; Klissouras, V.; Milic-Emili, J. Genetic differences in the ventilatory response to inhaled CO_2. J. Appl. Physiol. 36:6-11; 1974.
5. Astemborski, J.A.; Beaty, T.H.; Cohen, B.H. Variance components analysis of forced expiration in families. Am. J. Med. Genet. 21:741-53; 1985.
6. Berge, K.E.; Berg, K. No effect of insertion/deletion polymorphism at the *ACE* locus on normal blood pressure level or variability. Clin. Genet. 45:169-74; 1994.
7. Berge, K.E.; Berg, K. No effect of a Taq 1 polymorphism in DNA at the endothelin-1 (*EDN1*) locus on normal blood pressure level or variability. Clin. Genet. 41:90-95; 1992.
8. Bielen, E.C.; Fagard, R.H.; Amery, A.K. Inheritance of acute cardiac changes during bicycle exercise: An echocardiographic study in twins. Med. Sci. Sports Exerc. 23:1254-59; 1991.
9. Bielen, E.; Fagard, R.; Amery, A. The inheritance of left ventricular structure and function assessed by imaging and Doppler echocardiography. Am. Heart J. 121:1743-49; 1991.
10. Bloem, L.J.; Manatunga, A.K.; Tewksbury, D.A.; Pratt, J.H. The serum angiotensinogen concentration and variants of the angiotensinogen gene in white and black children. J. Clin. Invest. 95:948-53; 1995.
11. Bonnardeaux, A.; Davies, E.; Jeunemaître, S.; Fery, I.; Charru, A.; Clauser, E.; Tiret, L.; Cambien, F.; Corvol, P.; Soubrier, F. Angiotensin II type 1 receptor gene polymorphisms in human essential hypertension. Hypertension. 24:63-69; 1994.

12. Bonnardeaux, A.; Nadaud, S.; Charru, A.; Jeunemaître, X.; Corvol, P.; Soubrier, F. Lack of evidence for linkage of the endothelial cell nitric oxide synthase gene to essential hypertension. Circulation. 91:96-102; 1995.

13. Bouchard, C.; Malina, R.M.; Hollmann, W.; Leblanc, C. Submaximal working capacity, heart size and body size in boys 8 to 18 years. Eur. J. Appl. Physiol. 36:115-26; 1977.

14. Bouchard, T.J.; Likken, D.T.; McGue, M.; Segal, N.L.; Tellegen, A. Sources of human psychological differences: The Minnesota study of twins reared apart. Science. 250:223-28; 1990.

15. Brandfonbrener, M.; Landowne, M.; Shock, N.W. Changes in cardiac output with age. Circulation. 12:557-66; 1955.

16. Burke, W.; Motulsky, A.G. Molecular genetics of hypertension. In: Lusis, A.J.; Rotter, J.I.; Sparkes, R.S., eds. Molecular genetics of coronary artery disease: Candidate genes and processes in atherosclerosis. Basel: Karger; 1992:228-36.

17. Carter, C.L.; Kannel, W.B. Evidence of a rare gene for low systolic blood pressure in the Framingham Heart Study. Hum. Hered. 40:235-41; 1990.

18. Caulfield, M.; Lavender, P.; Farrall, M.; Munroe, P.; Lawson, M.; Turner, P.; Clark, A.J.L. Linkage of the angiotensinogen gene to essential hypertension. N. Engl. J. Med. 330:1629-33; 1994.

19. Caulfield, M.; Lavender, P.; Newell-Price, J.; Farrall, M.; Kamdar, S.; Daniel, H.; Lawson, M.; De-Freitas, P.; Fogarty, P.; Clark, A.J. Linkage of the angiotensinogen gene locus to human essential hypertension in African Caribbeans. J. Clin. Invest. 96:687-92; 1995.

20. Chen, Y.; Horne, S.L.; Rennie, D.C.; Dosman, J.A. Segregation analysis of two lung function indices in a random sample of young families: The Humboldt family study. Genet. Epidemiol. 13:35-47; 1996.

21. Christian, J.C. Association of haptoglobin with sodium sensitivity and resistance of blood pressure. Hypertension. 10:443-46; 1987.

22. Cohen, B.H. Chronic obstructive pulmonary disease: A challenge in genetic epidemiology. Am. J. Epidemiol. 112:274-88; 1980.

23. Collins, D.D.; Scoggin, C.H.; Zwillich, C.W.; Weil, J.V. Hereditary aspects of decreased hypoxic response. J. Clin. Invest. 21:105-10; 1978.

24. Cotch, M.F.; Beaty, T.H.; Cohen, B.H. Path analysis of familial resemblance of pulmonary function and cigarette smoking. Am. Rev. Resp. Dis. 142:1337-43; 1990.

25. Coultas, D.B.; Hanis, C.L.; Howard, C.A.; Skipper, B.J.; Samet, J.M. Heritability of ventilatory function in smoking and non-smoking New Mexico Hispanics. Am. Rev. Resp. Dis. 144:770-75; 1991.

26. Devor, E.J.; Crawford, M.H. Family resemblance for normal pulmonary function. Ann. Hum. Biol. 11:439-48; 1984.

27. Fagard, R.; van Den Broeke, C.; Bielen, E.; Amery, A. Maximum oxygen uptake and cardiac size and function in twins. Am. J. Cardiol. 60:1362-67; 1987.

28. Gerbase-DeLima, M.; DeLima, J.J.G.; Persou, L.B.; Silva, H.B.; Marcondes, M.; Bellotti, G. Essential hypertension and histocompatibility antigens: A linkage study. Hypertension. 14:604-9; 1989.
29. Ghio, A.J.; Crapo, R.O.; Elliott, C.G.; Adams, T.D.; Hunt, S.C.; Jensen, R.L.; Fisher, A.G.; Afman, G.H. Heritability estimates of pulmonary function. Chest. 96:743-46; 1989.
30. Gibson, J.B.; Martin, N.G.; Oakeshott, J.G.; Rowell, D.M. Lung function in an Australian population: Contributions of polygenic factors and the P_i locus to individual differences in lung function in a sample of twins. Ann. Hum. Biol. 10:547-56; 1983.
31. Goldbourt, U. Genetic variation and nutrition. In: Goldbourt, U.; de Faire, U.; Berg, K., eds. Genetic factors in coronary heart disease. Dordrecht, Netherlands: Kluwer Academic; 1994:397-408.
32. Goldman, R. Decline in organ function with aging. In: Rossman, I., ed. Clinical geriatrics. Philadelphia: Lippincott; 1979:23-59.
33. Grim, C.E.; Robinson, M.T. Blood pressure variation. In: Goldbourt, U.; de Faire, U.; Berg, K., eds. Genetic factors in coronary heart disease. Dordrecht, Netherlands: Kluwer Academic; 1994:153-77.
34. Hankins, D.; Drage, C.; Zamel, N.; Kronenberg, R. Pulmonary function in identical twins raised apart. Am. Rev. Resp. Dis. 125:119-21; 1982.
35. Havlik, R.J.; Garrison, R.J.; Fabsitz, R.; Feinleib, M. Variability of heart rate, P-R, QRS and Q-T durations in twins. J. Electrocardiol. 13:45-48; 1980.
36. Heise, E.R.; Moore, H.A.; Reid, Q.B.; Goodman, H.O. Possible association of MN locus haplotypes with essential hypertension. Hypertension. 9:634-40; 1987.
37. Higgins, M.W.; Keller, J. Familial occurrence of chronic respiratory disease and familial resemblance in ventilatory capacity. J. Chron. Dis. 28:239-51; 1975.
38. Hislop, A.; Reid, L. The similarity of the pulmonary artery branching system in siblings. Forensic Sci. Int. 2:37-52; 1973.
39. Hole, B.V.; Wasserman, K. Familial emphysema. Ann. Intern. Med. 63:1009-17; 1965.
40. Hubert, H.B.; Fabsitz, R.R.; Feinleib, M.; Gwinn, C. Genetics and environmental influences on pulmonary function. Am. Rev. Resp. Dis. 125:409-15; 1982.
41. Jeunemaître, X.; Lifton, R.P.; Hunt, S.C.; Williams, R.R.; Lalouel, J.M. Absence of linkage between the angiotensin converting enzyme locus and human essential hypertension. Nat. Genet. 1:72-75; 1992.
42. Jeunemaître, X.; Soubrier, F.; Kotelevtev, Y.V.; Lifton, R.P.; Williams, C.S.; Charru, A.; Hunt, S.C.; Hopkins, P.N.; Williams, R.R.; Lalouel, J.M.; Corvol, P. Molecular basis of human hypertension: Role of angiotensinogen. Cell. 71:169-80; 1992.
43. John, S.W.M.; Krege, J.H.; Oliver, P.M.; Hagaman, J.R.; Hodgin, J.B.; Pang, S.C.; Flynn, T.G.; Smithies, O. Genetic decreases in atrial natriuretic peptide and salt-sensitive hypertension. Science. 267:679-81; 1995.

44. Kawakami, Y.; Shida, A.; Yamamoto, H.; Yoshikawa, T. Pattern of genetic influence on pulmonary function. Chest. 87:507-11; 1985.

45. Kelly, D.P.; Strauss, A.W. Inherited cardiomyopathies. N. Engl. J. Med. 330:913-19; 1994.

46. Krieger, H.; Morton, N.E.; Rao, D.C.; Azevedo, E. Familial determinants of blood pressure in northeastern Brazil. Hum. Genet. 53:415-18; 1980.

47. Kupari, M.; Perola, M.; Koskinen, P.; Virolainen, J.; Karhunen, P.J. Left ventricular size, mass, and function in relation to angiotensin-converting enzyme gene polymorphism in humans. Amer. J. Physiol. 267:H1107-11; 1994.

48. Lahiri, S.; Delaney, R.G.; Brody, J.S.; Simpser, M.; Velasques, T.; Motoyama, E.K.; Polgar, C. Relative role of environmental and genetic factors in respiratory adaptation to high altitude. Nature. 261:133-35; 1976.

49. Landry, F.; Bouchard, C.; Dumesnil, J. Cardiac dimension changes with endurance training. JAMA. 254:77-80; 1985.

50. Larson, R.K.; Barman, M.L. The familial occurrence of chronic obstructive pulmonary disease. Ann. Intern. Med. 63:1001-8; 1965.

51. Lebowitz, M.D.; Knudson, R.J.; Burrows, B. Familial aggregation of pulmonary function measurements. Am. Rev. Resp. Dis. 129:8-11; 1984.

52. Lewitter, F.I.; Tager, I.B.; McGue, M.; Tishler, P.V.; Speizer, F.E. Genetic and environmental determinants of level of pulmonary function. Am. J. Epidemiol. 120:518-30; 1984.

53. Lockette, W.; Ghosh, S.; Farrow, S.; MacKenzie, S.; Baker, S.; Miles, P.; Schork, A.; Cadaret, L. α2-adrenergic receptor gene polymorphism and hypertension in blacks. Am. J. Hypertension. 8:390-94; 1995.

54. Luft, F.C.; Miller, J.Z.; Weinberger, M.H.; Christian, J.C.; Skrabal, F. Genetic influences on the response to dietary salt reduction, acute salt loading, or salt depletion in humans. J. Cardiovas. Pharmacol. 12:S49-55; 1988.

55. Malina, R.M.; Bouchard, C. Growth, maturation and physical activity. Champaign, IL: Human Kinetics; 1991.

56. Man, S.F.P.; Zamel, N. Genetic influence on normal variability of maximum expiratory flow curves. J. Appl. Physiol. 41:874-77; 1976.

57. Marian, A.J.; Yu, Q.T.; Workman, R.; Greve, G.; Roberts, R. Angiotensin-converting enzyme polymorphism in hypertrophic cardiomyopathy and sudden cardiac death. Lancet. 342:1085-86; 1993.

58. Mathers, J.A.L.; Osborne, R.H.; DeGeorge, F.V. Studies of blood pressure, heart rate, and the electrocardiogram in adult twins. Am. Heart J. 62:634-42; 1961.

59. Moller, P.; Heiberg, A. Atrioventricular conduction time—a heritable trait? I. Routine tracings in adult Norwegians. Clin. Genet. 18:450-53; 1980.

60. Mongeau, J.G. Heredity and blood pressure in humans: An overview. Pediatr. Nephrol. 1:69-75; 1987.

61. Morris, B.J.; Griffiths, L.R. Frequency in hypertensives of alleles for a RFLP associated with the renin gene. Biochem. Biophys. Res. Commun. 150:219-24; 1988.

62. Morris, B.J.; Zee, R.Y.L. Similarity of blood pressure for each genotype of the insertion/deletion polymorphism of the dipeptidyl carboxypeptidase-1 gene in different age groups of patients with severe, familial essential hypertension. Clin. Exp. Pharmacol. Physiol. 21:919-24; 1994.

63. Morton, N.E.; Gulbrandsen, C.L.; Rao, D.C.; Rhoads, G.G.; Kagan, A. Determinants of blood pressure in Japanese-American families. Hum. Genet. 53:261-66; 1980.

64. Motulsky, A.G. Genetics of coronary heart disease. Pediatrician. 6:366-70; 1977.

65. Mueller, W.H.; Chakraborty, R.; Barton, S.A.; Rothammer, F.; Schull, W.J. Genes and epidemiology in anthropological adaptation studies: Familial correlations in lung function in population residing at different altitudes in Chile. Med. Anthropol. 4:367-84; 1980.

66. Murphy, E.A. Genetics in hypertension. Circ. Res. 32-33:129-38; 1973.

67. Naftilan, A.J.; Williams, R.; Burt, D.; Paul, M. A lack of genetic linkage of renin gene restriction fragment length polymorphism with human hypertension. Hypertension. 14:614-18; 1989.

68. Nirmala, A.; Rice, T.; Chengal-Reddy, P.; Sreerama, K.; Krishna, P.; Ramana, V.; Rao, D.C. Commingling and segregation analysis of blood pressure in consanguineous and nonconsanguineous families from Andhra Pradesh, India. Am. J. Hum. Biol. 4:703-16; 1992.

69. Personen, E.; Norio, R.; Sarna, S. Thickening in the coronary arteries in infancy as an indication of genetic factors in coronary heart disease. Circulation. 51:218-25; 1975.

70. Pérusse, L.; Dériaz, O.; Dionne, F.T.; Bouchard, C. Association and linkage analyses of the alpha and beta genes of the sodium-potassium ATPase with age-related changes in blood pressure. Am. J. Hum. Genet. 55:A199; 1994.

71. Pérusse, L.; Leblanc, C.; Tremblay, A.; Allard, C.; Thériault, G.; Landry, F.; Talbot, J.; Bouchard, C. Familial aggregation in physical fitness, coronary heart disease risk factors, and pulmonary function measurements. Prev. Med. 16:607-15; 1987.

72. Pérusse, L.; Lortie, G.; Thériault, G.; Bouchard, C. Evidence of linkage between cardiac dimensions and genes located on chromosome 1. Med. Sci. Sports Exerc. 27:S41; 1995.

73. Pérusse, L.; Moll, P.P.; Sing, C.F. Evidence that a single gene with gender- and age-dependent effects influences systolic blood pressure determination in a population-based sample. Am. J. Hum. Genet. 49:94-105; 1991.

74. Pickering, T.G.; Gerin, W. Genetic factors, cardiovascular reactivity and blood pressure variability. In: Goldbourt, U.; de Faire, U.; Berg, K., eds. Genetic factors in coronary heart disease. Dordrecht, Netherlands: Kluwer Academic; 1994:385-96.

75. Rapp, J.P. Genetics of experimental and human hypertension. In: Genest, J.; Kuchel, O.; Hamet, P.; Cantin, M., eds. Hypertension: Physiopathology and treatment. New York: McGraw-Hill; 1983:582-98.

76. Rayment, I.; Holden, H.M.; Sellers, J.R.; Fananapazir, L.; Epstein, N.D. Structural interpretation of the mutations in the beta-cardiac myosin that have been implicated in familial hypertrophic cardiomyopathy. Proc. Natl. Acad. Sci. 92:3864-68; 1995.

77. Raynolds, M.V.; Bristow, M.R.; Bush, E.W.; Abraham, W.T.; Lowes, B.D.; Zisman, L.S.; Taft, C.S.; Perryman, M.B. Angiotensin-converting enzyme DD genotype in patients with ischemic or idiopathic dilated cardiomyopathy. Lancet. 342:1073-75; 1993.

78. Redline, S.; Tishler, P.V.; Lewitter, F.I.; Tager, I.B.; Munoz, A.; Speizer, F.E. Assessment of genetic and nongenetic influences on pulmonary function. Am. Rev. Resp. Dis. 135:217-22; 1987.

79. Rice, T.; Bouchard, C.; Borecki, I.B.; Rao, D.C. Commingling and segregation analysis of blood pressure in a French-Canadian population. Am. J. Hum. Genet. 46:37-44; 1990.

80. Rolfs, A.; Weber-Rolfs, I.; Regitz-Zagrosek, V.; Kallisch, H.; Riedel, K.; Fleck, E. Genetic polymorphisms of the angiotensin II type 1 (*AT1*) receptor gene. Eur. Heart J. 15:108-12; 1994.

81. Rybicki, B.A.; Beaty, T.H.; Cohen, B.H. Major genetic mechanisms in pulmonary function. J. Clin. Epidemiol. 43:667-75; 1990.

82. Saunders, N.A.; Leeder, S.R.; Rebuck, A.S. Ventilatory response to carbon dioxide in young athletes: A family study. Am. Rev. Resp. Dis. 113:497-502; 1976.

83. Schunkert, H.; Hans-Werner, H.; Holmer, S.R.; Stender, M.; Perz, S.; Keil, U.; Lorell, B.H.; Riegger, G.A.J. Association between a deletion polymorphism of the angiotensin-converting-enzyme gene and left ventricular hypertrophy. N. Engl. J. Med. 330:1634-38; 1994.

84. Scoggin, C.H.; Doekel, R.D.; Kryger, M.H.; Zwillich, C.W.; Weil, J.V. Familial aspects of decreased hypoxic drive in endurance athletes. Clin. Res. [Abstract]. 24:389A; 1976.

85. Shu-Chuan, L.; Carmelli, D.; Hunt, S.C.; Williams, R.R. Evidence for a major gene influencing 7-year increases in diastolic blood pressure with age. Am. J. Hum. Genet. 57:1169-77; 1995.

86. Siervogel, R.M. Genetic and familial factors in essential hypertension and related traits. In: Malina, R.M., ed. Yearbook of physical anthropology. New York: Alan R. Liss; 1983:37-63.

87. Sing, C.F.; Boerwinkle, E.; Turner, S.T. Genetics of primary hypertension. Clin. Exp. Hypertension. A8:623-51; 1986.

88. Solomon, S.D.; Wolff, S.; Watkins, H.; Ridker, P.M.; Come, P.; McKenna, W.J.; Seidman, C.E.; Lee, R.T. Left ventricular hypertrophy and morphology in familial hypertrophic cardiomyopathy associated with mutations of the beta-myosin heavy chain gene. J. Am. Coll. Cardiol. 72:498-505; 1993.

89. Soubrier, F.; Houot, A.M.; Jeunemaître, X.; Plouin, P.F.; Corvol, P. Molecular biology as a tool for genetic research in hypertension: Application to the renin gene. J. Cardiovas. Pharmacol. 12:S155-59; 1988.

90. Soubrier, F.; Jeunemaître, X.; Rigat, B.; Houot, A.M.; Cambien, F.; Corvol, P. Similar frequencies of renin gene restriction fragment length polymorphisms in hypertensive and normotensive subjects. Hypertension. 16:712-17; 1990.

91. Tager, I.B.; Rosner, B.; Tishler, P.V.; Speizer, F.E.; Kass, E.H. Household aggregation of pulmonary function and chronic bronchitis. Am. Rev. Resp. Dis. 114:485-92; 1976.

92. Takkunen, J. Anthropometric, electrocardiographic and blood pressure measurements in twins. Ann. Acad. Sci. Fenn. Suppl. 107:1-82; 1964.

93. Thériault, G.; Diano, R.; Leblanc, C.; Pérusse, L.; Landry, F.; Bouchard, C. The role of heredity in cardiac size: An echocardiographic study on twins, brothers and sisters, and sibs by adoption. Med. Sci. Sports Exerc. 18:S51; 1986.

94. Verhaaren, H.A.; Schieken, R.M.; Mosteller, M.; Hewitt, J.K.; Eaves, L.J.; Nance, W.E. Bivariate genetic analysis of left ventricular mass and weight in pubertal twins (the Medical College of Virginia Twin Study). Am. J. Cardiol. 68:661-68; 1991.

95. Ward, R. Familial aggregation and genetic epidemiology of blood pressure. In: Laragh, J.H.; Brenner, B.M., eds. Hypertension: Pathophysiology, diagnosis and management. New York: Raven Press; 1990:81-99.

96. Webster, P.M.; Lorimer, E.G.; Man, S.F.P.; Woolf, C.R.; Zamel, N. Pulmonary function in identical twins: Comparison of nonsmokers and smokers. Am. Rev. Resp. Dis. 119:223-28; 1979.

97. Weinberger, M.H.; Miller, J.Z.; Fineberg, N.S.; Luft, F.C.; Grim, C.E.; Christian, J.C. Association of haptoglobin with hypertension. Hypertension. 10:440-46; 1987.

98. Williams, R.R.; Hunt, S.C.; Hasstedt, S.J.; Berry, T.D.; Wu, L.L.; Barlow, G.K.; Stults, B.M.; Kuida, H. Definition of genetic factors in hypertension: A search for major genes, polygenes, and homogeneous subtypes. J. Cardiovas. Pharmacol. 12:S7-20; 1988.

99. Williams, R.R.; Hunt, S.C.; Hasstedt, S.J.; Hopkins, P.N.; Wu, L.L.; Berry, T.D.; Stults, B.M.; Barlow, G.K. Multigenic human hypertension: Evidence for subtypes and hope for haplotypes. J. Hypertension. 8:539-46; 1990.

100. Wilson, A.F.; Elston, R.C.; Tran, L.D.; Siervogel, R.M. Use of the robust sib-pair method to screen for single-locus, multiple-locus, and pleiotropic effects: Application to traits related to hypertension. Am. J. Hum. Genet. 48:862-72; 1991.

101. Wilson Cox, D.; Talamo, R.C. Genetic aspects of pediatric lung disease. Pediatric. Clin. N. Am. 26:467-80; 1979.

102. Zee, R.Y.L.; Bennett, C.L.; Schrader, A.P.; Morris, B.J. Frequencies of variants of candidate genes in different age groups of hypertensives. Clin. Exp. Pharmacol. Physiol. 21:925-30; 1994.

CHAPTER 13

Genetics of Metabolic Fitness Phenotypes and Other Risk Factors

Metabolic fitness is a recently recognized and important component of health-related fitness. It has been defined as resulting from adequate hormonal actions (particularly insulin), normal blood and tissue carbohydrate and lipid metabolism, and a favorable ratio of lipid to carbohydrate oxidized (25). This definition includes several aspects of health status on which physical activity could have beneficial effects. This chapter reviews the evidence of the role of genetic factors in observed variation in these factors. The role of genetic factors in the metabolic syndrome, a syndrome resulting from the clustering of several metabolic disturbances observed in the same individual and associated with an increased risk of cardiovascular disease and diabetes, is also addressed.

Life Span and Gender Variability

Blood lipids and lipoproteins

Blood lipids, lipoproteins, blood glucose and insulin levels change during a person's life span as discussed below.

Total serum cholesterol (CHOL), CHOL associated with high-density lipoprotein (HDL-CHOL) and with low-density lipoprotein (LDL-CHOL), and serum triglycerides (TG) increase rapidly during the first three years of life. Thereafter, mean values of blood lipids and lipoproteins remain relatively stable until puberty. During puberty, TG levels increase to reach adult values, while CHOL, LDL-CHOL, and HDL-CHOL show a transient decrease. The rise of TG and decline of HDL-CHOL associated with puberty is more pronounced in males than in females. During late adolescence, CHOL and LDL-CHOL increase, whereas HDL-CHOL tends to remain stable. CHOL associated with very-low-density lipoprotein (VLDL-CHOL) remains quite stable during growth. These changes in blood lipids and lipoproteins during growth suggest that males in particular, in contrast to females, progress toward an atherogenic lipid profile as a part of normal maturation. During adulthood, CHOL and TG levels increase. The age-associated increase in CHOL and TG levels is closely related to the age-associated changes observed in adiposity (16). Furthermore, there is evidence of a progressive accumulation of CHOL in extrahepatic tissues, particularly connective and adipose tissues, during the life span (37). Thus, aging appears to be associated with an expansion of both circulating and tissue pools of cholesterol.

Blood glucose and insulin

Fasting plasma glucose and insulin increase gradually with age, on average, in both sexes until puberty. During adolescence, glucose levels decline, more so in girls than in boys, while insulin increases through adolescence in girls but declines after about 13 to 14 years in boys. Mean glucose levels therefore tend to be higher in boys, while mean insulin levels tend to be higher in girls. Blood glucose levels remain relatively stable with age during adulthood. In healthy adults of both genders, fasting blood glucose levels change very slightly (approximately 1 mg · dl^{-1} · $decade^{-1}$) with age (101).

Similar age-related trends are observed for glucose and insulin metabolism. Following administration of a glucose load, the area under the glucose curve declines from childhood to adolescence and then increases into young adulthood. On the other hand, the area under the insulin curve increases from childhood through adolescence into adulthood. On average, mean areas under the glucose curve are higher in males, whereas mean areas under the insulin curves are higher in females, especially in adolescence and young adulthood, indicating that females probably secrete more insulin in response to a glucose load. It has been suggested that the average increase in blood glucose levels could be as high as 9.5 mg · dl^{-1} · $decade^{-1}$ and 5.3 mg · dl^{-1} · $decade^{-1}$ one hour and two hours, respectively, after a glucose load (41,108). These age-related changes in glucose and insulin metabolism suggest a markedly impaired utilization of glucose with aging associated with either reduced insulin secretion by the pancreas or reduced sensitivity to circulating insulin levels, that is, insulin resistance. The other pancreatic hormone responsible for glucose homeostasis, which has opposite effects of insulin, is glucagon. In general, fasting levels of glucagon are not influenced by age nor is the post-glucose-induced suppression of glucagon release. It thus appears that glucagon metabolism is spared of any significant changes with aging in contrast to the alterations observed in the insulin metabolism (101).

Other hormones

During growth, the weight of endocrine glands generally increases, whereas it appears to decrease with advancing age. Serum concentrations of most hormones are generally not influenced by age, although some hormones show reduced concentrations of their active form, and there is a trend toward a general decline in secretion rate with aging. The maintenance of normal basal levels in the presence of decreased secretion rates implies that clearance rates are reduced with age. The only exception to this general pattern is for norepinephrine and epinephrine, which show, respectively, a slight decrease and an increase with age. No systematic changes are observed in receptor binding capacity with age, but the intracellular response to hormones generally deteriorates with aging (101). Age-related changes observed in several specific hormone levels are described subsequently.

Growth hormone (GH), or somatotrophin, is the most abundant pituitary hormone and is produced throughout the life span. GH is secreted in a series of intermittent bursts during the day. Children have more bursts during 24 hours than adults, which explains the higher serum levels of the hormone in children than in adults. The average serum concentration of GH increases from childhood to adolescence, reaches a peak at late puberty, and subsequently declines into old age. Mean 24-hour GH concentrations relative to prepubertal years are 130% to 210% in late puberty, 60% in the third decade, 35% to 50% in the fourth to sixth decades, and 25% to 40% in the seventh decade and beyond (72). The age-related decline in GH concentration arises from both a decrease in GH production and an increase in clearance rate.

Serum levels of thyroid hormones (T_3 and T_4) are highest at birth and shortly thereafter and decline gradually during childhood and adolescence. Lowest values are attained in early adolescence. From adolescence through senescence, T_4 levels remain essentially unchanged, while T_3 decrease 25% to 40% after the sixth decade. Sex differences in thyroid hormone levels are not significant.

Parathyroid hormone (PTH) is secreted by the parathyroids, which are small glands located on the dorsal portion of the thyroid gland. PTH plays an important role in the regulation of calcium homeostasis and is secreted in response to hypocalcemia. The primary action of PTH is to raise the concentration of plasma calcium through action of the hormone on bone and the kidneys. Blood levels of PTH increase during growth and decrease with age except in cases of osteoporosis (58).

The steroid hormones secreted by the adrenal cortex are the mineralocorticoids, mainly aldosterone, and the glucocorticoids, mainly cortisol. The mineralocorticoids regulate electrolyte balance and extracellular fluid volume, while the glucocorticoids influence almost every organ and tissue of the body and, with others, are involved in the regulation of carbohydrate and fat metabolism and in coping with stress. Cortisol is often referred to as the stress hormone. Secretion of aldosterone and cortisol increases gradually with age during growth as a function of the age-associated increase in body size. Blood glucocorticoid levels show no significant difference between maturity and old age. This contrasts with aldosterone, basal levels of which appear to decline with aging. There tends to be a balanced reduction in the secretion and clearance rates of most adrenal hormones with aging (101). Despite the difference in autonomic activity between young and old people, there are no age-related changes in the secretion of epinephrine and norepinephrine by the adrenal medulla.

Sex hormones are synthesized by the male (testes) and female (ovaries) gonads and, in small amounts, by the adrenal cortex. They can also be produced by enzymatic conversion of steroid precursors in peripheral tissues. Testosterone in males and estradiol and progesterone in females are the major hormones synthesized by the gonads. The production of these hormones is regulated by two gonadotropic hormones released from the anterior pituitary: follicle-stimulating hormone (FSH) and luteinizing hormone (LH). During growth, serum concentrations of FSH and LH increase more in females than in males, especially during puberty. Testosterone and estradiol levels are low during childhood and rise markedly in midpuberty in males and females, respectively. Serum testosterone in girls and serum estradiol in boys also increase slightly during growth (89). Despite large interindividual differences, serum testosterone levels decline with age, especially around age 50. The capacity of the sex hormone binding globulin (SHBG) to bind testosterone increases with age, producing a decrease in the free testosterone fraction. This decrease in free testosterone appears more important than the decrease in total testosterone levels (101). Besides variation associated with the menstrual cycle, average serum levels of estrogens and progesterone remain constant with age until menopause, when a sharp drop occurs.

Blood Lipids, Lipoproteins, and Apolipoproteins

Lipid metabolism is an important factor of metabolic fitness. It is generally recognized that physical activity may favorably alter the lipid profile and contribute to the reduction of the risk for cardiovascular disease.

Genetic epidemiology studies

Because of their key role in the development of atherosclerosis and heart disease, the genetics of blood lipids, lipoproteins, and apolipoproteins have been extensively studied. Several in-depth reviews are available (30,31,50,52,88,118,137,139).

Several disorders of lipid metabolism have been attributed to single-gene defects. Various mutations in apolipoproteins, the protein component of lipoproteins, cellular receptors, or key enzymes of lipoprotein metabolism are responsible for these abnormalities (131-133). These genetic disorders are often classified on the basis of one of the major classes of lipoproteins found to be abnormal in the plasma or of the pattern of inheritance encountered in families where the gene is segregating. Familial hypercholesterolemia (FH) is probably the best known of these genetic disorders. FH is an autosomal disorder caused by various mutations in the LDL receptor gene. In heterozygotes, receptor activity is decreased by one-half, which results in about a twofold increase of plasma LDL levels. In homozygotes, there is little or no receptor activity, which results in LDL levels four to six times higher than normal. Subjects affected by this condition are at increased risk of premature atherosclerosis, which often leads to early coronary heart disease events in heterozygotes and early death, usually before adulthood, in homozygotes.

A large number of mutations in the LDL receptor gene have been characterized at the molecular level (74). However, in some populations, specific mutations appear to be more prevalent and account for the majority of cases reported. This is the case in French Canadians living in the province of Quebec, where the first discovered mutation—a 10-kb deletion in the 5' end of the LDL receptor gene—was found in about 60% of all FH patients (73). Although genetic dyslipidemias are often associated with dramatic changes in blood lipids and lipoprotein levels, they are relatively uncommon and therefore do not account for an important fraction of the variation in blood lipids and lipoproteins in the population at large.

The contribution of genetic and environmental factors in normolipemic individuals has been extensively studied in various populations. These studies have clearly established that genetic factors contribute significantly to interindividual differences in blood lipids and lipoproteins, with heritability estimates accounting for about 25% to 70% of the phenotypic variance, depending on the phenotype considered. Table 13.1 presents ranges of reported heritability estimates derived from several family studies (79,140). For all the phenotypes listed in table 13.1, the genetic effect was more important than environmental effects shared among family members, which accounted for 5% to 20% of the phenotypic variance. Table 13.1 indicates whether major gene effects have been identified using complex segregation analysis. Major gene effects have been reported for all the phenotypes (1,6,20,36,55,67,78,80,85,102, 103,106,109,146,153). Only one study reported evidence of a major gene segregating for triglyceride levels (90), but this was observed in families with high incidence of cardiovascular disease.

Table 13.1
Summary of Heritability Estimates and Evidence for Major Gene Effects for Several Phenotypes of Blood Lipid and Lipoprotein Metabolism

Phenotype	Heritability %	Major gene
Triglycerides	25-47	Yes
Total cholesterol (CHOL)	49-64	Yes
LDL-CHOL	39-67	Yes
HDL-CHOL	28-56	Yes
APOA I	43	Yes
APOA II	30	No
APOB	50-70	Yes
Lp(a)	70-95	Yes
LDL particles size	50	Yes

See text for references.

Genetic and environmental determinants of blood lipids and lipoproteins were also investigated in twins (70). Serum lipids and apolipoproteins were measured in 46 pairs of MZ twins reared apart, 67 pairs of MZ twins reared together, 100 pairs of DZ twins reared apart, and 89 pairs of DZ twins reared together. With this design, the authors were able to compare twins on the basis of zygosity and rearing status, which allows joint estimation of genetic and environmental effects on lipid levels. Results are summarized in table 13.2 by age and sex. The age and sex differences in genetic and environmental effects were evaluated by comparing models in which parameter estimates were constrained to be equal across subgroups to those in which estimates were allowed to differ. The absence of estimates in the "all twins" column indicates that estimates differed significantly across subgroups. No significant age

Table 13.2
Proportion of Phenotypic Variance in Levels of Lipids and Apolipoproteins Attributed to Genetic and Environmental Factors in Twins

	AGE			SEX		
Trait	**Younger twins (56-65 yr)**	**Older twins (66-86 yr)**	**All twins**	**Men**	**Women**	**All**
CHOL						
h^2	0.63	0.32	0.47	0.40	0.45	0.46
e^2	0.19	0.31	0.26	0.45	0.23	0.28
c^2	0.18	0.36	0.27	0.15	0.32	0.26
HDL-CHOL						
h^2	0.76	0.55	0.65	0.65	0.60	0.62
e^2	0.24	0.45	0.35	0.35	0.40	0.38
APOA I						
h^2	0.69	0.52	0.65	0.60	0.57	0.58
e^2	0.31	0.34	0.35	0.40	0.43	0.42
c^2	—	0.14	—	—	—	—
APOB						
h^2	0.78	0.51	—	0.53	0.63	—
e^2	0.22	0.31	—	0.47	0.19	—
c^2	—	0.18	—	—	0.18	—
TG						
h^2	0.72	0.28	—	0.45	0.55	0.52
e^2	0.28	0.72	—	0.55	0.45	0.48

Adapted from Heller et al. (70).

and sex differences in genetic and environmental sources of variance were observed for CHOL, HDL-CHOL, and APOA I. For these phenotypes, heritability (h^2) ranged between 46% and 65%, while nonshared environmental factors accounted from 26% to 42% of the variance. Shared environment contributed significantly to CHOL (around 25%), but not to HDL-CHOL and APOA I levels. Heritability estimates for APOB and TG ranged from 45% to 78%. However, significant age and sex differences were observed in the estimates, with higher heritabilities in younger twins and in females (only for APOB). Results of this study (70) show that genetic factors account for the majority of the variance in lipids and lipoproteins and that, except for CHOL, nonshared environmental factors account for the remainder of the phenotypic variance.

In an attempt to detect genetic variation in lipoproteins, a distinct class of lipoprotein particles, lipoprotein(a) or Lp(a), was identified in the early 1960s (12). Lp(a) is a lipoprotein composed of an LDL particle to which a single large glycoprotein, apolipoprotein(a) or apo(a) is attached. Lp(a) differs from LDL particles in electrophoretic mobility and in protein-lipid ratio, which is higher in Lp(a). Genetic studies subsequent to its discovery established that Lp(a) was genetically determined (15). The association between high levels of Lp(a) and coronary heart disease increased interest in the study of the genetics of this lipoprotein (11,18). Based on results of family studies (18), plasma concentrations of Lp(a) aggregate in families with heritability estimates in the range of 70% to 95%. Results from complex segregation analyses have also shown that Lp(a) levels are influenced by the segregation of a single gene that accounts for up to 73% of the phenotypic variance (66,105).

It now widely recognized that LDL particles are heterogeneous lipoproteins that can be classified into several subclasses based on differences in size, density, and chemical composition. Two distinct LDL subclass phenotypes, denoted A and B, have been described. LDL subclass phenotype A is characterized by a predominance of large LDL (average 26.3 nm) particles, while subjects with subclass phenotype B have a predominance of small (average 24.8 nm), dense LDL particles. It has been estimated that about 90% of subjects can be classified as either phenotype A or B, with the remainder having an intermediate phenotype. Subjects having small, dense LDL particles (phenotype B) are at increased risk for coronary heart disease (3). Twin studies indicate higher concordance rates in MZ (74%) than in DZ (32%) twins for this phenotype (3). Results of twin studies suggest that about 50% of the variation in the dichotomous LDL subclass phenotype or in the quantitative LDL peak particle diameter could be attributed to genetic factors (2). Complex segregation analyses also suggest a major gene effect for LDL subclass phenotypes (dichotomous and quantitative), with allele frequencies for the small, dense LDL particles ranging from about 0.2 to 0.3 (4,59). However, the mode of inheritance of this single gene has not been clearly established in these studies.

Molecular studies

Several early association studies of blood lipids undertaken with red blood cell protein polymorphisms showed that levels of serum CHOL were higher in persons of blood group A than in those of blood group O and that CHOL was also associated with the Lewis blood group (7,8,92,144). One of the most extensive of the earlier association studies is that of Sing and Orr (141), who analyzed 12 red blood cell polymorphisms in more than 3,000 males and 3,000 females. Significant differences in average CHOL levels between subjects of different phenotypes for the following four polymorphic systems were found: ABO, secretor, Gm, and haptoglobin. In another study, Børresen et al. (21) reported an association between haptoglobin and HDL-CHOL levels. This association between haptoglobin and HDL-CHOL may be explained by the presence of an enzyme involved in the metabolism of HDL-CHOL (lecithin-cholesterol acyltransferase, or LCAT) closely linked to the haptoglobin locus on the chromosomal region 16q22.1. Several candidate genes located on the long arm of chromosome 19 may explain the observed association between the secretor locus and cholesterol levels in the study of Sing and Orr (141), including apolipoprotein C-I, apolipoprotein C-II, and apolipoprotein E.

APOE is a constituent of VLDL and HDL and serves as ligand for recognition of lipoproteins by cellular receptors. Because of its important role in the regulation of cholesterol levels, the APOE gene is one of the most extensively studied genes of lipid metabolism. Its impact on plasma cholesterol levels has been evaluated in several populations around the world (56,63). The APOE gene is located on chromosome 19 and is polymorphic, with three common alleles designated ε2, ε3, and ε4, which code for three isoforms of the protein: E_2, E_3, and E_4. The APOE polymorphism has been consistently associated with plasma CHOL and LDL-CHOL levels (43,63) and more recently has also been associated with TG and HDL-CHOL levels (39,82,127). Several studies have also reported that the APOE polymorphism is associated with atherosclerosis (38,42,111).

A large number of association studies utilizing RFLP markers of the various apolipoprotein (APOA I, APOA II, APOA IV, APOB, APOC I, APOC II, APOC II, and APOE), LDL receptor, or cholesteryl ester transfer protein (CETP) genes have also been conducted, especially with dyslipoproteinemias, but also with serum levels of lipids, lipoproteins, and apolipoproteins (13,94). In many cases, the results are contradictory, and most of the positive findings have not been replicated. In general, genetic variation in the apolipoprotein genes is characterized by low levels of association with serum levels of the protein.

Glucose and Insulin

Genetic epidemiology studies

Few studies have been reported on the heritability and segregation characteristics of plasma glucose and insulin levels, glucose tolerance, insulin action on various tissues and organs, glucose disposal rate through oxidative or nonoxidative pathways, hepatic glucose production, insulin secretion, hepatic extraction of insulin, and related glucose/insulin phenotypes.

There is evidence of a strong familial component for phenotypes related to glucose and insulin metabolism (46,62,86,154). However, relatively few studies have reported on the heritability and segregation patterns of these phenotypes. Heritability estimates range from about 40% to 85% for fasting glucose and fasting insulin. Analyses of the distributions of fasting glucose (123) and fasting insulin (19) are compatible with a mixture of three distributions, which is consistent with the segregation of a major gene. However, contradictory results have been obtained from segregation analyses. Autosomal recessive, autosomal dominant, or codominant inheritance of a single gene with frequencies ranging from about 0.1 to 0.4 across studies or populations has been inferred (64). Recently, Rice et al. (123) examined the segregation patterns of fasting glucose in 1,969 individuals from 620 families in the Lipid Research Clinics family study. No evidence of a major gene effect was found. On the other hand, Schumacher et al. (135) suggested that fasting insulin levels were influenced by the contribution of a major locus segregating as an autosomal recessive allele with a frequency of 0.25 in a white sample in Utah. The variance in insulin levels accounted for by this major gene reached 33% with an additional polygenic component explaining 11% of the variance (135).

In the case of insulin-action phenotypes, a few reports have established that insulin sensitivity, assessed from intravenous glucose tolerance tests or insulin-mediated glucose disposal rate determined from clamp studies, was characterized by significant heritability. The familial aggregation of plasma fasting insulin, submaximal M values, and maximal M values reached 12%, 15%, and 34% of the phenotypic variance, respectively, in 45 sibships of Pima Indians in Arizona (86). Heritability values of about 35% to 40% for an insulin-mediated glucose disposal phenotype have been attained in 601 subjects from 155 nuclear families submitted to a glucose infusion test (81). Comparable results were obtained with a 3-hour intravenous glucose tolerance test of 183 nondiabetic individuals from 105 families that had two NIDDM parents (91).

There was no evidence of a major gene effect for the insulin-sensitivity phenotype in the study by Iselius et al. (81) after the data were subjected to complex segregation analysis. In contrast, the study of the white Utah sample suggested that 1-hour-postglucose insulin levels were characterized by the contribution of a major autosomal locus which accounted for 48% of the variance, with polygenic inheritance reaching 4% (135).

Molecular studies

A large number of candidate genes of glucose and insulin metabolism (insulin, insulin receptor, glucose transporters, glucokinase, glycogen synthase) have been investigated in relation to insulin resistance and the presence or absence of diabetes. Several reviews (9,51,71,119,138) suggest a failure of the candidate genes to be consistently associated with diabetes. When the associations are significant, they account for a relatively small fraction of the variance in the phenotypes or the risk of diabetes.

A few genetic loci involved in insulin response to a glucose load have been identified by linkage analyses. In one study of 123 siblings from 46 Pima Indian nuclear families, 11 polymorphic markers located on the long arm of chromosome 4 (4q26) were tested for linkage with fasting insulin and insulin-stimulated glucose uptake rates (117). Results from sib-pair linkage analysis revealed a strong linkage ($p < .001$) of maximal insulin action with markers of two closely linked loci on 4q26, intestinal fatty acid binding protein (FABP2), and the annexin V gene (*ANX5*). Fasting insulin levels were linked only to the *FABP2* gene locus. The acute insulin response (AIR) to an intravenous glucose load was also tested for linkage with 18 short-tandem-repeat polymorphism of the short arm of chromosome 1 in 175 sib pairs from the same population, all with normal glucose tolerance (145). The anonymous DNA marker D1S198 exhibited strong linkage ($p = .00005$) with AIR. Results of these two studies suggest that genes located on 1p and 4q may play a role in insulin action.

Substrate Oxidation

The relative proportion of carbohydrate and lipid oxidized can be indirectly determined by measuring the respiratory quotient (RQ) under standardized conditions. Few studies have addressed the role of genetic factors in substrate oxidation. Three lines of evidence suggest a role for inherited differences in the propensity to oxidize relatively more carbohydrates or lipids. First, twin studies show greater similarity among MZ twins than DZ twins in the RQ measured during submaximal exercise, suggesting a role for heredity in substrate oxidation (29). Second, experimental studies in which MZ twins were submitted to overfeeding or exercise training suggest significant within-pair resemblance in the pattern of substrate oxidation following a mixed meal or during a submaximal exercise test. In a study in which 12 pairs of MZ twins were submitted to a 1,000-kcal caloric surplus, 6 days a week for a period of 100 days under rigorously controlled conditions of energy intake and activity level (26), significant within-pair resemblance was observed in changes in the RQ at rest and for several hours following a mixed meal (unpublished observations). The results suggest that even when energy intake, macronutrient composition of the diet, and physical activity are standardized for more than three months, some individuals oxidize more fat than carbohydrate, and genetic factors appear to play a role in this phenomenon.

Evidence from a series of standardized endurance and high-intensity cycling exercise programs with MZ twins shows that improvement in a given submaximal power output or in maximal exercise is characterized by significant within-pair resemblance (23). Exercise training results in a decrease in RQ measured in relative steady state at a low-intensity level, but the changes observed in RQ are not randomly distributed among genotypes. Despite the fact that twins were all subjected to the same training protocol, individuals with the same genotype (within pairs) were more similar than unrelated individuals (between pairs) in the changes of RQ, suggesting that the pattern of substrate oxidation changes at low power output levels may be genetically determined.

The third line of evidence comes from the Quebec Family Study. Data at rest in a fasted state for adults from 121 nuclear families indicate an estimated familiality of about 30% for the RQ (Dériaz et al., unpublished results).

Blood Coagulation and Fibrinolytic Systems

Relatively little is known of the genetics of hemostatic function. Several proteins are involved in the processes of blood coagulation and fibrinolysis. Among them, coagulation factor VII, fibrinogen, and plasminogen activator inhibitor-1 (PAI-1) are associated with coronary heart disease (47,48,155).

Evidence for a role of the genotype in coagulation factor VII has been derived from the measured genotype approach. Variation in coagulation factor VII levels is apparently associated with an RFLP polymorphism resulting from the substitution of an arginine residue by a glutamine at position 353 of the gene locus in healthy men (61,77) and women (100).

Based on a path analysis of data from 170 families, Hamsten et al. (65) noted a significant genetic effect for plasma fibrinogen levels. Heritability of age- and sex-adjusted fibrinogen levels reached 51%, while cultural heritability was not significantly different from zero. A low correlation of 0.27 was observed for fibrinogen levels in 118 pairs of Norwegian MZ twins (14). More recently, Reed et al. (122) reported a heritability of

30% for plasma fibrinogen levels adjusted for age and 11 confounding variables associated with the risk of cardiovascular disease in 44 MZ and 39 DZ twin pairs. These moderate heritability estimates indicate that environmental factors have a stronger influence on plasma fibrinogen levels than do genetic factors. Genetic and environmental sources of variation in fibrinogen levels were recently studied in 82 randomly ascertained Israeli pedigrees (54). Significant familial resemblance was observed in age- and sex-adjusted fibrinogen levels. Complex segregation analysis of fibrinogen data provided evidence for a major environmental effect combined with polygenic effects that were age dependent. The single major environmental effect accounted for about 10% of the variance in fibrinogen levels at the age of 20 years up to about 65% at the age of 80 years. The polygenic effect accounted for about 50% of the variance at the age of 20 years and decreased to about 20% at the age of 80 years (54).

Fibrinogen is made of three polypeptide chains encoded by three different genes, α, β, and γ, grouped in a cluster of about 50 kb on the long arm of chromosome 4. The β chain appears to be the rate-limiting step in the synthesis of fibrinogen. Several polymorphisms of the fibrinogen gene cluster, especially β-fibrinogen, have been characterized and tested for association with plasma fibrinogen levels and risk of coronary heart disease. Despite some negative findings (14,35), results of most studies suggest that genetic variation in the fibrinogen gene is associated with fibrinogen levels (10,35,60,76,129) and also with coronary artery disease (10). The fibrinogen responses to acute exercise and chronic exercise training were also investigated in relation to polymorphism in the β-fibrinogen gene, and the acute rise in fibrinogen concentrations following intense exercise was found to be strongly genotype dependent (104).

Plasma PAI-1 levels vary considerably among individuals, and a significant fraction of this variation appears to be related to genetic factors (48). Mutations in the gene coding for PAI-1, located on chromosome 7 (7q21.3-q22), alter PAI-1 levels. In an Amish pedigree with PAI-1 deficiency, for example, plasma levels of PAI-1 of heterozygous individuals were intermediate, falling between those of affected homozygotes and normal individuals without the mutation (49).

Hormones

There are considerable interindividual differences in serum concentrations of most hormones in normal, healthy individuals, and both genetic and environmental factors account for this variation. The role of heredity in hormone levels has been investigated mainly with the twin model.

In a series of family and twin studies of steroid hormones, genetic factors were important for some hormones but not for others. For example, in a sample of 98 men from 66 families, Meikle et al. (96) reported higher between- than within-sibship variances for plasma levels of testosterone, SHBG, and estradiol. However, no familial resemblance was found for free testosterone, dihydrotestosterone (DHT, the most potent androgen derived from the enzymatic conversion of testosterone), and estrone levels. In another study (128), a heritability of 65% was reported for the serum level of dehydroepiandrosterone sulfate (DHEAS) in 178 individuals from 26 families.

In a study comparing 20 pairs of MZ and 20 pairs of DZ male twins, there was a significant genetic effect for plasma total and unbound cortisol levels, with heritability estimates of 45% and 51%, respectively; in contrast, no significant genetic effect was found for cortisol-binding globulin (98). In the same study, the heritability of DHEAS reached 58%. More recently, Linkowski et al. (87) studied the 24-hour cortisol profile of 11 MZ and 10 DZ male twin pairs; no significant genetic effect for the average 24-hour cortisol level was evident. However, the profile of cortisol secretion, especially the timing of the nocturnal nadir and the variability associated with pulsatility, was characterized by a significant genetic effect.

The contribution of genetic and nongenetic factors to variation of plasma sex steroid levels and action was investigated in 75 pairs of MZ twins and 88 pairs of DZ twins (95). Significant genetic effects were found for estradiol ($h^2 = 76\%$), LH (60%), estrone (46%), free testosterone (34%), FSH (28%), and total testosterone (26%). There was no significant genetic effect for DHT and SHBG levels, a finding confirmed by another report from the same group (17). Genetic regulation of tissue DHT formation is also suggested by the presence of a 48% genetic effect on the plasma content of 3α-androstanediol glucoronide, a marker of DHT formation (95). This was confirmed in a subsequent study by the same group showing that production rates of testosterone and DHT were strongly influenced by genetic factors, with heritability levels around 85% (97). The heritability of urinary adrenal androgen excretion rates was studied in 36 MZ and 21 DZ twins aged 6 to 17 years (116). Intraclass correlation coefficients for weight-adjusted excretion rates were 0.73 and 0.51 for MZ and DZ twins, respectively. The heritability reached 58%.

Variation in total and free thyroxine (T_4), thyronine (T_3), thyroid-binding globulin, and thyroid-stimulating hormone (TSH), after adjustment for the BMI, was compared in 15 MZ and 15 DZ pairs of male twins (99). The within-pair variance was two to six times higher within DZ pairs than within MZ pairs, but the within-pair estimate of genetic variance reached significance only for total and free T_4 levels. The within-pair resemblance for T_4 levels is shown in figure 13.1. Heritability estimates of 38% and 44% were obtained for total T_4 and free T_4, respectively.

The evidence for the role of heredity in circulating levels of several hormones is at times conflicting, which could be explained by the variation associated with measurement and with small sample sizes. In addition, many hormones are secreted in a pulsatile manner, so that 24-hour assays are essential. In other cases, plasma levels of the hormones do not reflect tissue levels of the hormones. The contribution of heredity appears to be stronger for production rate than for basal levels of the hormones. The genetic effect observed for sex steroids tends to be higher for plasma estrogens than for androgens.

Other Risk Factors

Risk factors, such as hyperhomocysteinemia and smoking, influence metabolic fitness. These will be discussed below.

Hyperhomocysteinemia

Homocysteine is an amino acid derived from the essential amino acid methionine. Homocystinuria represents a group of inborn errors of metabolism characterized by gross elevation of the concentration of homocysteine in the blood (hyperhomocysteinemia) accompanied by the presence of homocystine, the oxidized form of homocysteine, in the urine (136). Deficiency in the enzyme cysthationine β-synthase (CBS) is the most common cause of homocystinuria. This enzyme and other enzymes of homocysteine metabolism require vitamins B6 and B12 and folic acid as cofactors, and deficiencies in the intake of these vitamins are also associated with elevated blood levels of homocysteine.

Soon after the discovery of homocystinuria, the association with severe vascular disease was recognized and found to be the usual cause of premature death in these patients (34,130,149). Since recognition of this association, several studies of patients with premature vascular disease have noted a disproportionate number with mild elevation of plasma homocysteine, suggesting that hyperhomocysteinemia may increase the risk of cardiovascular disease (150). Despite normal fasting homocysteine levels, hyperhomocysteinemia may be unmasked by a methionine challenge, which involves administration of an oral methionine load followed by blood sampling for determination of plasma homocysteine levels, as in a glucose tolerance test. Patients with

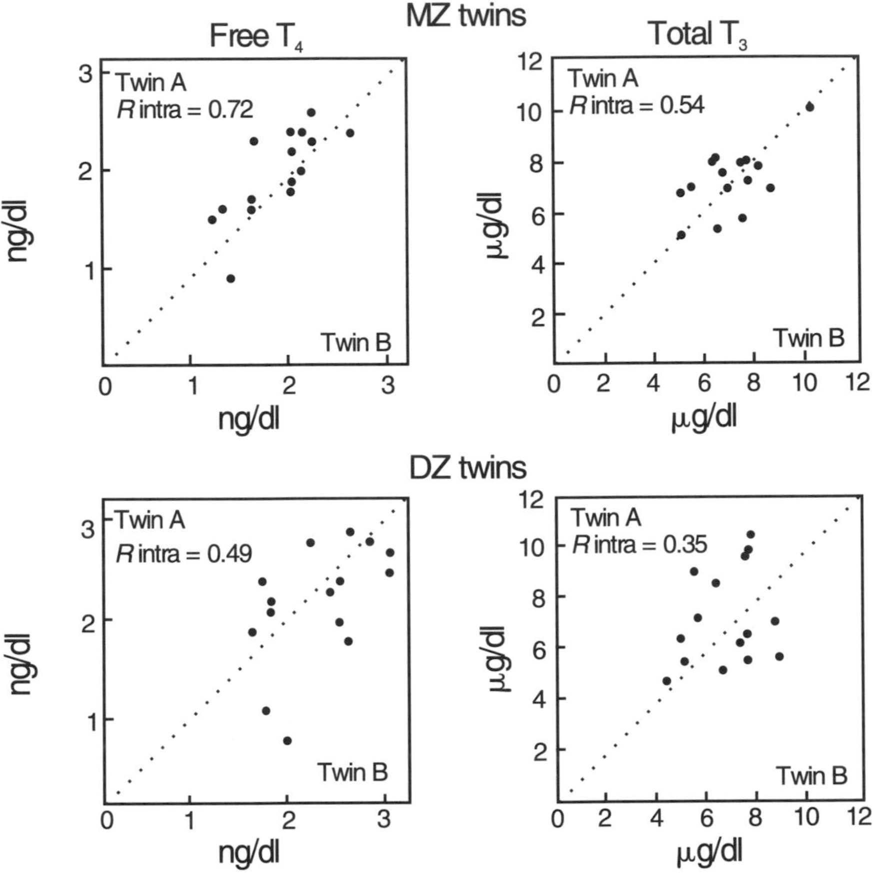

Figure 13.1 Plasma free (*left*) and total (*right*) T_4 concentrations in MZ twins (*upper panel*) and DZ twins (*lower panel*) after adjustment for body mass. Adapted from Meikle et al. (99).

premature coronary artery disease exhibit increased levels of plasma homocysteine after an oral methionine load (151). The association between elevated plasma homocysteine levels (even mild elevation) and increased risk of cardiovascular disease is now well established (126,150).

Three lines of evidence suggest that homocysteine levels are strongly influenced by genetic factors. First, CBS deficiency is a recessively inherited disorder, and several mutations of the CBS gene (21q22.3) have been identified (126). Second, MZ twins exhibit higher correlations than do DZ twins for plasma homocysteine levels (121). Third, plasma levels of homocysteine remain constant in individuals with vascular disease even when measured up to 14 years later (147). Despite the important role of genetic factors in the determination of homocysteine levels, environmental influences are also recognized (40).

Smoking

Cigarette smoking is a risk factor for coronary heart disease (CHD), and cessation of smoking is associated with a decline in the risk of CHD (5,53,57). The genetics of smoking behavior has been investigated in family and twin studies. A review of these studies suggests that the shared family environment, especially the environment shared by members of the same generation, is the major determinant of smoking behavior (84). Results from twin studies generally indicate that genetic factors contribute to the various aspects of smoking behavior: smoking initiation, quantity smoked, smoking dependence, and so on.

The degree of familial resemblance for smoking was studied in 18,073 subjects who participated in the Canada Fitness Survey (110). A spouse correlation of 0.61 was reported for smoking status. Parent-offspring and sibling correlations were 0.40 and 0.57, respectively. This pattern of correlations suggests that environmental factors shared by family members of the same generation and living in the same household are more important than genetic factors in explaining the smoking status.

Results from twins studies consistently show higher concordance rates in MZ twins than in DZ twins for smoking. Reported heritabilities range from 0.28 to 0.84 (75). A few studies based on large samples of twins suggest heritabilities around 50% for different aspects of smoking behavior. In a study of 5,044 adult male twins from the Finnish Twin registry (83), a factorial analysis derived score representing years of smoking, smoking status, and cigarettes smoked per day showed a heritability of 45%. In another study, Carmelli et al. (33) reported a heritability of 53% for number of cigarettes smoked, an estimate that was reduced to 35% after adjustment for alcohol and coffee use, occupation, and socioeconomic status. In a follow-up study of the same population, smoking status (never smoked, currently smoking, and quitting) was only moderately influenced by genetic factors (32). Among current smokers, concordance rates were higher in MZ twins than in DZ twins for light and heavy smoking, which suggests a genetic effect on the dependence on smoking. Smoking persistence has also been associated with genetic factors, with a heritability of 53% (69). Heath et al. (68) also reported heritabilities ranging from 33% to 67% for smoking initiation, depending on the population studied and the sex of the individual.

Koopmans et al. (84) studied smoking status in 574 pairs of MZ twins and 991 pairs of DZ twins, 13 to 22 years of age. Results derived from path analysis with age included in the model to estimate the relative contribution of genetic and environmental factors are summarized in figure 13.2. A model that included both genetic and shared environmental influences provided the best fit to the data. Under this model, 9% of the total variance was accounted for by individual-specific factors, 11% by age, 30% by additive genetic factors, and 50% by shared sibling environment. Results of this study thus suggest a stronger influence of shared environment than heredity for smoking.

The results of these studies support the findings of several other studies on the environmental determinants of smoking. They have shown that the risk of initiating smoking in youth is associated with smoking siblings, peer influences, and attitudes and beliefs about smoking (84).

Genotype-Environment Interaction

A large number of studies of genotype-environment interaction effects on metabolic fitness phenotypes have used diet and blood lipids and lipoproteins as phenotypes. The gene-diet interactions in lipid and lipoprotein metabolism have been reviewed by Dreon and Krauss (45) and were discussed in chapter 8. The response of body fat and fat distribution of MZ twins to the overfeeding and exercise training protocols, described in chapter 10, has also been considered for several other metabolic fitness phenotypes. Within-pair similarities in the response to either overfeeding (27,107,112) or exercise training (28,113) of

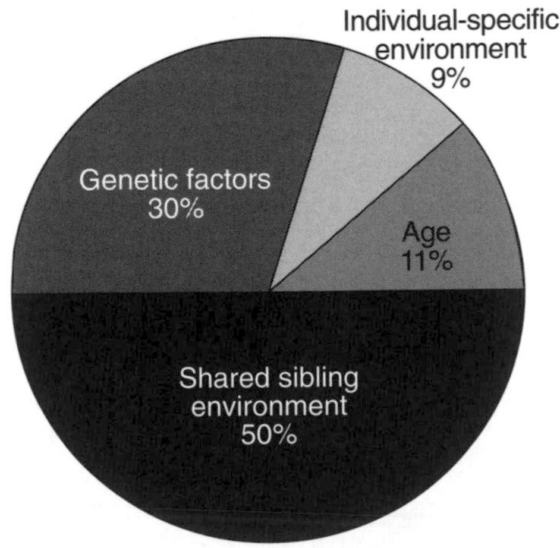

Figure 13.2 Genetic and environmental sources of variance in smoking. Adapted from Koopmans et al. (84).

fasting plasma levels of total cholesterol, triglycerides, glucose, insulin, glucagon, T_3 and T_4, and total areas under the curves of glucose, insulin, and glucagon after a 1,000-kcal meal test are summarized in table 13.3. Significant genotype-overfeeding interaction was noted for triglycerides ($r = 0.70$), fasting insulin ($r = 0.71$), fasting glucagon ($r = 0.53$), and areas under the curve of glucose ($r = 0.69$) and glucagon ($r = 0.67$). On the other hand, cholesterol, triglycerides, and T_4 responses to exercise training were apparently genotype dependent, with intraclass correlations of 0.61, 0.74, and 0.71, respectively.

The Metabolic Syndrome

Obesity, noninsulin dependent diabetes mellitus (NIDDM), atherogenic diseases, and hypertension are common disorders encountered in

Table 13.3
Within-Pair Similarity in the Response to Overfeeding and Exercise Training for Several Metabolic Fitness Phenotypes

	OVERFEEDING		EXERCISE TRAINING	
Variable	**F-ratio**	***r* intra**	**F-ratio**	***r* intra**
Cholesterol	2.7	0.45	4.1*	0.61
Triglycerides	5.4*	0.70	6.6*	0.74
Glucose	0.7	–0.19	0.6	–0.3
Insulin	5.9**	0.71	0.3	–0.5
Glucagon	3.2*	0.53	NA	NA
Glucose area	5.4**	0.69	NA	NA
Insulin area	1.8	0.29	NA	NA
Glucagon area	5.1**	0.67	NA	NA
T_3	1.4	0.10	1.4	0.17
T_4	1.2	0.10	6.0*	0.71

Results of overfeeding experiments are from short-term overfeeding (22 days) in 6 pairs of MZ twins for cholesterol, triglycerides, T_3 and T_4, and from long-term overfeeding (100 days) in 12 pairs of MZ twins for glucose, insulin, glucagon, and areas under the curves. Results from exercise-training experiments are from long-term (93 days) exercise training in 7 pairs of MZ twins for cholesterol, triglycerides, glucose, and insulin and from short-term (22 days) exercise training in 6 pairs of MZ twins for T_3 and T_4. NA = not available.

*p<.05

**p<.01

Adapted from Poehlman et al. (112, 113); Bouchard et al. (27); Oppert et al. (107).

a significant fraction of individuals living in industrialized societies. Reaven (120) was the first to propose that the occurrence of these disorders in the same individual could be attributable to hyperinsulinemia resulting from insulin resistance. It is now well accepted that a significant fraction of middle-aged individuals of both sexes have various metabolic disorders associated with insulin resistance and hyperinsulinemia, including impaired glucose tolerance, hypertension, dyslipidemia, obesity, and NIDDM. Because of the clustering of these metabolic disorders in the same individual, the syndrome is sometimes referred to as the metabolic syndrome. The number and the severity of the metabolic complications resulting from insulin resistance and the ensuing hyperinsulinemia vary among individuals, suggesting that genetic factors are probably important in determining susceptibility to this syndrome. Results from genetic epidemiology studies reveal that most of the causes and manifestations of the metabolic syndrome, taken individually, are influenced by genetic factors (22,24).

The genetic epidemiology studies reviewed in this chapter were based on a univariate approach, in which relevant phenotypes were considered individually in the genetic analyses. Relatively few studies have considered the role of genetic factors in the clustering of the various phenotypes associated with the metabolic syndrome, despite evidence of familial aggregation in the clustering of its components. For example, results from the San Antonio Heart Study reveal that hyperinsulinemia and CHD risk factors are encountered more often in nondiabetic subjects with a positive family history of NIDDM than in those without such a family history (62). A dyslipidemic profile characterized by high plasma triglyceride levels and low HDL cholesterol levels were observed in about 12% of patients with essential hypertension, and the condition was strongly aggregated in families (152). A strong familial aggregation was also reported for a dyslipidemic phenotype characterized by high triglyceride levels and low HDL cholesterol levels (142), and multivariate genetic analyses suggested that genetic factors may be important in determining the expression of the phenotype (93). These observations suggest that genetic factors are likely involved in the familial aggregation of the co-occurrence of the features of the metabolic syndrome.

Few studies have considered the role of shared genetic factors in the phenotypes associated with the metabolic syndrome. Such pleiotropic effects have already been reported among components of the metabolic syndrome. Rice et al. (125) showed that the covariation between upper-body fat (assessed by the ratio of trunk to extremity skinfolds) and blood pressure could be explained by shared genetic factors that could account for as much as 33% of the covariance between these two components of the metabolic syndrome. These results were compatible with those of

Schork et al. (134), who also reported significant genetic pleiotropy between blood pressure and various indices of body mass. Significant cross-trait familial resemblance between body fat and plasma insulin levels was noted, with a bivariate heritability estimate of 8% (124). Significant pleiotropic effects have also been found between body fat and blood lipids and lipoproteins, but these effects do not appear to be explained by shared genetic factors (Pérusse et al., unpublished data). Results of these multivariate genetic analyses thus suggest that genetic factors play a significant role in the covariation of metabolic phenotypes encountered in the metabolic syndrome.

The search for molecular markers associated or linked with these phenotypes or responsible for the covariation among them has just begun. Polymorphisms of genes involved in lipoprotein metabolism influence the relationships between various components of the metabolic syndrome. For example, Pouliot et al. (115) related the APOE polymorphism and plasma lipoproteins to body fat and fat distribution in 63 premenopausal women and found that correlations between indicators of body fat or fat distribution and lipoprotein levels were influenced by the APOE phenotypes. The association between abdominal visceral fat, assessed by computed tomography, and plasma lipoprotein levels was also altered by genetic variation in the APOB-100 gene (114). The role of APOE in modulating the association between body fat and blood lipids and lipoproteins was also noted in children (143). The covariation between insulin and plasma triglyceride levels was also influenced by the APOE polymorphism (44). Recently, a tumor necrosis factor-β polymorphism was associated with higher triglyceride levels in NIDDM patients than in control subjects (148). The limited research thus suggests that genetic variation in the apolipoprotein genes may influence phenotypes related to the metabolic syndrome.

Summary

The genetics of phenotypes associated with metabolic fitness, especially blood lipids and lipoproteins, has been extensively studied. Research on the inherited disorders of lipid metabolism indicates that genes are strong determinants of blood levels of lipids and lipoproteins. Besides the genetic disorders of lipid metabolism, several genetic epidemiology studies have shown a significant contribution of genetic factors to plasma levels of lipids and lipoproteins encountered in normolipemic individuals. Heritabilities ranging from about 25% to 70% are reported, depending on the phenotype considered. In addition to this polygenic effect, major gene effects as well as genotype-environment effects are commonly observed for blood lipids and lipoproteins.

Compared with blood lipids and lipoproteins, less is known regarding the influence of genetic factors on other metabolic fitness phenotypes. Heritabilities ranging from about 40% to 85% have been reported for fasting levels of blood glucose and insulin. The available evidence also suggests that insulin sensitivity is influenced by genetic factors. Significant genetic effects have also been reported for blood coagulation and fibrinolytic systems, but environmental factors appear to be more important than genetic factors. The evidence regarding the role of heredity on hormone levels is sometimes conflicting. However, results from twin studies generally indicate higher plasma hormonal level resemblance in MZ twins than in DZ twins, suggesting a significant genetic effect. Evidence of genotype-training interaction effects has been reported for several metabolic fitness phenotypes, including total cholesterol, triglycerides, and T_4 levels. Finally, a syndrome characterized by the clustering of several metabolic disorders in the same individual was described and shown to be influenced by genetic factors.

References

1. Amos, C.L.; Elston, R.C.; Srinivasan, S.R.; Wilson, A.F.; Cresanta, J.L.; Ward, L.J.; Berenson, G.S. Linkage and segregation analysis of apolipoproteins A1 and B, and lipoprotein cholesterol levels in a large pedigree with excess coronary heart disease: The Bogalusa Heart Study. Genet. Epidemiol. 4:115-28; 1987.
2. Austin, M.A. Genetics of low-density lipoprotein subclasses. Curr. Opin. Lipidol. 4:125-32; 1993.
3. Austin, M.A. Low density lipoprotein subclass phenotypes. In: Goldbourt, U.; de Faire, U.; Berg, K., eds. Genetic factors in coronary heart disease. Dordrecht, Netherlands: Kluwer Academic; 1994:105-13.
4. Austin, M.A.; Jarvik, G.P.; Hokanson, J.E.; Edwards, K. Complex segregation analysis of LDL peak particle diameter. Genet. Epidemiol. 10:599-604; 1993.
5. Bartecchi, C.E.; Mackensie, T.D.; Schrier, R.W. The human cost of tobacco use. N. Engl. J. Med. 330:907-11; 1994.
6. Beaty, T.H.; Kwiterovich, P.O.; Khoury, M.J.; White, S.; Bachorik, P.S.; Smith, H.H.; Teng, B.; Sniderman, A.D. Genetic analysis of plasma sitosterol, apoprotein B, and lipoproteins in a large Amish pedigree with sitosterolemia. Am. J. Hum. Genet. 38:492-504; 1986.
7. Beckman, L.; Olivecrona, T. Serum-cholesterol and ABO and Lewis blood-groups. Lancet. 1:1000; 1970.
8. Beckman, L.; Olivecrona, T.; Hernell, O. Serum lipids and their relation to blood groups and serum alkaline phosphatase isozymes. Hum. Hered. 20:569-79; 1970.

9. Beck-Nielsen, H.; Groop, L.C. Metabolic and genetic characterization of prediabetic states. Sequence of events leading to non-insulin-dependent diabetes mellitus. J. Clin. Invest. 94:1714-21; 1994.

10. Behague, I.; Poirier, O.; Nicaud, V.; Evans, A.; Arveiler, D.; Luc, G.; Gambou, J.P.; Scarabin, P.Y.; Bara, L.; Green, F.; Cambien, F. β-fibrinogen gene polymorphisms are associated with plasma fibrinogen and coronary artery disease in patients with myocardial infarction. The ECTIM Study. Circulation. 93:440-49; 1996.

11. Berg, K. Lp(a) lipoprotein: A monogenic risk factor for cardiovascular disease. In: Goldbourt, U.; de Faire, U.; Berg, K., eds. Genetic factors in coronary heart disease. Dordrecht, Netherlands: Kluwer Academic; 1994:275-87.

12. Berg, K. A new serum type system in man: The Lp system. Acta Pathol. Microbiol. Immunol. Scand. 59:369-82; 1963.

13. Berg, K. From random genetic markers to candidate genes in association and linkage studies of coronary heart disease and its risk factors. In: Goldbourt, U.; de Faire, U.; Berg, K., eds. Genetic factors in coronary heart disease. Dordrecht, Netherlands: Kluwer Academic; 1994:301-8.

14. Berg, K.; Kierulf, P. DNA polymorphisms at fibrinogen loci and plasma fibrinogen concentration. Clin. Genet. 36:229-35; 1989.

15. Berg, K.; Mohr, J. Genetics of the Lp system. Acta Genet. Med. Gemellol. 13:349-60; 1963.

16. Bierman, E.L. Arteriosclerosis and aging. In: Finch, C.E.; Schneider, E.L., eds. Handbook of the biology of aging. New York: Van Nostrand Reinhold; 1985:842-58.

17. Bishop, D.T.; Meikle, A.W.; Slattery, M.L.; Stringham, J.D.; Ford, M.H.; West, D.W. The effect of nutritional factors on sex hormone levels in male twins. Genet. Epidemiol. 5:43-59; 1988.

18. Boerwinkle, E. Genetics of plasma lipoprotein (a) concentrations. Curr. Opin. Lipidol. 3:128-36; 1992.

19. Bogardus, C.; Lillioja, S.; Nyomba, B.L.; Zurlo, F.; Swinburn, B.; Eposito-Delpuente, A.; Knowler, W.C.; Ravussin, E.; Mott, D.M.; Bennet, P.H. Distribution of in vivo insulin action in Pima Indians as a mixture of three normal distributions. Diabetes. 38:1423-32; 1989.

20. Borecki, I.B.; Laskarzewski, P.; Rao, D.C. Genetic factors influencing apolipoprotein AI and AII levels in kindred with premature coronary heart disease. Genet. Epidemiol. 5:393-406; 1988.

21. Børresen, A.L.; Leren, T.; Berg, K.; Solaas, M.H. Effect of haptoglobin subtypes on serum lipid levels. Hum. Hered. 37:150-56; 1987.

22. Bouchard, C. Genetics and the metabolic syndrome. Int. J. Obes. 19:S52-59; 1995.

23. Bouchard, C.; Dionne, F.T.; Simoneau, J.A.; Boulay, M.R. Genetics of aerobic and anaerobic performances. Exerc. Sport Sci. Rev. 20:27-58; 1992.

24. Bouchard, C.; Pérusse, L. Genetics of causes and manifestations of the

metabolic syndrome. In: Crepaldi, G.; Tiengo, A.; Mamzato, E., eds. Diabetes, obesity and hyperlipidemia. V. The plurimetabolic syndrome. Amsterdam: Elsevier Science; 1993:67-74.

25. Bouchard, C.; Shephard, R.J. Physical activity, fitness, and health: The model and key concepts. In: Bouchard, C.; Shephard, R.J.; Stephens, T., eds. Proceedings and consensus statement. Physical activity, fitness, and health. Champaign, IL: Human Kinetics; 1994:77-88.
26. Bouchard, C.; Tremblay, A.; Després, J.P.; Nadeau, A.; Lupien, P.J.; Thériault, G.; Dussault, J.; Moorjani, S.; Pineault, S.; Fournier, G. The response to long-term overfeeding in identical twins. N. Engl. J. Med. 322:1477-82; 1990.
27. Bouchard, C.; Tremblay, A.; Després, J.P.; Poehlman, E.T.; Thériault, G.; Nadeau, A.; Lupien, P.; Moorjani, S. Sensitivity to overfeeding: The Quebec experiment with identical twins. Prog. Food Nutr. Sci. 12:45-72; 1988.
28. Bouchard, C.; Tremblay, A.; Després, J.P.; Thériault, G.; Nadeau, A.; Lupien, P.J.; Moorjani, S.; Prud'homme, D.; Fournier, G. The response to exercise with constant energy intake in identical twins. Obes. Res. 2:400-410; 1994.
29. Bouchard, C.; Tremblay, A.; Nadeau, A.; Després, J.P.; Thériault, G.; Boulay, M.R.; Lortie, G.; Leblanc, C.; Fournier, G. Genetic effect in resting and exercise metabolic rates. Metabolism. 38:364-70; 1989.
30. Breslow, J.L. Apolipoprotein genes and atherosclerosis. Clin. Invest. 70:377-84; 1992.
31. Breslow, J.L. Apolipoprotein genetic variation and human disease. Physiol. Rev. 65:85-132; 1988.
32. Carmelli, D.; Swan, G.E.; Robinette, D.; Fabsitz, R.R. Genetic influence on smoking—A study of male twins. N. Engl. J. Med. 327:829-33; 1992.
33. Carmelli, D.; Swan, G.E.; Robinette, D.; Fabsitz, R.R. Heritability of substance use in the NAS-NRC twin registry. Acta Genet. Med. Gemellol. 39:91-98; 1990.
34. Carson, N.A.J.; Dent, C.E.; Field, C.M.B.; Gaull, G.E. Homocystinuria: Clinical and pathological review of cases. J. Pediatr. 66:563-83; 1965.
35. Connor, J.M.; Fowkes, F.G.R.; Wood, J.; Smith, F.B.; Donnan, P.T.; Lowe, G.D.O. Genetic variation at fibrinogen loci and plasma fibrinogen levels. J. Med. Genet. 29:480-82; 1992.
36. Coresh, J.; Beaty, T.H.; Kwiterovich, P.J. Inheritance of plasma apolipoprotein B levels in families of patients undergoing coronary arteriography at an early age. Genet. Epidemiol. 10:159-76; 1993.
37. Crouse, J.R.; Grundy, S.M.; Ahrens, E.H., Jr. Cholesterol distribution in the bulk tissues of man: Variation with aging. J. Clin. Invest. 51:1292-96; 1972.
38. Dallongeville, J. Apolipoprotein E polymorphism and atherosclerosis risk. In: Goldbourt, U.; de Faire, U.; Berg, K., eds. Genetic factors in coronary heart disease. Dordrecht, Netherlands: Kluwer Academic; 1994:289-97.

39. Dallongeville, J.; Lussier-Cacan, S.; Davignon, J. Modulation of plasma triglycerides levels by apo E phenotype: A meta-analysis. J. Lipid. Res. 33:447-54; 1992.

40. Daly, L.; Robinson, K.; Tan, K.S.; Graham, I.M. Hyperhomocysteinaemia: A metabolic risk factor for coronary heart disease determined by both genetic and environmental influences? Q. J. Med. 86:685-89; 1993.

41. Davidson, M.B. The effect of aging on carbohydrate metabolism: A review of the English literature and a practical approach to the diagnosis of diabetes mellitus in the elderly. Metabolism. 28:688-705; 1979.

42. Davignon, J. Apolipoprotein E polymorphism and atherosclerosis. In: Born, G.V.R.; Schwartz, C.J., eds. New horizons in coronary heart disease: Current science. London: Science Press; 1993:5.1-5.21.

43. Davignon, J.; Gregg, R.E.; Sing, C.F. Apolipoprotein E polymorphism and atherosclerosis. Arteriosclerosis. 8:1-21; 1988.

44. Després, J.P.; Verdon, M.F.; Moorjani, S.; Pouliot, M.-C.; Nadeau, A.; Bouchard, C.; Tremblay, A.; Lupien, P.J. Apolipoprotein E polymorphism modifies relation of hyperinsulinemia to hypertriglyceridemia. Diabetes. 42:1474-81; 1993.

45. Dreon, D.M.; Krauss, R.M. Gene-diet interactions in lipoprotein metabolism. In: Lusis, A.J.; Rotter, J.I.; Sparkes, R.S., eds. Molecular genetics of coronary artery disease. Candidate genes and processes in atherosclerosis. Monographs in human genetics, Vol. 14. Basel: Karger; 1992:325-49.

46. Erickson, J.; Franssila-Kalluni, A.; Ekstrand, A.; Saloranta, C.; Widen, E.; Schalin, C.; Groop, G. Early metabolic defects in persons at increased risk for non-insulin-dependent diabetes mellitus. N. Engl. J. Med. 321:337-43; 1989.

47. Ernst, E. The role of fibrinogen as a cardiovascular risk factor. Atherosclerosis. 100:1-12; 1993.

48. Fay, W.P.; Ginsburg, D. Fibrinogen, factor VII, and plasminogen activator inhibitor-1. In: Goldbourt, U.; de Faire, U.; Berg, K., eds. Genetic factors in coronary heart disease. Dordrecht, Netherlands: Kluwer Academic; 1994:125-38.

49. Fay, W.P.; Shapiro, A.D.; Shih, J.L.; Schleef, R.R.; Ginsburg, D. Complete deficiency of plasminogen-activator inhibitor type 1 due to a frame-shift mutation. N. Engl. J. Med. 327:1729-33; 1992.

50. Ferrell, R.E. Genetics of the apolipoprotein and the contribution of allelic variation to quantitative variation in lipid and lipoprotein levels in the population. Curr. Opin. Lipidol. 3:122-27; 1992.

51. Ferrell, R.E.; Lyengar, S. Molecular studies of the genetics of non-insulin-dependent diabetes mellitus. Am. J. Hum. Biol. 5:415-24; 1993.

52. Fisher, E.A. Gene polymorphisms and variability of human apolipoproteins. Ann. Rev. Nutr. 9:139-60; 1989.

53. Frank, E. Benefits of stopping smoking. West. J. Med. 159:83-87; 1993.

54. Friedlander, Y.; Elkana, Y.; Sinnreich, R.; Kark, J.D. Genetic and environmental sources of fibrinogen variability in Israeli families: The Kibbutzim Family Study. Am. J. Hum. Genet. 56:1194-1206; 1995.

55. Friedlander, Y.; Kark, J.D.; Stein, Y. Complex segregation analysis of low levels of plasma high-density lipoprotein cholesterol in a sample of nuclear families in Jerusalem. Genet. Epidemiol. 3:285-97; 1986.
56. Gerdes, L.U.; Klausen, I.C.; Sihm, I.; Faergeman, O. Apolipoprotein E polymorphism in a Danish population compared to findings in 45 other study populations around the world. Genet. Epidemiol. 9:155-67; 1992.
57. Glantz, S.A.; Parmley, W.W. Passive smoking and heart disease: Epidemiology, physiology and biochemistry. Circulation. 83:1-11; 1991.
58. Goldman, R. Decline in organ function with aging. In: Rossman, I., ed. Clinical geriatrics. Philadelphia: Lippincott; 1979:23-59.
59. Graaf, J.D.; Swinkels, D.W.; Haan, A.F.J.D.; Demacker, P.N.M.; Stalenhoef, A.F.H. Both inherited susceptibility and environmental exposure determine the low-density lipoprotein-subfraction pattern distribution in healthy Dutch families. Am. J. Hum. Genet. 51:1295-1310; 1992.
60. Green, F.; Hamsten, A.; Blombäck, M.; Humphries, S. The role of β-fibrinogen genotype in determining plasma fibrinogen levels in young survivors of myocardial infarction and healthy controls from Sweden. Thromb. Haemost. 70:915-20; 1993.
61. Green, F.; Kelleher, C.; Wilkes, H.; Temple, A.; Meade, T.; Humphries, S. A common genetic polymorphism associated with lower coagulation factor VII levels in healthy individuals. Arterioscler. Thromb. 11:540-46; 1991.
62. Haffner, S.M.; Stern, M.P.; Hazuda, H.P.; Mitchell, B.D.; Patterson, J.K. Increased insulin concentrations in non-diabetic offspring of diabetic parents. N. Engl. J. Med. 319:1297-1301; 1988.
63. Hallman, D.M.; Boerwinkle, E.; Saha, N.; Sandholzer, C.; Menzel, H.J.; Csazar, A.; Utermann, G. The apolipoprotein E polymorphism: A comparison of allele frequencies and effects in nine populations. Am. J. Hum. Genet. 49:338-49; 1991.
64. Hamman, R.F. Genetic and environmental determinants of non-insulin-dependent diabetes mellitus (NIDDM). Diabetes Metab. Rev. 8:287-338; 1992.
65. Hamsten, A.; de Faire, U.; Iselius, L.; Blombäck, M. Genetic and cultural inheritance of plasma fibrinogen concentration. Lancet. 31:988-90; 1987.
66. Hasstedt, S.J.; Williams, R.R. Three alleles for quantitative Lp(a). Genet. Epidemiol. 3:53-55; 1986.
67. Hasstedt, S.J.; Wu, L.; Williams, R.R. Major locus inheritance of apolipoprotein B in Utah pedigrees. Genet. Epidemiol. 4:67-76; 1987.
68. Heath, A.C.; Cates, R.; Martin, N.G.; Mexer, J.; Hewitt, J.K.; Neale, M.C.; Eaves, L.J. Genetic contribution to risk of smoking initiation: Comparisons across birth cohorts and across cultures. J. Subst. Abuse. 5:221-46; 1994.
69. Heath, A.C.; Martin, N.G. Genetic models for the natural history of smoking: Evidence for a genetic influence on smoking persistence. Addict. Behav. 18:19-34; 1993.

70. Heller, D.A.; de Faire, U.; Pedersen, N.L.; Dahlen, G.; McClean, G.E. Genetic and environmental influences on serum lipid levels in twins. N. Engl. J. Med. 328:1150-56; 1993.

71. Hitman, G.A.; Melcalfe, K.A. The genetics of diabetes—An update. In: Home, P.D.; Marshall, S.M.; Alberti, K.G.M.M.; Krall, L.P., eds. The diabetes annual. Amsterdam: Elsevier Science; 1993:1-16.

72. Ho, K.K.Y.; Hoffman, D.M. Aging and growth hormone. Horm. Res. 40:80-86; 1993.

73. Hobbs, H.H.; Brown, M.S.; Russel, D.W.; Davignon, J.; Goldstein, J.L. Deletion in the gene for the low-density-lipoprotein receptor in a majority of French Canadians with familial hypercholesterolemia. N. Engl. J. Med. 317:734-37; 1987.

74. Hobbs, H.H.; Russel, D.W.; Brown, M.S.; Goldstein, J.L. The LDL receptor locus in familial hypercholesterolemia: Mutational analysis of a membrane protein. Annu. Rev. Genet. 24:133-70; 1990.

75. Hugues, J.R. Genetics of smoking: A brief review. Behav. Ther. 17:335-45; 1986.

76. Humphries, S.E.; Cook, M.; Dubowitz, M.; Stirling, Y.; Meade, T.W. Role of genetic variation at the fibrinogen locus in determination of plasma fibrinogen concentrations. Lancet. 1:1452-55; 1987.

77. Humphries, S.E.; Lane, A.; Green, F.R.; Cooper, J.; Miller, G.J. Factor VII coagulant activity and antigen levels in healthy men are determined by interaction between factor VII genotype and plasma triglyceride concentration. Arterioscler. Thromb. 14:193-98; 1994.

78. Iselius, L. Complex segregation analysis of hypertriglyceridemia. Hum. Hered. 31:222-26; 1981.

79. Iselius, L. Genetic epidemiology of common diseases in humans. In: Weir, B.S.; Eisen, E.J.; Goodman, M.M.; Namkoong, G., eds. Proceedings of the 2nd International Conference on Quantitative Genetics. Sunderland: Sinauer Associates; 1988:341-52.

80. Iselius, L.; Carlson, L.A.; Morton, N.E.; Effendic, S.; Lindsten, J.; Luft, R. Genetic and environmental determinants for lipoprotein concentrations in blood. Acta. Med. Scand. 217:161-70; 1985.

81. Iselius, L.; Lindsten, J.; Morton, N.E.; Effendic, S.; Cerasi, E.; Haegermark, A.; Luft, R. Genetic regulation of the kinetics of glucose-induced insulin release in man. Studies in families with diabetic and non-diabetic probands. Clin. Genet. 28:8-15; 1985.

82. Kamboh, M.L.; Aston, C.E.; Hamman, R.F. The relationship of Apo E polymorphism and cholesterol levels in normoglycemic and diabetic subjects in a biethnic population from the San Luis Valley, Colorado. Atherosclerosis. 112:145-59; 1995.

83. Kaprio, J.; Koskenvuo, M.; Sarna, S. Cigarette smoking, use of alcohol, and leisure-time physical activity among same-sexed adult male twins. In: Gedda, L.; Parisse, P.; Nance, W.E., eds. Twin research 3. Part C. Epidemiological and clinical studies. New York: Alan R. Liss; 1981:37-46.

84. Koopmans, J.R.; Van Doornen, L.J.P.; Boomsma, D.I. Smoking and sports participation. In: Goldbourt, U.; de Faire, U.; Berg, K., eds. Genetic factors in coronary heart disease. Dordrecht, Netherlands: Kluwer Academic; 1994:217-35.
85. Lambert-Prenger, V.; Beaty, T.H.; Kwiterovich, P.O. Genetic determination of high-density lipoprotein-cholesterol and apolipoprotein A-1 plasma levels in a family study of cardiac catheterization patients. Am. J. Hum. Genet. 51:1047-57; 1992.
86. Lillioja, S.; Mott, D.M.; Zawadzki, J.K.; Young, A.A.; Abbott, W.G.H.; Knowler, W.C.; Bennett, P.H.; Moll, P.P.; Bogardus, C. In vivo insulin action is familial characteristic in nondiabetic Pima Indians. Diabetes. 36:1329-35; 1987.
87. Linkowski, P.; Van Onderbergen, A.; Kerkhofs, M.; Bosson, D.; Mendlewicz, J.; Van Cauter, E. Twin study of the 24-h cortisol profile: Evidence for genetic control of the human circadian clock. Amer. J. Physiol. 264:E173-81; 1993.
88. Lusis, A.J. Genetic factors affecting blood lipoproteins: The candidate gene approach. J. Lipid. Res. 29:397-429; 1988.
89. Malina, R.M.; Bouchard, C. Growth, maturation and physical activity. Champaign, IL: Human Kinetics; 1991.
90. Marazita, M.L.; Elston, R.C.; Namboodiri, K.K.; Hames, C.G. Genetic analysis of serum lipid levels and blood pressure in a large kindred. Am. J. Med. Genet. 26:511-19; 1987.
91. Martin, B.C.; Warram, J.H.; Rosner, B.; Rich, S.S.; Soeldner, J.S.; Krolewski, A.S. Familial clustering of insulin sensitivity. Diabetes. 41:850-54; 1992.
92. Mayo, O.; Fraser, G.R.; Stamatoyannopoulos, G. Genetic influences on serum cholesterol in two Greek villages. Hum. Hered. 19:86-99; 1969.
93. McGue, M.; Rao, D.C.; Reich, T.; Laskarzewski, P.; Glueck, C.J.; Russel, J.M. The Cincinnati Lipid Research Clinic family study. Bivariate path analyses of lipoprotein concentrations. Genet. Res. Camb. 42:117-35; 1983.
94. Mehrabian, M.; Lusis, A.J. Genetic markers for studies of atherosclerosis and related risk factors. In: Lusis, A.J.; Rotter, J.I.; Sparkes, R.S., eds. Molecular genetics of coronary artery disease. Basel: Karger; 1992:363-418.
95. Meikle, A.W.; Bishop, D.T.; Stringham, J.D.; West, D.W. Quantitating genetic and non-genetic factors that determine plasma sex steroid variation in normal male twins. Metabolism. 35:1090-95; 1986.
96. Meikle, A.W.; Stanish, W.M.; Taylor, N.; Edwards, C.Q.; Bishop, C.T. Familial effects on plasma sex-steroid content in man: Testosterone, estradiol and sex-hormone binding globulin. Metabolism. 31:6-9; 1982.
97. Meikle, A.W.; Stringham, J.D.; Bishop, D.T.; West, D.W. Quantitating genetic and non-genetic factors influencing androgen production and clearance rates in man. J. Clin. Endocrinol. Metab. 67:104-9; 1988.
98. Meikle, A.W.; Stringham, J.D.; Woodward, M.G.; Bishop, D.T. Heritability of variation of plasma cortisol levels. Metabolism. 37:514-17; 1988.
99. Meikle, A.W.; Stringham, J.D.; Woodward, M.G.; Nelson, J.C. Hereditary

and environmental influences on the variation of thyroid hormones in normal male twins. J. Clin. Endocrinol. Metab. 66:588-92; 1988.

100. Meilahn, E.; Ferrell, R.; Kiss, J.; Temple, A.; Green, F.; Humphries, S.; Kuller, L. Genetic determination of coagulation factor VIIc levels among healthy middle-aged women. Thromb. Haemost. 73:623-25; 1995.

101. Minaker, K.L.; Meneilly, G.S.; Rowe, J.W. Endocrine systems. In: Finch, C.E.; Schneider, E.L., eds. Handbook of the biology of aging. 2nd ed. New York: Van Nostrand Reinhold; 1985:433-56.

102. Moll, P.P.; Michels, V.M.; Weidman, W.H.; Kottke, B.A. Genetic determination of plasma apolipoprotein AI in a population-based sample. Am. J. Hum. Genet. 44:124-39; 1989.

103. Moll, P.P.; Sing, C.F.; Lussier-Cacan, S.; Davignon, J. An application of a model for a genotype-dependent relationship between a concomitant (age) and a quantitative trait (LDL cholesterol) in pedigree data. Genet. Epidemiol. 1:301-14; 1984.

104. Montgomery, H.E.; Clarkson, P.; Nwose, O.M.; Mikailidis, D.P.; Jagroop, I.A.; Dollery, C.; Moult, J.; Benhizia, F.; Deanfield, J.; Jubb, M.; World, M.; McEwan, J.R.; Winder, A.; Humphries, S. The acute rise in plasma fibrinogen concentration with exercise is influenced by the G-453-A polymorphism of the β-fibrinogen gene. Arterioscler. Thromb. 16:386-91; 1996.

105. Morton, N.E.; Berg, K.; Dahlen, G.; Ferrell, R.E.; Rhoads, G.G. Genetics of the Lp lipoprotein in Japanese-Americans. Genet. Epidemiol. 2:113-21; 1985.

106. Morton, N.E.; Gulbrandsen, C.L.; Rhoads, G.G.; Kagan, A.; Lew, R. Major loci for lipoprotein concentrations. Am. J. Hum. Genet. 30:583-89; 1978.

107. Oppert, J.M.; Nadeau, A.; Tremblay, A.; Després, J.P.; Thériault, G.; Dériaz, O.; Bouchard, C. Plasma glucose, insulin, and glucagon before and after long-term overfeeding in identical twins. Metabolism. 44:96-105; 1995.

108. O'Sullivan, J.B.; Mahan, C.M.; Freedlander, A.E.; Williams, R.F. Effect of age on carbohydrate metabolism. J. Clin. Endocrinol. Metab. 33:619-23; 1971.

109. Pairitz, G.; Davignon, J.; Mailloux, H.; Sing, C.F. Sources of interindividual variation in the quantitative levels of apolipoprotein B in pedigrees ascertained through a lipid clinic. Am. J. Hum. Genet. 43:311-21; 1988.

110. Pérusse, L.; Leblanc, C.; Bouchard, C. Familial resemblance in lifestyle components: Results from the Canada Fitness Survey. Can. J. Public Health. 79:201-5; 1988.

111. Plump, A.S.; Breslow, J.L. Apolipoprotein E and the apolipoprotein E–deficient mouse. Ann. Rev. Nutr. 15:495-518; 1995.

112. Poehlman, E.T.; Tremblay, A.; Fontaine, E.; Després, J.P.; Nadeau, A.; Dussault, J.; Bouchard, C. Genotype dependency of the thermic effect of a meal and associated hormonal changes following short-term overfeeding. Metabolism. 35:30-36; 1986.

113. Poehlman, E.T.; Tremblay, A.; Nadeau, A.; Dussault, J.; Thériault, G.; Bouchard, C. Heredity and changes in hormones and metabolic rates with short-term training. Am. J. Physiol. 250:E711-17; 1986.

114. Pouliot, M.C.; Després, J.P.; Dionne, F.T.; Vohl, M.C.; Moorjani, S.; Prud'homme, D.; Bouchard, C.; Lupien, P.J. Apo B-100 gene EcoR1 polymorphism. Relations to plasma lipoprotein changes associated with abdominal visceral obesity. Arterioscler. Thromb. 14:527-33; 1994.

115. Pouliot, M.C.; Després, J.P.; Moorjani, S.; Lupien, P.J.; Tremblay, A.; Bouchard, C. Apolipoprotein E polymorphism alters the association between body fatness and plasma lipoproteins in women. J. Lipid. Res. 31:1023-30; 1990.

116. Pratt, J.H.; Manatunga, A.K.; Li, W. Familial influences on the adrenal androgen excretion rate during the adrenarche. Metabolism. 43:186-89; 1994.

117. Prochazka, M.; Lillioja, S.; Tait, J.F.; Knowler, W.C.; Mott, D.M.; Spraul, M.; Bennet, P.H.; Bogardus, C. Linkage of chromosomal markers on 4q with a putative gene determining maximal insulin action in Pima Indians. Diabetes. 42:514-19; 1993.

118. Rader, D.J.; Brewer, H.B., Jr. Lipids, apolipoproteins and lipoproteins. In: Goldbourt, U.; de Faire, U.; Berg, K., eds. Genetic factors in coronary heart disease. Dordrecht, Netherlands: Kluwer Academic; 1994:83-103.

119. Raffel, L.J.; Shohat, T.; Rotter, J.I. Diabetes and insulin resistance. In: Goldbourt, U.; de Faire, U.; Berg, K., eds. Genetic factors in coronary heart disease. Dordrecht, Netherlands: Kluwer Academic; 1994:203-15.

120. Reaven, G.M. Role of insulin resistance in human disease. Diabetes. 37:1595-1607; 1988.

121. Reed, T.; Malinow, M.R.; Christian, J.C.; Upson, B. Estimates of heritability for plasma homocyst(e)ine with progression of symptomatic peripheral arterial disease. J. Vasc. Surg. 13:128-36; 1991.

122. Reed, T.; Tracy, R.P.; Fabsitz, R.R. Minimal genetic influences on plasma fibrinogen level in adult males in the NHLBI twin study. Clin. Genet. 45:71-77; 1994.

123. Rice, T.; Laskarzewski, P.M.; Rao, D.C. Commingling and complex segregation analysis of fasting plasma glucose in the Lipid Research Clinics family study. Am. J. Med. Genet. 44:399-404; 1992.

124. Rice, T.; Nadeau, A.; Pérusse, L.; Bouchard, C.; Rao, D.C. Familial correlations in the Quebec family study: Cross-trait familial resemblance for body fat with plasma glucose and insulin. Diabetologia. 39: 1357-64; 1996.

125. Rice, T.; Province, M.; Pérusse, L.; Bouchard, C.; Rao, D.C. Cross-trait familial resemblance for body fat and blood pressure: Familial correlations in the Québec family study. Am. J. Hum. Genet. 55:1019-29; 1994.

126. Robinson, K.; Tan, K.S.; Graham, I.M. Homocysteine. In: Goldbourt, U.; de Faire, U.; Berg, K., eds. Genetic factors in coronary heart disease. Dordrecht, Netherlands: Kluwer Academic; 1994:139-52.

127. Robitaille, N.; Cormier, G.; Couture, R.; Bouthillier, D.; Davignon, J.; Pérusse, L. Apolipoprotein E polymorphism in a French Canadian population of northeastern Quebec: Allele frequencies and effects on blood lipids and lipoproteins. Hum. Biol. 68:357-70; 1996.

128. Rotter, J.I.; Wong, F.L.; Lifrak, E.T.; Parker, L.N. A genetic component to the variation of dehydroepiandrosterone sulfate. Metabolism. 34:731-36; 1985.

129. Scarabin, P.Y.; Bara, L.; Ricard, S.; Poirier, O.; Camboum, J.P.; Arveiler, D.; Luc, G.; Evans, A.E.; Samana, M.M.; Cambien, F. Genetic variation at the β-fibrinogen locus in relation to plasma fibrinogen concentrations and risk of myocardial infarction. The ECTIM Study. Arterioscler. Thromb. 13:886-91; 1993.

130. Schmike, R.N.; McKusick, V.A.; Huang, T.; Pollack, A.D. Homocystinuria: Studies of 20 families with 38 affected members. JAMA. 193:711-19; 1965.

131. Schonfeld, G. Disorders of lipid transport: Relationship to abnormalities of apoproteins, enzymes and cellular receptors. In: Elston, R.C.; Kuller, L.K.; Feinleib, M.; Carter, C.; Havlik, R., eds. Genetic epidemiology of coronary heart disease: Past, present and future. New York: Alan R. Liss; 1984:375-402.

132. Schonfeld, G. Inherited disorders of lipid transport. Am. J. Physiol. Endocr. Metab. 19:229-57; 1990.

133. Schonfeld, G.; Krul, E.S. Genetic defects in lipoprotein metabolism. In: Goldbourt, U.; de Faire, U.; Berg, K., eds. Genetic factors in coronary heart disease. Dordrecht, Netherlands: Kluwer Academic; 1994:239-66.

134. Schork, N.J.; Weder, A.B.; Trevisan, M.; Laurenzi, M. The contribution of pleiotropy to blood pressure and body-mass index variation: The gubbio study. Am. J. Hum. Genet. 54:361-73; 1994.

135. Schumacher, M.C.; Hasstedt, S.J.; Hunt, S.C.; Williams, R.R.; Elbein, S.C. Major gene effect for insulin levels in familial NIDDM pedigrees. Diabetes. 41:416-23; 1992.

136. Scriver, C.R.; Beaudet, A.L.; Sly, W.S.; Valle, D. The metabolic basis of inherited disease, Vols. I and II. 6th ed. New York: McGraw-Hill; 1989.

137. Segal, P.; Rifkind, B.M.; Schull, W.J. Genetic factors in lipoprotein variation. Epidemiol. Rev. 4:127-60; 1982.

138. Shohat, T.; Raffel, L.F.; Vadheim, C.M.; Rotter, J.I. Diabetes mellitus and coronary heart disease genetics. In: Lusis, A.J.; Rotter, J.I.; Sparkes, R.S., eds. Molecular genetics of coronary artery disease. Candidate genes and processes in atherosclerosis. Monographs in human genetics, Vol. 14. Basel: Karger; 1992:272-310.

139. Sing, C.F.; Boerwinkle, E.A. Genetic architecture of inter-individual variability in apolipoprotein, lipoprotein and lipid phenotypes. In: Bock, G.; Collins, G.M., eds. Molecular approaches to human polygenic disease. New York: Wiley; 1987:99-127.

140. Sing, C.F.; Boerwinkle, E.; Moll, P.P.; Templeton, A.R. Characterization of genes affecting quantitative traits in humans. In: Weir, B.S.; Eisen, E.J.; Goddman, M.M.; Namkoong, G., eds. Proceedings of the Second International Conference on Quantitative Genetics. Sunderland: Sinauer Associates; 1988:250-69.

141. Sing, C.F.; Orr, J.D. Analysis of genetic and environmental sources of variation in serum cholesterol in Tecumseh, Michigan. III. Identification

of genetic effects using 12 polymorphic genetic marker systems. Am. J. Hum. Genet. 28:453-64; 1976.

142. Sprecher, D.L.; Feigelson, H.S.; Laskarzewski, P.M. The low HDL cholesterol/high triglyceride trait. Arterioscler. Thromb. 13:495-504; 1993.

143. Srinivasan, S.R.; Ehnholm, C.; Wattigney, W.A.; Berenson, G.S. Relationship between obesity and serum lipoproteins in children with different apolipoprotein E phenotypes: The Bogalusa Heart Study. Metabolism. 43:470-75; 1994.

144. Talbot, S.; Wakley, E.J.; Langman, M.J.S. A19, A29, B, and O blood-groups, Lewis blood-groups, and serum triglyceride and cholesterol concentrations in patients with venous thromboembolic disease. Lancet. 1:1152-54; 1972.

145. Thompson, D.B.; Janssen, R.C.; Ossowski, V.M.; Prochazka, M.; Knowler, W.C.; Bogardus, C. Evidence for linkage between a region on chromosome 1p and the acute insulin response in Pima Indians. Diabetes. 44:478-81; 1995.

146. Tiret, L.; Steinmetz, J.; Herbeth, B.; Visvikis, S.; Rakotovao, R.; Ducimetière, P.; Cambien, F. Familial resemblance of plasma apolipoprotein B: The Nancy Study. Genet. Epidemiol. 7:187-97; 1990.

147. Ueland, P.; Refsum, H.; Brattstrom, L. Plasma homocysteine and cardiovascular disease. In: Francis, R.J., Jr., ed. The hemostatic system, endothelial function, and cardiovascular disease. New York: Dekker; 1992:183-236.

148. Vendrell, J.; Gutierrez, C.; Pastor, R.; Richart, C. A tumor necrosis factor–polymorphism associated with hypertriglyceridemia in non-insulin-dependent diabetes mellitus. Metabolism. 44:691-94; 1995.

149. White, H.H.; Rowland, L.P.; Araki, S.; Thompson, H.L.; Cowen, D. Homocystinuria. Arch. Neurol. 13:455-70; 1965.

150. Wilcken, D.E.L.; Dudman, N.P.B. Homocystinuria and atherosclerosis. In: Lusis, A.J.; Rotter, J.I.; Sparkes, R.S., eds. Molecular genetics of coronary artery disease. Candidate genes and processes in atherosclerosis. Monographs in human genetics, Vol. 14. Basel: Karger; 1992:311-24.

151. Wilcken, D.E.L.; Wilcken, B. The pathogenesis of coronary artery disease: A possible role for methionine metabolism. J. Clin. Invest. 57:1079-82; 1976.

152. Williams, R.R.; Hunt, S.C.; Hopkins, P.N.; Stults, B.M.; Wu, L.L.; Hasstedt, S.J.; Barlow, G.K.; Stephenson, S.H. Familial dyslipidemic hypertension. Evidence from 58 Utah families for a syndrome present in approximately 12% of patients with essential hypertension. JAMA. 259:3579-86; 1988.

153. Williams, W.R.; Lalouel, J.M. Complex segregation analysis of hyperlipidemia in a Seattle sample. Hum. Hered. 32:24-36; 1982.

154. Wing, R.R.; Matthews, K.R.; Kuller, L.H.; Smith, D.; Becker, D.; Plantinga, P.L.; Meilahn, E.N. Environmental and familial contributions to insulin levels and change in insulin levels in middle-aged women. JAMA. 268: 1890-95; 1992.

155. Yang, X.C.; Jing, T.Y.; Resnick, L.M.; Phillips, G.B. Relation of hemostatic risk factors to other risk factors for coronary heart disease and to sex hormones in men. Arterioscler. Thromb. 13:467-71; 1993.

CHAPTER 14

Genetics and Motor Development

Motor development is the process through which an individual acquires competence in a variety of movements. Motor learning refers to the development of proficiency or skill in specific movements, usually in the context of specific instruction and/or practice. Skill implies accuracy, economy, precision of movement, and in turn performance.

The process of motor development involves the interaction of several factors, including rate of neuromuscular maturation, prior movement experiences, current movement experiences, and of course, a variety of environmental factors that influence and mediate movement experiences. Additional factors are physical characteristics, opportunity to move, and specific practice. Growth of specific body segments and tissues is differential, so that body proportions and composition change. These changes occur quite rapidly during the first 5 or 6 years of life, the time when basic movement patterns are developing. However, relationships between changes in proportions and in body composition and the development of movement patterns have not been satisfactorily addressed. In contrast, interest in the issue of specific practice of motor skills at young ages has a long history, which is considered later in the chapter.

Changes During the Life Span

Motor development begins prenatally and continues postnatally. With advances in technology for fetal monitoring, specifically ultrasonography, early movement behavior of the fetus can be more accurately documented. Earliest movements occur about the seventh to eighth week postfertilization, that is, toward the end of the embryonic period. The movement repertoire of the fetus is quite diverse (34). During the second half of gestation, fetal movements, on average, increase in number through the 34th week and then decrease in an inverted-U pattern. However, there are substantial individual differences that are relatively stable (7). Continuity of prenatal and postnatal movement activity is often assumed, but studies of the interrelationship of prenatal movement activity and subsequent postnatal movement activity and motor development are not available.

Motor development during the first year or so of postnatal life is aimed largely at the development of walking, that is, upright posture and independent locomotion. Commonly used scales for the assessment of early motor development [e.g., the motor scales of the Bayley Scales of Infant Development (1) or the Denver Developmental Screening Test (10)] include a variety of gross and fine motor items. The focus of this discussion is primarily on gross motor achievements. With the onset of independent walking, the development of proficiency in a variety of fundamental movement patterns is a major developmental task of the next 6 or 7 years. These include running, jumping, hopping, skipping, throwing, kicking, striking, catching, and others. Development of these basic movement skills progresses rapidly during early childhood and continues into middle childhood. Sex differences are small, with the exception of throwing and kicking (earlier in boys) and hopping and skipping (earlier in girls) (20). On average, mature patterns of most of the fundamental motor skills are developed by 6 to 7 years of age. Subsequently, they are refined through practice and perhaps instruction and integrated into more complex movement sequences such as those required for specific games and sports. The quality and quantity of performance thus improve.

Beyond the acquisition of fundamental movement patterns early in middle childhood, the motor development literature does not ordinarily extend into older ages. Emphasis is more on the outcomes of performance, especially how they are influenced by growth and maturation and then by changes associated with aging. These are briefly considered in chapter 15.

Genetics and Early Motor Development

Most of the available research that deals with the role of inherited characteristics in motor development has been done on twins with the motor scale of the Bayley Scales of Infant Development. MZ and DZ twins do not differ significantly from each other in motor achievements on the developmental scales, nor does the first born twin differ from the second born (6). Twins often differ in size at birth and are often born prematurely. Nevertheless, intrapair differences in the motor development of MZ and DZ twins dissimilar in birth weight are not significant during the first two years of life (12,45). On the other hand, MZ and DZ twins tend to show a consistent lag in motor development compared to singletons at least through 5 years of age (4,6,45).

Concordance in early motor development is higher in MZ than in DZ twins (11,38,45). Although intrapair similarities differ between twin types, within-pair correlations for DZ twins are quite high compared to those of MZ twins (table 14.1). Longitudinal observations also indicate higher concordance for the overall profile of motor development in MZ twins during the first year; however, between 12 and 18 months, MZ twins are no more concordant than DZ twins in the rate of gain in motor development (45). The preceding findings thus suggest the presence of shared environmental influences, for example, environmental pressure for similarity or mutual imitation. Additionally, difficulties in testing young children at these ages (stranger anxiety, crying, etc.) must be considered.

Table 14.1
Within-Pair Correlations for Scores on the Motor Scale of the Bayley Scales of Infant Development

Age (months)	MZ (n = 71-93)[a]	DZ (n = 77-91)[a]	MZ (n = 39)[b]	DZ (n = 21)[b]
3	0.50	0.41		
6	0.87	0.75		
8			0.93	0.82
9	0.84	0.61		
12	0.75	0.63		
18	0.70	0.77		

[a]Adapted from Wilson and Harpring (45), mixed longitudinal.
[b]Adapted from Saudino and Eaton (38).

Data from the Colorado Adoption Project indicate significant, though low (< 0.2), correlations between self-reported parental athletic ability and Bayley motor scores at 12 and 24 months of age (33). At 12 months, parent-child correlations between self-reported athletic ability of the mother are similar for biological and adoptive mothers, which suggests genetic and environmental influences. At 24 months, parent-offspring correlations are not significant in adopted children but are higher in corresponding correlations for parents and their biological offspring, which may suggest some genetic influence.

Data dealing with specific movements are less extensive. Early observations indicate greater concordance among MZ than DZ twins for first efforts at walking (3,44). The development of other movement patterns during early childhood, such as crawling, running, stair climbing, and so on, generally shows greater concordance among MZ twins (13), but much of the data is based upon descriptive comparisons of single pairs of twins (14,15,31,32,41).

Information about the early motor development of biological siblings is extremely limited. One study indicates variable heritabilities for ages at turning, sitting up, standing alone, and walking in siblings from 66 families (25). Estimates ranged from zero for age at turning over to 0.42 for age at walking.

Genetics and Movement Patterns

A good deal of research on fundamental motor skills, such as running, jumping, and throwing, is based on cinematographic and video evaluation of different aspects of the movement patterns. Little is known about similarities or dissimilarities in movement patterns between twins and siblings. One exception is a comprehensive comparison of the kinematic structure of the dash in twins 11 to 15 years of age (40). The data are based on cinematographic analysis of specific aspects of running style during the performance of a 60-m dash (table 14.2). Within-pair variances for stride length, tempo, and various limb and trunk angles are consistently smaller in MZ twins, and in male more so than in female twins. This suggests a significant genotypic contribution to variation in the kinematic structure of the dash. However, the smaller differences between female MZ and DZ twins suggests that the running performance of girls is more amenable to environmental influences, including social and motivational factors for an all-out performance.

More recently, Goya et al. (16,17) reported results of a motion analysis (cinematographic and video) of the dash, throw, and swimming crawl in 10 pairs of MZ and DZ twins followed longitudinally for 6 to 9 years. Intrapair differences in kinematic variables of the dash were smaller in MZ than in DZ twins, while intrapair differences in related variables for the throw and swimming crawl were similar in MZ and DZ twins. The results for the dash are consistent with those of Sklad (40), suggesting a genotypic contribution to variation in the kinematic structure of the dash. However, the results for the throw and swimming crawl indicate a more important role for environmental influences, especially practice and experience.

Experience from horse breeding suggests that pacing (two legs on the same side of the body move forward together) and trotting (diagonally opposite legs move forward together) movement patterns are character-

Table 14.2
Within-Pair Variances for Certain Kinematic Aspects of a 60-m Dash on 11- to 15-Year-Old Twins

	MALES		FEMALES	
	V_{MZ} (n = 15)	V_{DZ} (n = 11)	V_{MZ} (n = 12)	V_{DZ} (n = 12)
Stride length take-off	7.6	160.1	1.8	14.3
Mean stride length	9.5	106.3	2.3	13.4
Tempo[a]	0.7	2.7	0.5	1.3
Take-off, right leg:				
Angles between				
R arm and forearm	26.5	247.7	91.5	370.2
R arm and trunk	6.3	27.5	4.7	18.2
Trunk and L thigh	27.4	49.5	44.3	38.0
L thigh and leg	46.5	323.4	43.5	145.6
R and L thighs	25.5	136.7	37.6	47.8

[a]Tempo = speed and number of strides over 60 m.
Adapted from Sklad (40).

ized by a small genetic component (5,23). The specific genetic contribution to gait, however, is not known. In an electrophoretic study of 13 protein loci and 10 blood group loci in 600 trotters and 1,227 pacers, the two breeds shared common alleles at 20 of the 23 loci; there were, however, significant differences in allele frequencies at 21 of the 23 loci (5). The results emphasize the need for linkage analysis between the specific loci and gait characteristics.

Motor Development in Individuals With Genetic Anomalies

Advances in molecular biology and clinical genetics provide important insights into the molecular genetic basis of many diseases. In turn, genotypic influences on motor development may be inferred from clinical phenotypes of individuals with inherited anomalies. The behavior of infants and young children born with inherited anomalies is routinely screened, and the screening tests characteristically include developmental scales such as those described earlier. Quite often, it is a developmental delay—for example, a delay in attaining motor development milestones such as head control or sitting alone—that brings the child to a clinic for more detailed examination. The scales include a variety of mental, language, social, fine motor, and gross motor items that are largely schedules of sensory and motor achievements. Some scales use the label psychomotor, while others use the label motor, emphasizing the interrelatedness of early development in a variety of behavioral domains.

Inborn errors of metabolism

Hereditary metabolic disorders may disturb the normal concentrations of important metabolites and in turn compromise the developing brain, peripheral nervous system, skeletal muscles, or other systems, and thus directly affect motor development. The catalog of inborn errors of metabolism is extensive and increasing with advances in molecular genetics (see 39). Several inborn errors of metabolism that occur in childhood and that interfere with motor development are summarized in table 14.3. The list of inborn errors that delay psychomotor development in childhood is more extensive (24). The current state of knowledge of many inborn errors of metabolism is largely at the protein level; however, progress is being made in identifying processes at the level of gene expression. For example, the genetic basis of Lesch-Nyhan disease, an X-linked recessive disorder due to a defect in the enzyme hypoxanthine-guanine phosphoribosyltransferase (HPRT) is well known. The gene encoding for HPRT is located near the end of the long arm of the X chromosome (Xq26-Xq27.2), is 44 kb in length, contains nine exons, and encodes for 1.6-kb mRNA (30). Boys with Lesch-Nyhan syndrome (girls are rarely affected) present severely delayed motor development during the first year of life (42).

Muscular dystrophies

The muscular dystrophies are a clear example of an inherited condition that markedly influences motor development, among other features. They are a group of "genetically determined, progressive, and degenerative primary disorders of muscle" (19, p. 2870). Duchenne muscular dystrophy, the best-known and most serious of the muscular dystrophies, is X-linked. It is determined by a single locus on the X chromosome in the p21 region. Early clinical features include late walking and overall developmental delay. The majority of boys with Duchenne muscular dystrophy are not walking independently by 18 months of age.

Table 14.3
Inborn Errors of Metabolism Interfering With Normal Motor Development

Name	Heredity[a]	Enzyme defect
Carbohydrate metabolism		
Glycogen storage disease, type II (Pompe), juvenile forms	AR	α-1,4-glucosidase
Glycogen storage disease, type III (Cori)	AR	Amylo-1,6-glucosidase
Glycogen storage disease, type IV (Andersen)	AR	Amylo-1,4→1,6-transglucosidase
Lipid metabolism (lipidoses)		
Metachromatic leukodystrophy (sulfatide lipidosis), infantile onset type, juvenile onset type	AR	(Exebroside sulfatase) (Arylsulfatase A)
Juvenile (G_{M1}) ganglioside storage disease	AR	
Oligosaccharide and glycoprotein metabolism (formerly mucolipidoses)		
Farber's lipogranulomatous disease	AR	Acid ceramidase
Disorders of amino acid or organic acid metabolism		
Nonketotic hyperglycinemia	AR	Glycine decarboxylase
Serum carnosinase deficiency	AR	Negatively charged serum carnosinase
Disorders due to malfunction of endocrine glands		
Group of entities caused by defects in homeostasis, synthesis, storage, and utilization of thyroid hormone	AR	Enzyme defects, unknown in most instances
Disorders of pyrine and pyrimidine metabolism		
Lesch-Nyhan disease	XR	Hypoxanthine-guanine phosphoribosyltransferase

[a]AR = autosomal recessive; XR = X-linked recessive.

Adapted from Leroy (24).

Changes in ploidy

Delayed motor development is characteristic of individuals with changes in ploidy, for example, trisomy 21 (Down syndrome) (8), 47,XXY and 47,XYY males (37), 45,X and 47,XXX females (2,36), the fragile X syndrome (22), and mosaic-trisomy 8 (21). The degree of delay varies among conditions and individuals and in some instances is seemingly mild. Late walking is an especially common characteristic, often occurring after 15 months of age. Infants with Down syndrome show a reduction in complex movement sequences related to kicking, which is associated with their later ages at independent walking (43).

The impaired motor development of children with changes in ploidy often persists through childhood and adolescence. It is commonly reflected in poor muscle tone, clumsiness, and fine and gross motor dysfunction. For example, 45,X and 46,XXX girls and 47,XXY and 47,XYY boys show generally poorer gross and fine motor performance on the Bruininks-Oseretsky test of motor proficiency and mild-to-moderate impairment of sensory-motor integration (36,37). Children with Down syndrome also perform significantly below control children on a variety of strength and motor tests (9).

Idiopathic late walking

Lundberg (26) has described a condition of dissociated motor development associated with very late walking (see also 18). It is characterized by marked delay in the gross motor domain without any abnormal neurological signs, but normal fine motor development. Among 65 children who could not walk at 17 months, 79% had dissociated sitting behavior, that is, they sat unsupported without using arms at the expected normal age but were very delayed in sitting actively without help; 51% showed a prewalking progression that involved shuffling (in a sitting position with or without using the arms and hands) in contrast to crawling or creeping. Late walking was idiopathic in 35 (54%), and of these, 13 (37%) presented a family history of shuffling, which would suggest a genetic influence. Subsequent study of six idiopathic late walkers (27,29) indicated reduced muscle fiber size, especially for type II fibers, and lowered concentrations of ATP, phosphocreatine (PCr), and glycogen compared to controls of the same age. These characteristics may be related to physical inactivity. Similar trends occur with atrophy, while training can increase fiber size and concentration of metabolic substrates. Nevertheless, reduced concentrations of stored metabolic substrates or impairment of substrate mobilization, which occur in several metabolic diseases, may have implications for motor development. For example, children with celiac disease also have reduced muscle concentrations of ATP, PCr, and glycogen and are also delayed in motor development; with treatment, levels of the muscle substrates and motor development are similar to control subjects (28).

In contrast to the reduced size of type II fibers in six children with idiopathic late walking, Qazi et al. (35) recently described three boys with a syndrome of extreme motor developmental delay, among other clinical features. Upon biopsy of the vastus lateralis, all three boys had a predominance of type II fibers, which were also considerably larger than type I fibers. Although the etiology of the syndrome is unknown, some of its features are similar to a disorder described as congenital fiber-type disproportion, whose inheritance is also uncertain (19).

Summary

Very little is known about the genetics of motor development. Early motor development shows more similarity among MZ than DZ twins, but the presence of shared environmental effects is also apparent. Data for the development of specific movement patterns beyond walking are lacking, although the pattern of sprint running in children shows more similarity among pairs of MZ twins than among pairs of DZ twins.

In contrast to the lack of data for healthy children, genotypic influences on motor development may be inferred from phenotypes of individuals with inherited anomalies, many of whom present delayed attainment of motor milestones, especially postural control and independent walking.

References

1. Bayley, N. Manual for the Bayley Scales of Infant Development. Berkeley, CA: Psychological Corp.; 1969.
2. Bender, B.; Puck, M.; Salbenblatt, J.; Robinson, A. Cognitive development of unselected girls with complete and partial monosomy. Pediatrics. 73:175-82; 1984.
3. Bossik, L.J. K voprosu o roli nasledstvennosti i sredi v fiziologii i patologii detskovo vovrasta. Trudy Medytsinsk of Biologi Institut. 3:33; 1934. As cited by R. Kovar. Human variation in motor abilities and its genetic analysis. Prague: Charles University 1981, 73-74.
4. Cook, C.F.; Broadhead, G.D. Motor performance of pre-school twins and singletons. Phys. Educ. 41:16-20; 1984.
5. Cothran, E.G.; MacCluer, J.W.; Weitkamp, L.R.; Bailey, E. Genetic differentiation associated with gait within American Standardbred horses. Anim. Genet. 18:285-96; 1987.

6. Dales, R.J. Motor and language development of twins during the first three years. J. Genet. Psychol. 114:263-71; 1969.
7. Eaton, W.O.; Saudino, K.J. Prenatal activity level as a temperament dimension? Individual differences and developmental functions in fetal movement. Inf. Behav. Develop. 15:57-70; 1992.
8. Epstein, C.H. Down syndrome (trisomy 21). In: Scriver, C.R.; Beaudet, A.L.; Sly, W.S.; Valle, D., eds. The metabolic basis of inherited disease. 6th ed. New York: McGraw-Hill; 1989:291-326.
9. Francis, R.J.; Rarick, G.L. Motor characteristics of the mentally retarded. Cooperative research monograph no. 1. Washington, DC: U.S. Department of Health, Education, and Welfare; 1960.
10. Frankenburg, W.K.; Dodds, J.B. The Denver Developmental Screening Test. J. Pediatr. 71:181-91; 1967.
11. Freedman, D.G.; Keller, B. Inheritance of behavior in infants. Science. 140:196-98; 1963.
12. Fujikura, T.; Froehlich, L.A. Mental and motor development in monozygotic co-twins with dissimilar birth weights. Pediatrics. 53:884-89; 1974.
13. Gesell, A. The ontogenesis of infant behavior. In: Carmichael, L., ed. Manual of child psychology. New York: Wiley; 1954:335-73.
14. Gesell, A.; Thompson, H. Learning and growth in identical infant twins: An experimental study by the method of co-twin controls. Genet. Psychol. Monogr. 6:1-120; 1929.
15. Gifford, S.; Murawski, B.J.; Brazelton, T.B.; Young, G.C. Differences in individual development within a pair of identical twins. Int. J. Psychoanal. 47:261-68; 1966.
16. Goya, T.; Amano, Y.; Hoshikawa, T.; Matsui, H. Longitudinal study on selected sports performance related with the physical growth and development in twins. XIIIth International Congress on Biomechanics book of abstracts. Perth: University of Western Australia; 1991:139-41.
17. Goya, T.; Amano, Y.; Hoshikawa, T.; Matsui, H. Longitudinal study on the variation and development of selected sports performance in twins: Case study for one pair of female monozygous (MZ) and dizygous (DZ) twins. Sport Sci. 14:151-68; 1993.
18. Haidvogl, M. Dissociation of maturation: A distinct syndrome of delayed motor development. Dev. Med. Child Neurol. 21:52-57; 1979.
19. Harper, P.S. The muscular dystrophies. In: Scriver, C.R.; Beaudet, A.L.; Sly, W.S.; Valle, D., eds. The metabolic basis of inherited disease, II. 6th ed. New York: McGraw-Hill; 1989:2869-2902.
20. Haubenstricker, J.; Seefeldt, V. Acquisition of motor skills during childhood. In: Seefeldt, V., ed. Physical activity and well-being. Reston, VA: American Alliance for Health, Physical Education, Recreation, and Dance; 1986:41-102.
21. Kautza, M.; Schwanitz, G.; Hosenfeld, D.; Grote, W.; Hunze-Fuhrmann, D.; Brandt, I.; Schleiermacher, E.; Gellissen, K.; Bopp, E.; Zerres, K.

Psychomotor development of three children with mosaic-trisomy 8 and literature review. Acta Med. Auxol. 23:215-26; 1991.

22. Largo, R.H.; Schinzel, A. Developmental and behavioral disturbances in 13 boys with fragile X syndrome. Eur. J. Pediatr. 143:269-75; 1985.
23. Laskey, J.F. Genetics of livestock improvement. Englewood Cliffs, NJ: Prentice Hall; 1978.
24. Leroy, J.G. Heredity, development, and behavior. In: Levine, M.D.; Carey, W.B.; Crocker, A.C.; Gross, R.T., eds. Developmental behavioral pediatrics. Philadelphia: Saunders; 1983:315-45.
25. Livshits, G. Environmental and sibling resemblance components of variance and covariance in traits of early child development. Anthropol. Anz. 46:41-50; 1988.
26. Lundberg, A.E. Dissociated motor development: Developmental patterns, clinical characteristics, causal factors and outcome, with special reference to late walking children. Neuropediatrics. 10:161-82; 1979.
27. Lundberg, A.E. Normal and delayed walking age: A clinical and muscle morphological and metabolic study. In: Berg, K.; Eriksson, B.O., eds. Children and exercise IX. Baltimore: University Park Press; 1980:23-31.
28. Lundberg, A.E.; Eriksson, B.O.; Jansson, G. Muscle abnormalities in coeliac disease: Studies on gross motor development and muscle fibre composition, size and metabolic substrates. Eur. J. Pediatr. 130:93-103; 1979.
29. Lundberg, A.E.; Eriksson, B.O.; Mellgren, G. Metabolic substrates, muscle fibre composition and fibre size in late walking and normal children. Eur. J. Pediatr. 130:79-92; 1979.
30. Martin, J.B. Molecular genetics: Applications to the clinical neurosciences. Science. 238:765-72; 1987.
31. McGraw, M.B. Growth: A study of Johnny and Jimmy. New York: Appleton-Century; 1935.
32. McGraw, M.B. Later development of children specially trained during infancy: Johnny and Jimmy at school age. Child Dev. 10:1-19; 1939.
33. Plomin, E.; DeFries, J.C. Origins of individual differences in infancy: The Colorado Adoption Project. New York: Academic Press; 1985.
34. Prechtl, H.R.F. Prenatal motor development. In: Wade, M.G.; Whiting, H.T.A., eds. Motor development in children: Aspects of coordination and control. Dordrecht, Netherlands: Martinus Nijhoff; 1986:53-64.
35. Qazi, Q.H.; Markovizoa, D.; Rao, C.; Sheikh, T.; Beller, E.; Kula, R. A syndrome of hypotonia, psychomotor retardation, seizures, delayed and dysharmonic skeletal maturation, and congenital fibre type disproportion. J. Med. Genet. 31:405-9; 1994.
36. Salbenblatt, J.A.; Meyers, D.C.; Bender, B.G.; Linden, M.G.; Robinson, A. Gross and fine motor development in 45,X and 47,XXX girls. Pediatrics. 84:678-82; 1989.

37. Salbenblatt, J.A.; Meyers, D.C.; Bender, B.G.; Linden, M.G.; Robinson, A. Gross and fine motor development in 47,XXY and 47,XYY males. Pediatrics. 80:240-44; 1987.
38. Saudino, K.J.; Eaton, W.P. Infant temperament and genetics: An objective twin study of motor activity level. Child Dev. 62:1167-74; 1991.
39. Scriver, C.R.; Beaudet, A.L.; Sly, W.S.; Valle, D. The metabolic basis of inherited disease, Vols. I and II. 6th ed. New York: McGraw-Hill; 1989.
40. Sklad, M. Similarity of movement in twins. Wychowanie Fizycznie i Sport. 16:119-41; 1972.
41. Smith, N.W. Twin studies and heritability. Hum. Dev. 19:65-68; 1976.
42. Stout, J.T.; Caskey, C.T. Hypoxanthine phosphoribosyltransferase deficiency: The Lesch-Nyhan syndrome and gouty arthritis. In: Scriver, C.R.; Beaudet, A.L.; Sly, W.S.; Valle, D., eds. The metabolic basis of inherited disease. 6th ed. New York: McGraw-Hill; 1989:1007-28.
43. Ulrich, B.D.; Ulrich, D.A. Spontaneous leg movements of infants with Down syndrome and nondisabled infants. Child Dev. 66:1844-55; 1995.
44. Verschuer, O. Studien an 102 eineligen und 45 gleichgeschlechtlichten zweieligen Zwillings- und an 2 Drillingspaaren. Ergebnisse der Inneren Medizin und Kinderheilkunde. 31:35-120; 1927.
45. Wilson, R.S.; Harpring, E.B. Mental and motor development in infant twins. Dev. Psychol. 7:277-87; 1972.

CHAPTER 15

Genetics and Performance

Physical performance is a comprehensive term that includes activities carried out under a variety of conditions and contexts, for example, the play and games of children and youths, highly organized athletic competitions, and activities of daily living. There are many performance phenotypes: strength, endurance, flexibility, motor abilities, perceptual-motor characteristics, and aerobic and anaerobic outcomes, all of which can be measured under specified and standardized conditions. Application of these phenotypes to tasks of daily living, play, games, sports, and leisure pursuits, among others, is highly variable among individuals.

Life Span and Gender Variability

Performance in strength and motor tasks improves, on average, during childhood and adolescence, and boys generally perform better than girls on tasks requiring strength, speed, and power. Lower-back flexibility (sit-and-reach) decreases, on average, during childhood into adolescence and then increases. Boys also show adolescent growth spurts in strength, speed, power, and flexibility. With the exception of strength, performances of girls do not show clearly defined adolescent spurts, though longitudinal data are very limited. Performances of males continue to improve until the mid- or late 20s, while those of females are more variable. In both sexes, performances on strength, motor, and flexibility tasks tend to decline with advancing age beginning in the 30s, although there is variation by specific tasks (2,3,49,58,86,96). Several examples of differences in strength, motor performance, and flexibility through the life span are illustrated in figures 15.1*a* through 15.3*d*.

Changes in perceptual-motor characteristics across the life span generally follow a pattern similar to that for strength and motor tasks. Performances improve during childhood and adolescence. Longitudinal data are not available during adolescence, so the presence of adolescent spurts is not certain. Perceptual-motor performances tend to decline,

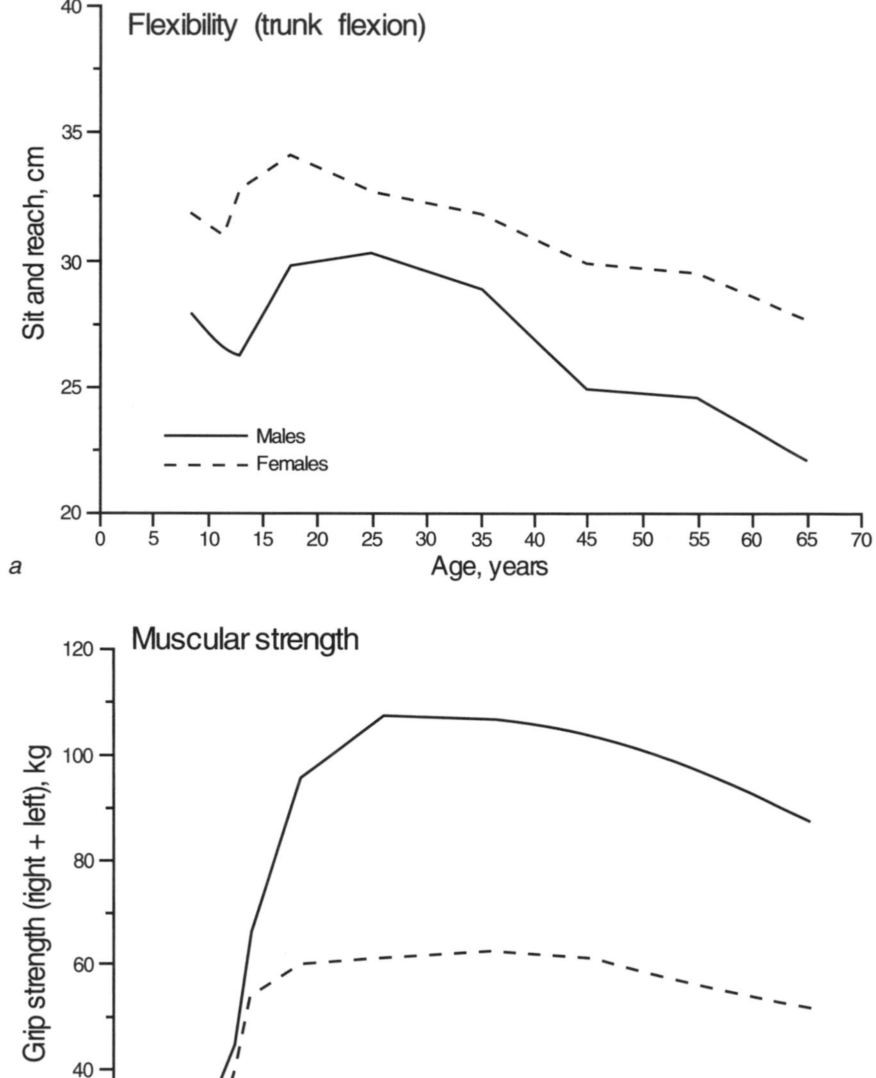

(continued)

Figure 15.1 Changes in (*a*) flexibility, (*b*) muscular strength, and (*c* and *d*) muscular endurance with age in Canadians 7 to 69 years old. Data are means from the Canada Fitness Survey (16).

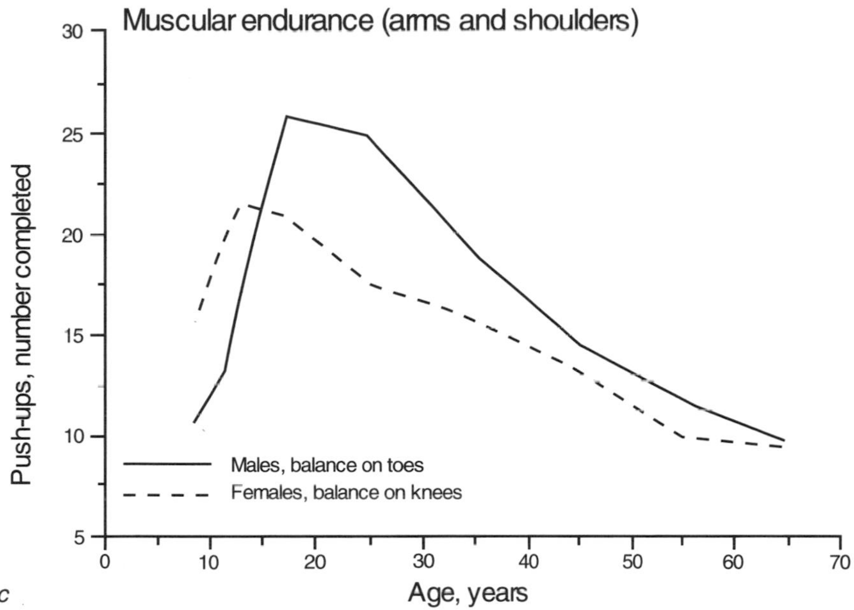

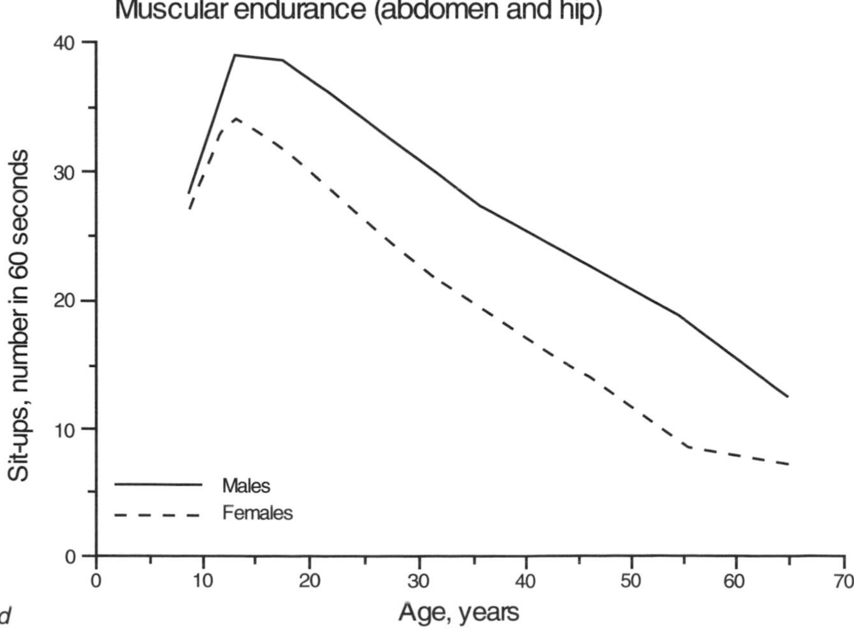

Figure 15.1 *(continued)*

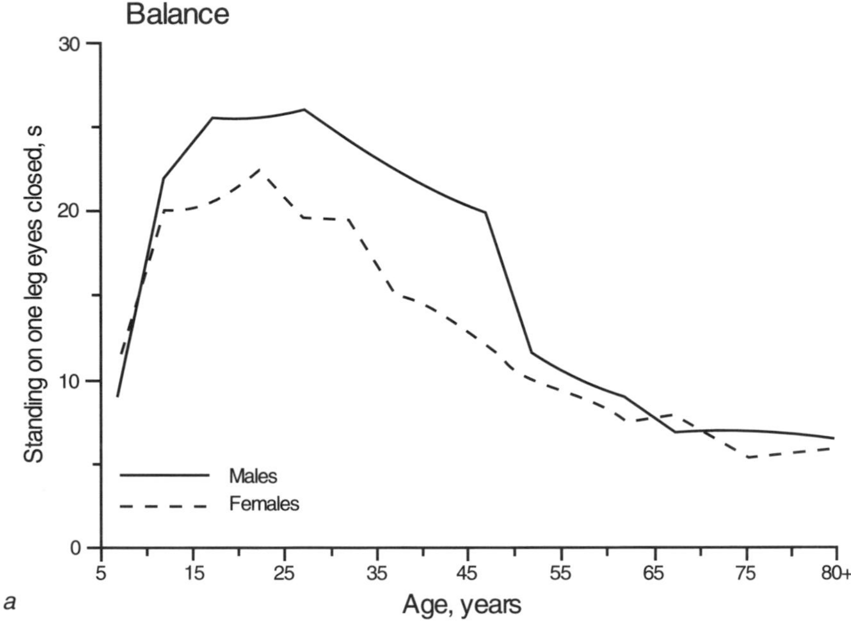

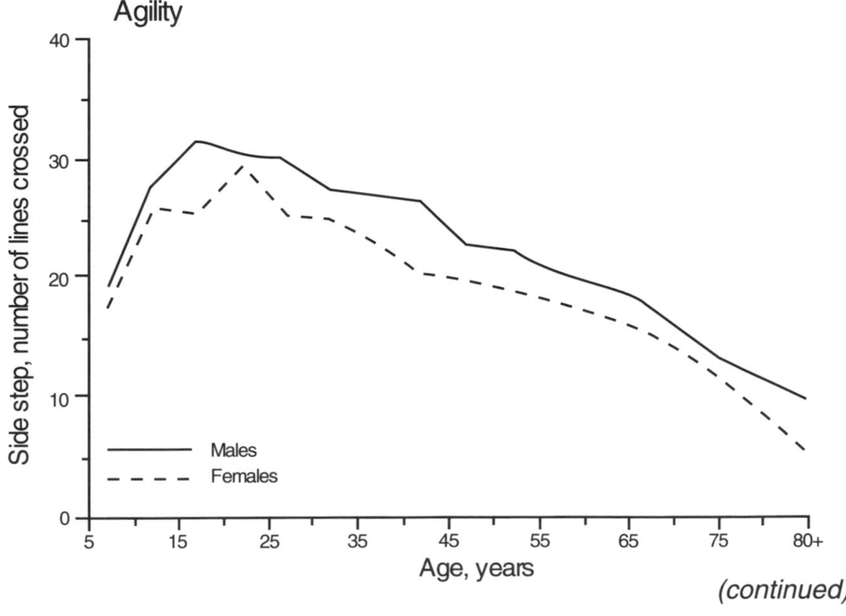

(continued)

Figure 15.2 Changes in (*a*) balance, (*b*) agility, (*c*) power, and (*d*) strength with age in Taiwanese 9 to 80+ years old. Data are redrawn after Kuo (49).

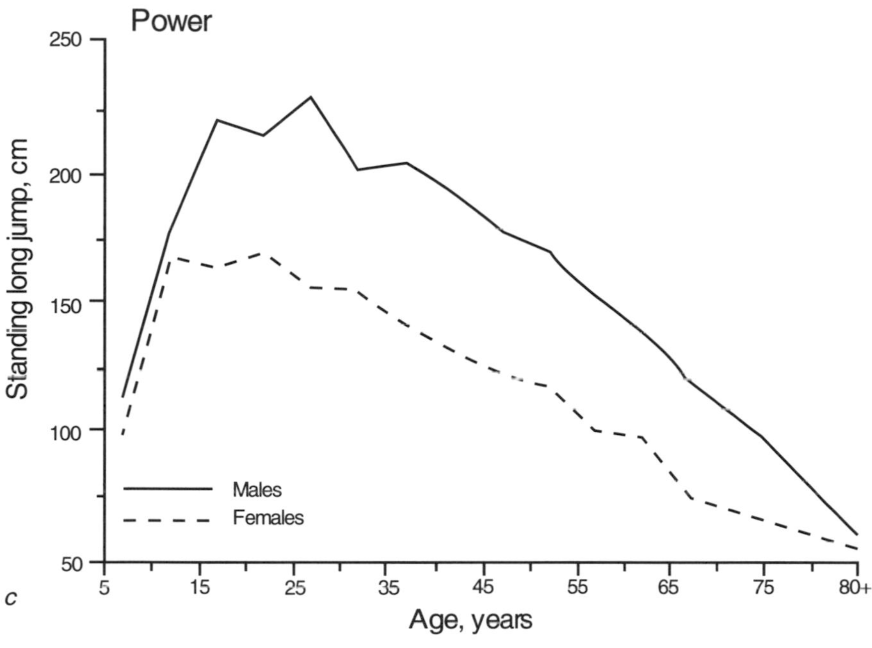

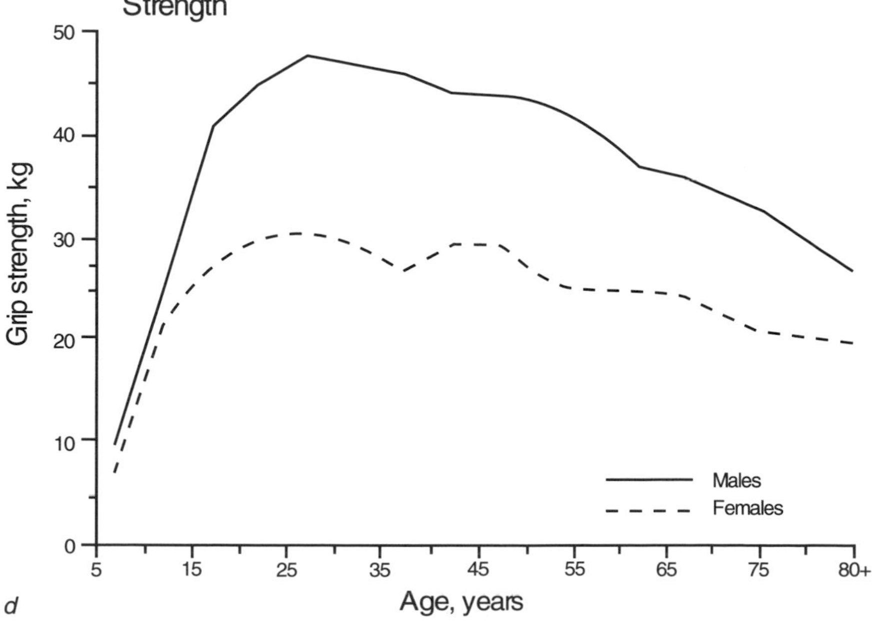

Figure 15.2 *(continued)*

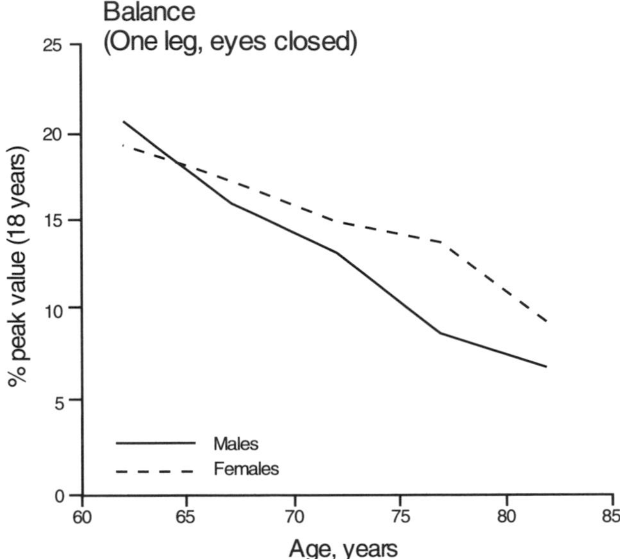

a

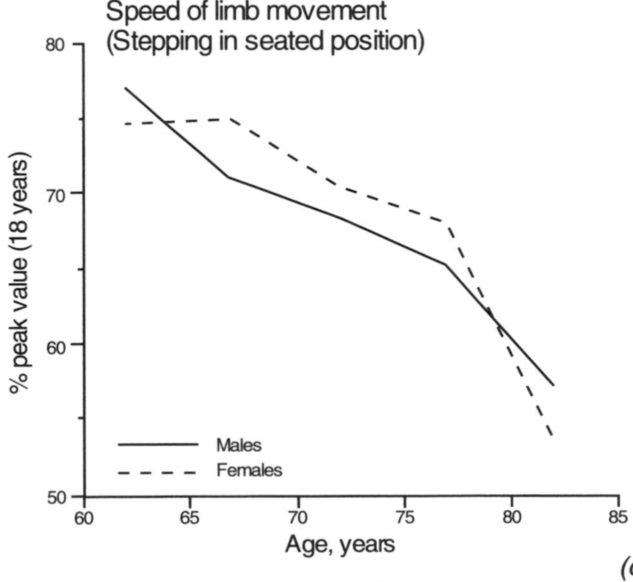

b

(continued)

Figure 15.3 Relative changes in (*a*) balance, (*b*) speed of limb movement, (*c*) power, and (*d*) strength with age in Japanese 60 to 80+ years old. Data are percentages of peak values in late adolescence/young adulthood from Kimura et al. (42).

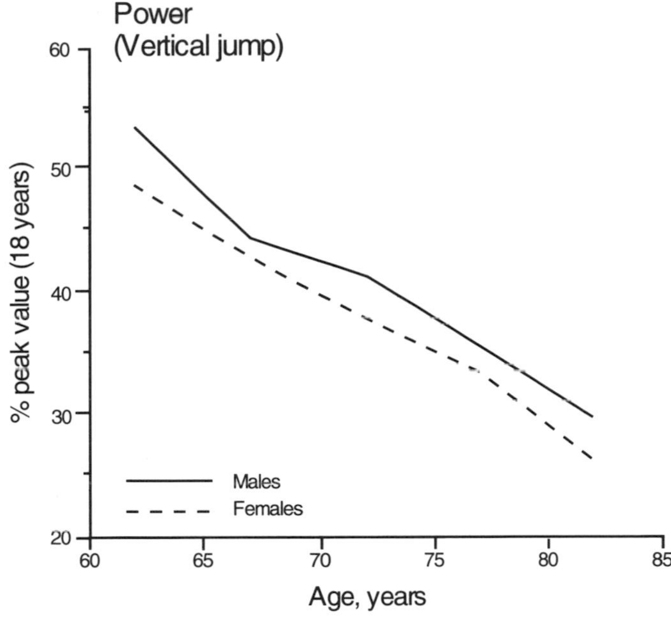

c

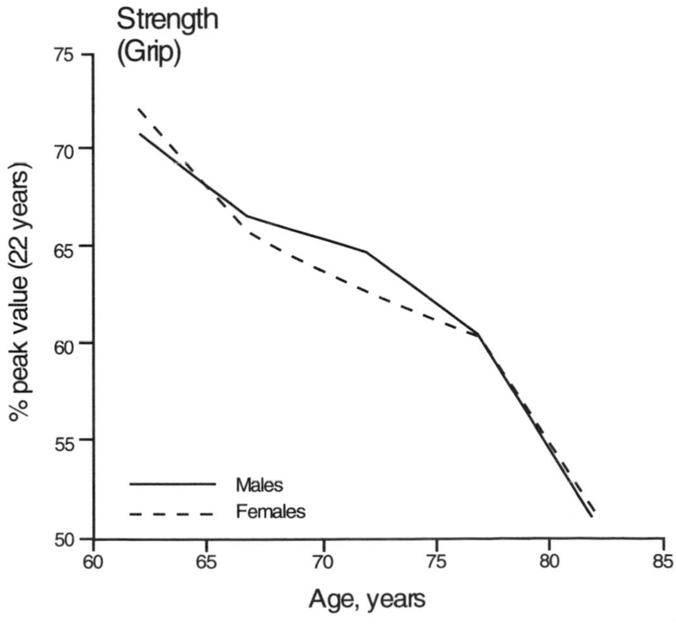

d

Figure 15.3 *(continued)*

on average, beginning in the 30s. Greater performance decrements occur in more complex coordination tasks, especially those requiring rapid changes in a series of different actions (3,70,86,97). Changes in response speed to a discrimination reaction time apparatus with age are shown in figure 15.4. The pattern of age- and sex-associated variation is similar to that for strength, balance, agility, and power shown in figures 15.1*a* through 15.2*d*. The relative decline in a variety of coordination tests with one hand and functional tasks requiring two hands between 20 and 80 years of age is shown in figure 15.5. The age-associated decrease varies among tasks (75). Some of the decline in functional tasks with age is related to age-associated changes in muscle strength, and a decrease in the estimated number of motor units appears to be a major factor in the loss of contractile strength with age (26).

Submaximal exercise capacity and maximal aerobic power show similar age trends and sex differences during childhood and adolescence. Both sexes show clearly defined adolescent spurts in absolute

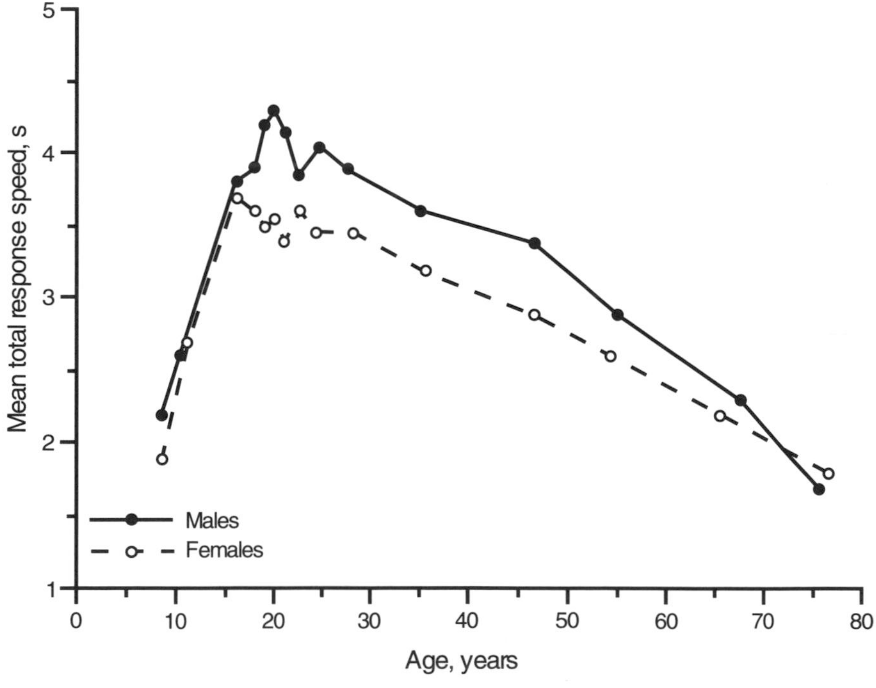

Figure 15.4 Mean total response speed on a discrimination reaction time apparatus with age. Each point is based on data for 20 subjects averaged over an entire practice period. Redrawn from Noble et al. (70).

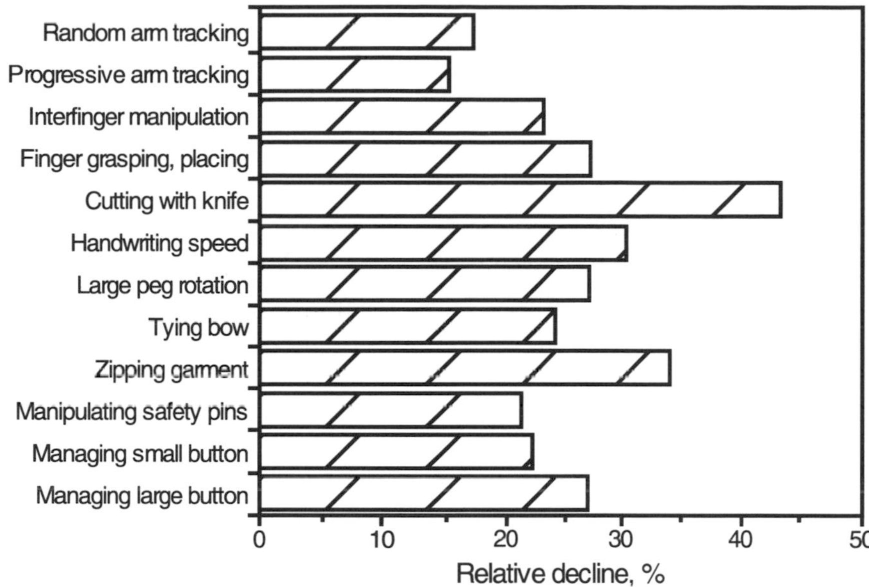

Figure 15.5 Relative decline in performance on several coordination tasks requiring one hand and functional tasks requiring two hands between 20 and 80 years of age. Drawn from data reported by Potvin et al. (75).

maximal aerobic power and submaximal power output (58). Absolute submaximal performance and maximal aerobic power increase to the midteens in females and into the mid-20s in males; subsequently, average performances decline with age (figure 15.6, *a* and *b*). Relative aerobic power (expressed per unit body mass) tends to decline with age in girls and to remain stable or slightly increase with age in boys during childhood and adolescence. During adulthood, relative aerobic power declines with age in both sexes.

Habitual physical activity is a factor that may influence performance. The data for children and youths are, however, variable. Cross-sectional surveys that incorporate strength and motor tasks indicate better performance levels in active children, while longitudinal comparisons of active and nonactive children are equivocal. In contrast, both cross-sectional and longitudinal data indicate better aerobic power in active children (55,56). During adulthood and aging, the rate of decline with age is affected by habitual physical activity and training. Although performances decline, on average, with age, the decreases in a variety of functions tend to be slower in more active individuals (3,28,40,79,87).

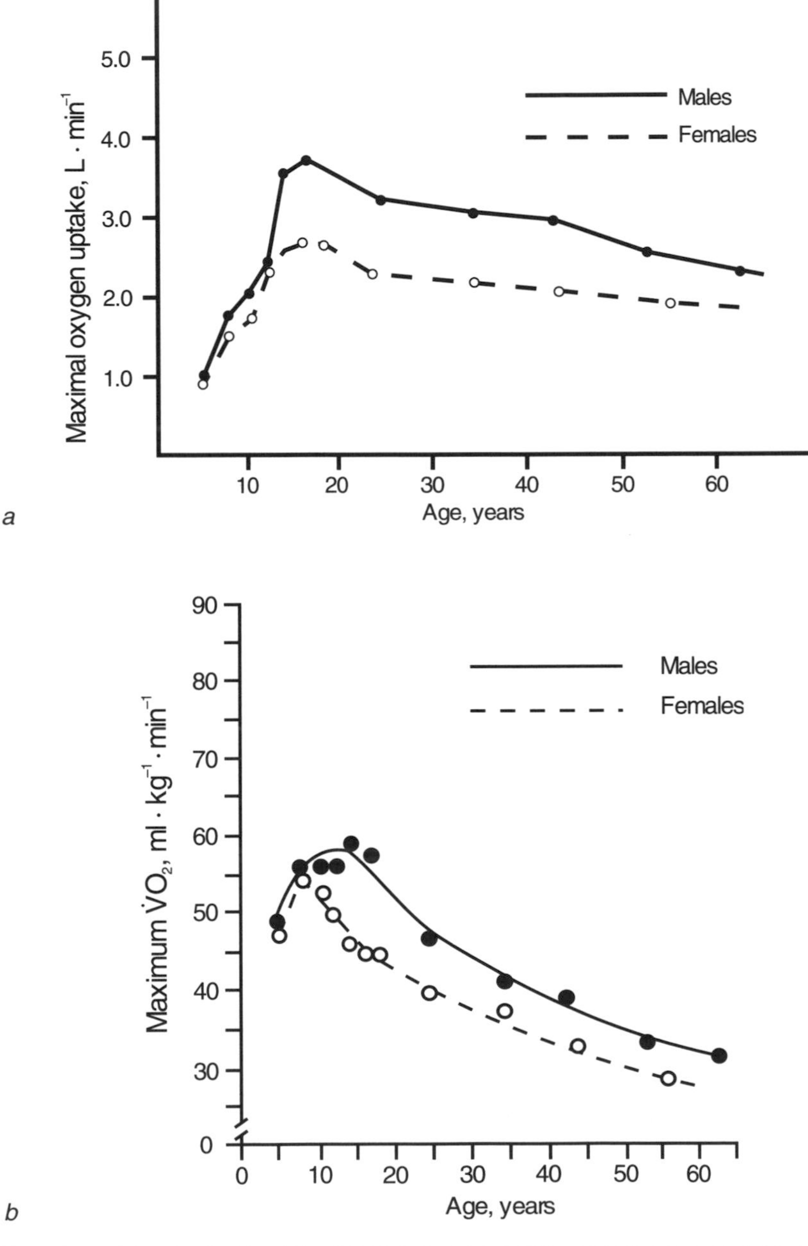

Figure 15.6 Changes in (*a*) absolute and (*b*) relative aerobic power with age in Swedish subjects 1 to 65 years old. Redrawn after Åstrand and Rodahl (1).

Overview of Genetic Studies

Studies of the genotypic contribution to performance phenotypes are reasonably extensive. However, they are only of the genetic epidemiology type and are limited largely to comparisons of twins, siblings, and parent-offspring pairs. Hence, they have limited utility in defining the genetic contribution to performance phenotypes. More comprehensive studies utilizing different multivariate models to estimate additive and nonadditive genetic and shared and unshared environmental sources of variation are not extensive, and the results are not entirely consistent. Moreover, there are only a few papers that have considered the contribution of specific genes or markers to performance phenotypes and their responses to training.

A variety of all-out tests are used to measure the components of motor fitness, for example, dashes, jumps, and endurance runs. Many of these tests are not suitable for older adult samples. Hence, studies of the genetic contribution to motor fitness are limited largely to samples of school-age children and youths and young adults. In contrast, corresponding data for aerobic tasks are largely derived from adult samples.

Muscular Strength and Endurance

Strength is an expression of muscular force, or the capacity of the individual to exert force against some external resistance. Two kinds of strength measurements are commonly used: (1) static or isometric, in which force is exerted without any change in muscle length, and (2) explosive, in which maximal force is released in the shortest possible time. The latter is often referred to as muscular power and is frequently measured by tests of jumping (vertical jump, standing long jump) and throwing (throws for distance).

Muscular endurance refers to the ability to exert strength over a period of time. It is often measured by the maximum number of push-ups performed consecutively without a time limit, the maximum number of sit-ups performed in one minute, or the bent (flexed) arm hang (the duration that the chin can be held above a pull-up bar with the body suspended off the ground). Depending upon the instructions for these tests, they are sometimes labeled *dynamic strength tests* since they require repeated shortening (concentric contraction) and lengthening (eccentric contraction) of muscle groups.

This section considers measures of static strength (dynamometric tests) and muscular endurance and/or dynamic strength (push-ups, sit-ups, flexed arm hang). Indicators of muscular power are discussed in the section on motor performance.

Three early Japanese studies of primary- through high-school-age twins in the late 1940s and 1950s expressed strength differences within twin pairs as relative deviation scores (39,41,67). Tasks included dynamometric measures of grip, shoulder, and back strength; push-ups; and pull-ups. Although results varied somewhat among studies, the general trends were consistent with estimates of heritability in samples of European and American twins and siblings to be described.

Twin and sibling studies

Studies of twins (46,47,84,95) and siblings (59,89) from childhood through young adulthood are reported as correlations and estimated heritabilities. The latter vary among studies and strength tests (0 to 0.83), are often higher for boys than girls, and are generally higher in studies of twins than of siblings (54,57). Heritability estimates for strength of specific muscle groups, strength scores summed over several muscle groups, or strength expressed per unit of body weight tend to be similar. Age variation within most samples is not ordinarily considered. The lower estimated heritabilities for females suggest environmental sources of variation.

Two twin studies merit separate discussion. Komi et al. (45) used several measures of isometric, concentric, and eccentric muscle force in small samples of male and female MZ (15 pairs) and DZ (14 pairs) twins 10 to 14 years of age. None of the intrapair variances between MZ and DZ twins differed significantly, and female MZ twins showed as much variability in muscular forces as female DZ twins. Among male twins 18 to 19 years of age, Engstrom and Fischbein (27) used a composite strength score (grip, arm flexion, and knee extension, with each differently weighted) expressed per unit stature. Intrapair correlations were about 0.83 and 0.47 for 39 MZ and 55 DZ pairs, respectively (values interpolated from a graph). However, in a subset of twins who were concordant for the amount of leisure-time physical activity, the intraclass correlation for 22 MZ twin pairs did not change, while that for 21 DZ pairs decreased to about 0.33. The results thus indicate a potential role of habitual physical activity in influencing within–twin pair similarities and the need to control for activity in attempting to quantify the genotypic contribution to variance in strength and probably performance.

Relatively large-scale parent-offspring studies of strength in Poland are reported only as correlations: parents and offspring 7 to 39 years old from 570 three-generation rural families (104), and parents and offspring 3 to 42 years old from 347 primarily one-generation urban families (89,91). Parent-offspring correlations were variable, ranging from –0.24 to +0.62. A generally similar pattern of correlations was reported for grip and back strength between Czechoslovak parents and their 16- to 17-year-old sons (48).

Strength data from the Tecumseh (Michigan) community health study were analyzed in a different manner (69). The aggregation of siblings into tertiles by age and sex was noted for several measurements of strength: sum of right and left grip, arm strength using both arms simultaneously, and an index based on the two measurements corrected for body size and fatness. The results indicated a significant degree of sibling similarity in strength, but the magnitude of the resemblance was not quantified. Parent-offspring resemblances were evaluated in a similar manner. Resemblances were significant, and there was no age effect, that is, resemblances between parents and older offspring (16-39 years) were similar to those between parents and younger offspring (10-15 years). However, female offspring tended to resemble their parents more than male offspring.

The preceding studies are generally limited to one type of biological relative: twins only, parents and offspring only, and so on. Studies that include a variety of biological and nonbiological relatives are few. Correlations between different kinds of relatives for muscular strength and endurance are summarized in table 15.1. The highest correlations are obtained for MZ twins. However, the presence of significant correlations in nongenetically related individuals (foster parents–adopted child) suggest that environmental factors and lifestyle shared by family members living within the same household can contribute to the observed familial resemblance (71,72). Mother-child and father-child correlations for muscular strength and endurance did not differ (73).

Spouses show significant resemblance in a variety of muscular strength and endurance tasks (table 15.2). Most correlations range between 0.01 and 0.30. Although generally low, spouse similarities in muscular strength and endurance emphasize the role of assortative mating for the traits or some covariate and/or the effects of common lifestyle (e.g., patterns of habitual physical activity or exercise habits).

Motor Performance

Tests of sprinting, jumping, and throwing are perhaps the most commonly used in assessing gross motor performance. The standing long jump, vertical jump, and distance throws are measures of explosive strength (power) and coordination. Dashes or sprints are primarily anaerobic. Dashes, jumps, and throws are generally reliable and do not require elaborate equipment. Estimated heritabilities from twins

Table 15.1
Correlations in Muscular Strength and Endurance in Relatives by Descent and Adoption

Relatives	Strength (quadriceps isometric)[a]	Endurance (sit-ups)[a]	Strength (grip/weight)[b]	Strength (push-ups)[b]	Endurance (sit-ups)[b]
Parent-child	0.32	0.23	0.20	0.25	0.24
Father-child	0.31	0.22			
Mother child	0.33	0.25			
Sibs	0.28	0.37	0.29	0.38	0.34
DZ twins	0.39	0.19			
MZ twins	0.76	0.71			
Uncle/aunt–nephew/niece	–0.12	–0.07	–0.02	0.51	0.54
First-degree cousins	0.02	0.00			
Unrelated sibs	0.08	0.03			
Foster parents–adopted child	0.12	0.15			

[a]Pérusse et al. (72, 73), French Canadian families in Quebec City area.
[b]Pérusse et al. (71), Canada Fitness Survey.

Table 15.2
Spouse Correlations for Measures of Muscular Strength and Endurance

Source, *n* pairs	Test	*r*
Kovar (48), 60	Grip	0.26
	Back	0.26
Szopa (89), 347 families	Grip, R/L	0.15/0.26
	Grip/weight	0.23
	Arm	0.17
	Arm/weight	0.17
Devor and Crawford (24), 53	Dominant grip	0.01
Pérusse et al. (72, 73), 295-319	Quadriceps, isometric	0.22
	Sit-ups	0.30
Pérusse et al. (71)[a], 1,590-3,183	Grip/weight	0.08
	Sit-ups	0.28
	Pull-ups	0.15
Maes (52)[a], 84-97	Arm pull	0.03
	Sit-ups	0.32
	Leg lifts	0.26
	Flexed arm hang	0.17

[a]Maximum likelihood estimates.

(46,47,84) and siblings (59,89) 3 through 42 years of age range from 0.14 to 0.91. Sex differences in heritabilities are not as apparent as for strength tests. Heritabilities for the dashes vary somewhat with distance: 20 m, 0.83; 30 m, 0.62, 0.81; 60 m, 0.45, 0.72, 0.80, 0.91 (57). These heritabilities do not support the conclusion of that heritability is highest for shorter dashes and decreases with length of the dash (105).

Data from the three Japanese twin studies, described earlier for strength, are consistent with the preceding findings and suggest a moderate genetic influence. MZ twins generally show smaller within-pair deviations than DZ twins in a dash, two jumps, and a throw (39,41,67).

Parent-offspring data for explosive strength tasks are quite limited, probably reflecting the task demands and age-associated decrements in performance on such tasks during adulthood (see figures 15.1*a*-15.3*d*). Parent-offspring correlations for the vertical jump in a study of an urban Polish sample ranged from 0.17 to 0.54 (89,91). Midparent-offspring correlations were higher than parent-specific correlations in each of three age groups: 3-10, 11-17, and 18-42 years. However, in the oldest group, sons were more similar to their fathers (0.46), while daughters were more similar to their mothers (0.48). In addition, vertical jump performance showed significant spouse resemblance (0.35) in this sample, which indicates positive assortative mating for the trait or a covariate and/or the effects of common lifestyle.

An interesting study of parent-offspring similarity in motor performance compared the performances of 24 college-age men with that of their fathers when the latter were of college age 34 years earlier (21). Father-son correlations were 0.86 for the running long jump and 0.59 for the 100-yard dash. Hence, fathers and sons attained reasonably similar performances in these speed and power tasks when they were about the same age in young adulthood.

Balance

Balance is a skill that requires a combination of gross and fine motor control in the maintenance of equilibrium in static and dynamic tasks. Heritability estimates for various balance tests in samples of twins 8 to 18 years range from 0.27 to 0.86 (84,94,101,102). The estimates for a ladder climb and stabilometer (dynamic balance) are lower than those for a rail balance (static balance). Parent-offspring correlations for a beam-walking task are low (–0.01 to +0.21) and similar to those for a timed turning-balance test (0.10 to 0.23) (104). Clearly, the genetic or familial contribution to balance tasks is quite variable and in most instances quite low. Spouse resemblance data for balance tasks are not extensive. Spouses do not resemble each other in a beam-walking task (–0.05) but show some similarity in a timed turning-balance test (0.25) (103).

Psychomotor skills

The movement tasks considered in this section are heterogeneous. Some require rapid movements of the limbs, while others place a premium on fine motor control, that is, they require precision of movement, often with speed. As a group, they are often labeled *psychomotor*. Perceptual-motor characteristics, such as spatial abilities, perceptual speed, and others, are often indicated as important in skillful performance (100), although their relationship to motor skill has not been thoroughly specified.

Speed of arm and leg tapping show moderate to moderately high heritabilities, 0.62 to 0.90, in a sample of twins 8 to 15 years of age (84). Corresponding estimates for speed and accuracy of arm movement and manipulative dexterity in the 8- to 15-year-old twins and maze tracing in male twins 11 to 25 years old are lower, 0.13 to 0.64 (47,84). The data suggest slightly lower heritabilities for female twins in all tasks except speed and accuracy of arm movement (84).

Estimated heritabilities for fine motor skills such as mirror drawing, tweezer dexterity, hand steadiness, card sorting, and rotary pursuit with the preferred hand are of similar magnitude in same-sex twins of high school age, 0.37 to 0.71. Corresponding estimates for the nonpreferred hand are more variable, 0.05 to 0.71, and generally lower for most of the tasks (94). Data for Japanese twins primary- through high-school-age are consistent with the preceding findings but suggest age-associated variation in similarities for tapping speed, manual dexterity, and finger dexterity within pairs of twins (41).

Sibling similarities are also apparent in manual dexterity tasks, with intraclass correlations of 0.23 to 0.44 for tapping and 0.27 to 0.41 for a dot-filling task (17). The sibling similarities persisted in the dot-filling task when laterality was controlled, but not in the tapping task. Data for parents and their 16- to 17-year-old sons indicate higher mother-son correlations than father-son correlations for tapping speed (0.26 vs. 0.12) and errors in tracing a maze (0.24 vs. 0.10), but no difference in father-son and mother-son correlations for time to trace a maze (0.24 vs. 0.22) (48).

Reaction time is an individual characteristic that is often indicated as an important factor in some motor performances. Results of studies with school-age twins, however, yield variable heritability estimates for reaction time to a light stimulus, 0.22 (94) and 0.55 (84). In contrast, data for 16-year-old twins indicate a high heritability of 0.77 for peripheral nerve conduction velocity (78).

Parent-offspring correlations for several perceptual-motor tasks are quite variable in two Polish samples, ranging from –0.10 to +0.75 in a rural sample (104) and from –0.01 to +0.41 in an urban sample (90). These analyses, however, are limited to correlations, and age was not controlled.

Correlations between different kinds of relatives for reaction time and movement time in two studies are summarized in table 15.3. In the more extensive French Canadian sample, the highest correlations are obtained for MZ twins (0.38, 0.39). However, the presence of significant correlations in nongenetically related individuals for movement time (foster parents–adopted child, 0.17; unrelated siblings, 0.28; spouses, 0.15) suggests a role in the observed familial resemblance in this task for environmental factors and lifestyle shared by family members living within the same household. The genetic influence may be stronger for reaction time, since the correlations of genetically unrelated individuals approximate zero (73). The Mennonite data are consistent for siblings but show greater resemblance between uncles/aunts and nephews/nieces than between parents and offspring in three of the four tasks.

Table 15.3
Correlations for Several Psychomotor Tasks in Different Kinds of Relatives in Two Studies

	MENNONITES[a]				FRENCH CANADIANS[b]	
Relatives	**Reaction time**	**Movement time**	**Eye-hand coordination**	**Hand steadiness**	**Reaction time**	**Movement time**
Parent-child	0.13	0.09	0.04	0.05	0.13	0.14
Sibs	0.37	0.33	0.32	0.31	0.23	0.24
DZ twins					0.19	0.15
MZ twins					0.38	0.39
Uncle/aunt–nephew/niece	0.20	0.18	0.23	–0.10	0.06	0.02
First-degree cousins					0.03	0.20
Grandparent-grandchild	0.00	–0.22	0.12	–0.10	—	—
Unrelated sibs					0.11	0.28
Foster parents–adopted child					0.03	0.17
Spouses					0.01	0.15

[a]Devor and Crawford (24).
[b]Pérusse et al. (73).

Spouse correlations for several psychomotor tasks are variable. Among Czechoslovak spouses, correlations are 0.07 and 0.44 for maze tracing and 0.25 for tapping speed (46). Corresponding correlations for reaction time and movement time in French Canadian spouses are 0.01 and 0.15, respectively (72) and in Mennonite spouses are –0.10 for hand steadiness, –0.26 for hand-eye coordination, 0.10 for reaction time, and 0.27 for movement time (24). In contrast to these generally low spouse correlations, corresponding correlations for a rural Polish population are high, reaching 0.64 and 0.52 for reaction time measures (103).

Individual differences in perceptual-motor characteristics, such as spatial abilities, perceptual speed, perception of direction, intersensory integration, and so on, have a significant unspecified genetic component (62,80,93,94). The role of such perceptual characteristics in skillful performance, though often assumed, needs further study.

Flexibility

Flexibility, or suppleness of movement at a joint, tends to be joint specific. Flexibility of the lower back and upper thighs as measured in the sit-and-reach test is indicated as a component of health-related fitness. Note, however, flexibility may be more of a morphological characteristic (i.e., a feature of connective tissue) rather than of performance- or health-related fitness.

Flexibility data for biological relatives are not extensive. Estimated heritability for lower-back flexibility in a sample of male twins 11 to 15 years old was 0.69 (47), while heritabilities in a combined sample of male and female twins 12 to 17 years old were 0.84, 0.70, and 0.91 for trunk, hip, and shoulder flexibility, respectively (46). In contrast, estimated heritability of the sit-and-reach in Indian twins of both sexes 10 to 27 years of age was only 0.18; controlling for age and several anthropometric indicators of body size raised the estimated heritability to 0.50 (19).

Correlations for lower-back flexibility in biological siblings and parent-offspring pairs were, respectively, 0.43 and 0.29 in a Mennonite community (24) and 0.36 and 0.26 in a nationally representative sample of the Canadian population (71). Interestingly, among the Mennonite community, grandparent-grandchild and uncle/aunt–nephew/niece correlations were of similar magnitude, 0.37 and 0.30, respectively (24). Although the data are limited, the preceding findings may suggest somewhat more genetic influence in flexibility than in strength and motor tasks. In addition, spouse resemblance in lower-back flexibility is quite low, –0.09 in the Mennonite (24) and 0.10 in the Canadian (71) samples.

Aerobic performance

Aerobic performance tests are both maximal and submaximal, and both procedures have been used in estimates of genetic effects in performance. Two maximal aerobic performances, however, must be distinguished. Maximal aerobic *power,* or maximal oxygen uptake, is measured under standardized laboratory conditions, ordinarily on a cycle ergometer or a treadmill. In some studies, maximal aerobic power is predicted from submaximal power output and of course has the limitations associated with predicted parameters. Maximal aerobic power is generally expressed per unit body weight. Maximal aerobic *capacity* refers to the total capacity for prolonged work, and there is no commonly accepted procedure for its measurement. A test requiring 90 minutes of work on a cycle ergometer at the highest sustainable intensity has been developed (13).

Submaximal aerobic performance is generally measured as the power output at a specified heart rate, for example, at a heart rate of 150 or 170 beats per minute. The term *physical work capacity* is ordinarily used, and it is expressed as power output per kilogram of body weight (e.g., W/kg). These submaximal estimates of aerobic performance generally correlate well with measures of maximal aerobic power and maximal aerobic capacity.

Most studies of the genetic effect in aerobic performance are based upon the twin model, with less data available for other types of relatives. Intraclass correlations from twin studies are summarized in table 15.4. The data vary in test protocol (measured or predicted, maximal or submaximal), numbers of twin pairs, uncontrolled age or sex effects, and differences in mean or variance between twin types (8,10). The available twin studies are thus of unequal quantitative value and yield heterogeneous and sometimes contradictory results. Estimated heritabilities range from near zero to more than 90%. The studies with the largest sample sizes, and with no mean and variance differences within twin types, give reasonably similar results. In these studies, MZ intraclass correlations reach about 0.6 to 0.7, while DZ correlations reach about 0.3 to 0.5. The latter are compatible with those generally reported for biological siblings (see later discussion). The study of Sundet et al. (88) is derived from a population-based twin panel of conscripts. Although the data are based on predicted $\dot{V}O_2$max values that were subsequently transformed to categorical scores from low to high maximal aerobic power, the intraclass correlations for the categorical scores are similar to those of other twin studies with larger sample sizes (table 15.4).

The study of Engstrom and Fischbein (27) is significant among the twin studies. The sample is limited to male twins of the same age (i.e., no sex and age effects); in addition, an estimate of the amount of habitual

Table 15.4
Intraclass Correlations From Twin Studies of Aerobic Performance

Source	*n* PAIRS MZ	*n* PAIRS DZ	Test	MZ	DZ
Klissouras (43) Males	15	10	$\dot{V}O_2max/kg$	0.91	0.44
Klissouras et al. (44) Males and females	23	16	$\dot{V}O_2max/kg$	0.95	0.36
Komi et al.(45) Males and females	15	14	PWC_{205}/kg	0.83[a]	0.43[a]
Engstrom and Fischbein (27) Males	39 22	55 21	PWCmax/kg	~ 0.70[b] ~ 0.74	~ 0.28[b] ~ 0.53
Bouchard et al. (9) Males and females	54	56	PWC_{150}/kg	0.60	0.41
Bouchard et al. (11) Males and females	53	33	$\dot{V}O_2max/kg$	0.71	0.51
Fagard et al. (29)	29	19	$\dot{V}O_2max/kg$	0.77	0.04
Maes et al. (53) Males and females	41	50	$\dot{V}O_2max/kg$	0.85	0.56/0.59[c]
Sundet et al. (88) Males	436	622	$\dot{V}O_2max/kg$, predicted[d]	0.62	0.29

[a]Calculated from data in the original paper. Data were missing for one pair of MZ twins.

[b]Values interpolated from a graph. Second set of correlations is for a subset of twins concordant for physical activity.

[c]The second correlation is for the same-sex DZ twins.

[d]Maximal aerobic power was predicted from a nomogram, and the predicted $\dot{V}O_2max$ was subsequently transformed to a categorical score from 1 to 9. The intraclass correlations are based upon the categorical scores.

physical activity was available for a subset of the twins—control of an aspect of lifestyle. Using the maximum amount of work performed during 6 minutes on a cycle ergometer (PWC_{max}/kg), there was approximately twice as much within-pair variance in DZ as in MZ pairs. However, in the subsample in which both twins in each pair were concordant for leisure-time physical activity, MZ pairs were as variable as DZ pairs. Thus, the estimated genetic influence was significantly reduced by habitual physical activity. In twins of both sexes 10 to 27 years of age, intrapair variances for an index of cardiovascular fitness derived from the Harvard step test did not differ significantly between MZ and DZ twins (19).

Relatively few studies of aerobic performance in various family members and relatives are available. In the Tecumseh community health study, heart rate response to a step test (20.3-cm bench, 24 steps per min, 3 min, energy expenditure of about 5 METs) was measured in parents and their offspring (69). There were significant parent-child similarities in the heart rate response to the step test. Using measured or predicted (for older fathers) maximal aerobic power adjusted for age, weight, and fatness, the father-son correlation was 0.34 (68). The relationship was stronger, 0.65, when only fathers below 40 years of age were considered. These results emphasize the differential effect of environmental factors associated with ages of fathers and sons. In contrast, the brother-brother correlation in maximal aerobic power was low, 0.19, and not significant.

Resemblances between different kinds of biological (e.g., by descent) and nonbiological (i.e., by adoption) relatives in maximal and submaximal aerobic performances are summarized in table 15.5. The data are derived primarily from the Quebec Family Study. Submaximal and predicted maximal aerobic performances were derived during work on a cycle ergometer (9,51,72,73). Maximal aerobic power ($\dot{V}O_2max$) was measured on a cycle ergometer (11) or treadmill (50) in several subsamples. Supplementary estimates of submaximal power output derived from a step test in a nationally representative sample of the Canadian population in the Canada Fitness Survey are also included (71).

Submaximal power output is characterized by significant familial resemblance; however, correlations among relatives are apparent not only among biological relatives, but also among relatives by adoption. Although less extensive, correlations for predicted and measured maximal aerobic power also indicate significant familial resemblance. However, in the only study that measured $\dot{V}O_2max$ in parents and offspring (50), the parent-child correlation is low and not significantly different from zero. Interestingly, the mother-child correlation is higher than that for father-child pairs, which may suggest a possible maternal effect for this phenotype. Similar differential parent-child correlations are not evident in predicted maximal aerobic power (51). Nevertheless, familial resemblance in predicted and measured maximal aerobic power is seen primarily in first-degree relatives of the same generation, that is, in brothers and sisters as well as twins. When concomitants of aerobic performance (e.g., fatness, smoking, habitual physical activity, and socioeconomic status) are statistically controlled, variation in submaximal and maximal aerobic power is about two to three times greater between families than within families (4).

Little data are available for endurance performance (i.e., performance near maximal level for an extended period of time). When defined as the

Table 15.5
Familial Correlations for Aerobic Performance in Relatives by Descent and Adoption

Source: / Test:	Lesage et al. (50) $\dot{V}O_2$max, measured	Bouchard et al. (11) $\dot{V}O_2$max, measured	Lortie et al. (51) $\dot{V}O_2$max, predicted	Bouchard et al. (9) PWC 150/kg	Pérusse et al. (72, 73) PWC 150/kg	Pérusse et al. (71) PWC 150/kg
Parent-child	0.03		0.17		0.14	0.47
Father-child	–0.10		0.17		0.13	
Mother-child	0.28		0.17		0.16	
Sibs	0.19		0.33	(0.25)	0.25	0.26
Brother-brother		0.55 (0.41)	0.30	0.31		
Sister-sister			0.63	0.45		
Brother-sister			0.24	0.12		
Uncle/aunt-nephew/niece					–0.21	0.68
First-degree cousins				(0.14)	–0.11	
Foster parent–adopted child					0.20	
MZ		0.85 (0.70)		(0.60)	0.61	
DZ		0.74 (0.51)		(0.46)	0.46	
Unrelated sibs–adoptive sibs				(0.00)	0.08	

Correlations in parentheses are intraclass.
All data except Pérusse et al. (71) are based on French Canadian families in the Quebec City area. Pérusse et al. (71) is based on families in the Canada Fitness Survey.

total amount of high-intensity work performed during a 90-min cycle ergometer task, there was more than three times as much variance among DZ twins than among MZ pairs; the respective intraclass correlations were 0.47 and 0.82 (11). The estimated genetic contribution to endurance performance in this sample (about 70%) was greater than that for maximal aerobic power (40%).

Aerobic performances also show significant resemblance between spouses (table 15.6). Such similarity is probably associated with positive assortative mating for the trait or a relevant covariate, that is, deviations from random mating and/or a common lifestyle associated with cohabitation (e.g., similar levels of habitual energy expenditure or exercise habits).

Table 15.6
Spouse Correlations for Aerobic Performance

Source	Test	*r*
Montoye and Gayle (68)	$\dot{V}O_2$max, measured or estimated, adjusted for age, weight, fatness	0.18
Lortie et al. (51)	$\dot{V}O_2$max, predicted	0.33
Lesage et al. (50)	$\dot{V}O_2$max, measured	0.22
Maes (52)	$\dot{V}O_2$max, L/O_2 ml/min/kg	0.42[a] 0.35[a]
Bouchard et al. (9)	PWC_{150}/kg	(0.19)[b]
Pérusse et al. (72, 73)	PWC_{150}/kg	0.21(0.25)
Pérusse et al. (71)	PWC_{150}/kg	0.17

[a]Maximum likelihood correlations.

[b]Correlations in parentheses are intraclass.

Anaerobic performance

Anaerobic performance refers to the maintenance of maximal effort in tasks of short duration. Sprints, which require an all-out effort, are primarily anaerobic. As discussed earlier, the genetic contribution to running in a variety of short dashes is quite variable. Data for twins, biological siblings, and adopted siblings indicate that anaerobic performance of short duration (maximal power output in a cycling task for 10 seconds) is characterized by significant familial resemblance, which is not observed among relatives by adoption. Intraclass correlations for anaerobic performance per unit body weight or FFM, respectively, were as follows: 0.80 and 0.77 in MZ twins, 0.58 and 0.44 in DZ twins, 0.46 and

0.38 in biological siblings, and –0.01 and 0.06 in adoptive siblings (82). The estimated heritability for short-term anaerobic performance (total power output in 10 seconds per unit body weight) was >50% (8). Corresponding data for maximal performance with a large anaerobic component lasting more than 10 seconds are not presently available.

Modeling Twin and Familial Resemblance

More recent analytical strategies have been applied to estimate genetic and environmental sources of variation in various performance phenotypes. For example, Fagard et al. (29) did a path analysis of $\dot{V}O_2$max in 29 MZ and 19 DZ pairs. With the data adjusted for body mass, skinfold thickness, and sports participation, the resulting heritability estimate was 66%. More recently, Maes et al. (53) applied univariate maximum likelihood estimation in path analysis to a variety of strength, motor, and aerobic measures in 10-year-old twins: 21 female MZ, 20 male MZ, 13 female DZ, 20 male DZ, and 17 unlike-sex DZ pairs. Within-pair correlations were higher in MZ pairs, but these did not significantly differ from those of DZ twins, with the exception of static arm strength and the vertical jump. Within-pair correlations for same-sex and unlike-sex DZ twins did not differ. Two path-analytic models were used, a simple genotype-environment model and one incorporating common and specific environmental factors. Estimates of the variance explained by genetic factors were consistently higher with the simple model, more so for health-related than performance-related fitness tasks: maximal aerobic power, 83% vs. 52%; lower-back flexibility, 82% vs. 57%; abdominal strength (leg lifts), 65% vs. 60%; muscular endurance (flexed arm hang), 63% vs. 27%; static arm strength, 72% vs. 72%; vertical jump, 63% vs. 63%; speed and agility (shuttle run), 71% vs. 40%; balance (flamingo stand), 51% vs. 30%; and speed of limb movement (plate tapping), 42% vs. 43% (53). The estimated heritabilities are higher than those derived from traditional twin studies and from similarly analyzed familial data (see later discussion).

Other studies have used different models and different types of relatives. Two have used the TAU path analysis model (24,71), while one has used the BETA path analysis model (72). The TAU model assumes that a phenotype P may be partitioned as $P = E + T$, where T represents the transmission of both genetic and cultural factors from parent to offspring and E denotes all other environmental influences that are not transmissible. The TAU model does not differentiate between the transmissible component of phenotypic variance that is genetic and that which is cultural. The BETA model assumes that a phenotype P can be partitioned as $P = A + B + E$, where A and B denote additive genetic and cultural factors, respectively, transmitted from parent to offspring, and E represents all other environmental effects not transmitted between generations.

Application of the TAU model to six neuromuscular traits in Mennonite families showed variability in the degree of parent-offspring transmission (24). All six traits showed substantial residual sibling resemblance due to common environmental effects. Transmissibility was highest for lower-back flexibility, 66%, and was not evident for grip strength, <1%. It was low for four psychomotor traits: reaction time, 25%; hand-eye coordination, 16%; movement time, 11%; and hand steadiness, 7%. Subsequent analysis suggested commingling in the distributions of four of the six traits studied: grip strength, hand steadiness, reaction time, and movement time (23). However, it was not possible to estimate whether the effect was the result of genetic factors, environmental processes, or both.

Use of the TAU model in a large, nationally representative sample of Canadian families resulted in the following estimates of intergenerational transmissibility: grip strength, 37%; muscular endurance (sit-ups), 37%; muscular endurance (push-ups), 44%; lower-back flexibility, 48%; and PWC_{150}/kg derived from a step test, 28% (71). An attempt to detect maternal or paternal effects in familial transmission did not markedly alter the estimates of transmissibility, and there was no difference between maternal and paternal transmission, except for push-ups.

Application of the BETA model to French Canadian relatives by descent, including MZ and DZ twins, and by adoption indicated the following trends: PWC_{150}/kg, total transmissible variance 22%, all of which was cultural; muscular strength and endurance, 63% and 55%, respectively, of which about one-half was genetic; reaction time, 27%, of which about 75% was genetic; and movement time, 18%, of which all was cultural (73). Note, however, that most of the variance in PWC, movement time, and reaction time (73%-82%) was nontransmissible, so the environment accounted for most of the variation. In contrast, less than 50% of the variance in muscular strength and endurance was nontransmissible.

Results of the three path analyses of familial data (23,24,71,73) suggest that biological variation observed in performance phenotypes is mainly associated with nontransmissible environmental factors and that the contribution of heredity is moderate and clearly lower than previously reported for twins, siblings, and parents and offspring, and for path analyses of twin data. Discrepancies among estimates should thus be interpreted with caution. There may also be variation in measurement of the different phenotypes. Estimates of transmissibility derived from models of path analysis represent only the approximate transmissible effect in a given population and can be affected by sampling procedures and sample sizes. Estimates of transmissibility are also specific to the population from which they are derived, and for a given phenotype, the contribution of environmental factors may be quite different from one population to another. For example, the Kansas Mennonite community studied by Devor and Crawford (23,24) was formed by religious immigrants among whom the use of alcohol and smoking is prohibited. This plus a largely agricultural lifestyle may affect the relative contribution of transmissible and nontransmissible factors for a given phenotype and makes any comparison difficult.

Responses to Training

Performance capacities, in general, readily respond to specific training programs; that is, performance in strength, motor, aerobic, and anaerobic tasks are trainable phenotypes. The response of these phenotypes to training, or their trainability, is not the same at all ages. This is especially apparent in children, in whom changes associated with growth and maturation and those associated with training move in the same direction. Hence, it is difficult to partition growth- and maturation-related changes from those associated with training. With the exception of several motor tasks, the role of genotype in the response to training has not been systematically studied in children and youths. On the other hand, the role of genotype in the response to strength, aerobic, and anaerobic training has been addressed in young adults of both sexes.

Motor learning

Motor learning implies improvement in performance with practice over time. Children and youths learn new motor skills or sequences of skills as they grow, with instruction (as in physical education or youth sports) and with practice associated with everyday activity. Learning of motor skills continues through adulthood, although changes in response to specific motor practice associated with aging need to be established (86).

Planned instructional programs can enhance the development of basic skills in preschool children and more complex skills in school-age children. Guided instruction by specialists or trained parents, appropriate motor task sequences, and adequate time for practice are essential components of successful instructional programs (36). Responses to the specific practice of motor skills (i.e., training) are analogous to responses of tissues and systems to regular physical training. Thus, a question that needs study is: Are gains in motor skills associated with practice and learning dependent upon genetic characteristics?

Studies of motor learning in twins have a relatively long history. The issue of maturation versus learning, or nature versus nurture, was the focus of several early studies of special practice on motor development per se and on the development of specific motor skills in infancy and early childhood, for example, the co-twin control studies of Gesell and Thompson (34), Hilgard (37), McGraw (63), and Mirenva (66). One twin was given specific practice or training in one or several movement activities, while the other twin received no special training or practice. Although the studies showed some improvement with training, maturational processes, which were viewed as genotypically mediated, were seemingly more important than the special training and practice during infancy and early childhood. In the classic study of Johnny and Jimmy (63,64), one twin (Johnny) was stimulated daily in a variety of activities, while the other twin (Jimmy) was left in his crib with few toys and routine care. After 22 months, Jimmy received 2-1/2 months of intensive practice in the same activities to which Johnny was exposed earlier. Jimmy quickly caught up to Johnny, but the quality of his performances was less. The initial results suggested that

> there is no one age period or developmental stage which clearly demarcates an earlier state of immaturity during which the child is incapable of improving through practice from the subsequent state in which improvement with practice becomes feasible. (64, p. 1)

The results were interpreted as evidence for the presence of critical periods that were dependent upon the state of neuromuscular maturation. Johnny and Jimmy were evaluated 4 years later, with the general observation that the trained twin (Johnny) maintained his advantage in motor coordination (64). Results of this and the other early studies are of historical value in demonstrating attempts to tease out genotypic bases for motor development and learning motor skills early in life.

Results of five experimental studies with twins during childhood and adolescence, dating from 1933 to 1980, are summarized in table 15.7. The experimental tasks tend to be fine motor skills that stress manual dexterity and precision of movement and more gross movements of isolated body segments. The stabilometer is an exception. The results generally indicate that the rate of learning is more similar in MZ than in DZ twins. However, estimates of the genetic contribution to the learning of motor skills vary from task to task, emphasizing the specificity of motor learning. Estimates of the genetic contribution also vary over a series of practice trials or training sessions.

The study of Sklad (83) is unique in that learning curves were mathematically fitted to the performances of the twins. Three parameters for

Table 15.7
Summary of Experimental Studies of the Genetics of Motor Learning in Twins

Source/Sample	Task	Learning/Training effects
McNemar (65) Junior-high-school-age males: 47 MZ, 48 DZ pairs	Pursuit rotor, spool packing, car sorting	Seven practice segments did not alter similarity between twins but increased similarity between DZ twins in two tasks; changes in heritabilities: 0.78 to 0.67 on pursuit rotor, 0.30 to 0.00 on spool packing, 0.44 to 0.43 on car sorting.
Brody (15) 8- to 14-year-old males: 29 MZ, 33 DZ pairs, reanalyzed by Wilde (99)	Mechanical ability	Six practice trials, greater increase in MZ than DZ similarity; heritability increased from 0.35 on first trial to 0.59 on sixth trial.
Sklad (83) 9- to 13-year-old males and females in like-sex pairs: 24 MZ, 22 DZ pairs	Plate tapping with hand, one foot tapping, mirror tracing, ball toss for accuracy	Performance curves over 10-14 days were more similar in MZ than DZ, more similar in MZ females than MZ males, especially for mirror tracing and ball toss.
Marisi (61) 11- to 18-year-olds, both sexes: 35 MZ, 35 DZ pairs	Pursuit rotor	Heritability decreased from 0.96 on trial 1 to 0.45 after 30 trials; after rest period, increased to 0.85 and decreased to 0.58 after 20 additional trials.
Williams and Gross (101) 11- to 18-year-olds, both sexes: 22 MZ, 41 DZ pairs	Stabilometer	Heritability low on first day of practice, 0.27; increased to 0.69 on second day and remained close to this level over next four days.

each skill were obtained: level of learning, rate of learning, and final level. The level and rate of learning were more similar in MZ twins. Although there was some variation among tasks, intrapair correlations were generally higher in male than in female MZ twins. The third parameter of the learning curve, the final level, was quite variable between twin types and sexes and among tasks.

Factors that are potentially capable of influencing sensitivity to learning (e.g., age, prior experience, and current phenotype) are not considered in any of the studies. One study suggests possible sex differences. Although the genotype is apparently an important determinant of the ease or difficulty with which new motor tasks are learned, or of the improvements in performance that occur with practice, there is a need for more comprehensive study of the individuality of responses to specific training or regular practice of motor tasks.

The potential role of physical activity–induced changes in gene expression influencing learning was considered in two strains of mice using a spatial learning task (32). Moderate physical activity is associated with enhanced spatial learning in rats (33) and humans (81). The transcription regulatory factor, zif 268, which is related to neuronal plasticity, was studied. Levels of zif 268 mRNA in the hippocampus and overlying cortex using *in situ* hybridization were studied after chronic and acute physical activity in a water maze. Chronic activity enhanced spatial learning in both strains but suppressed basal expression of zif 268 in only one strain of mice. In contrast, acute physical activity increased zif 268 mRNA levels in both strains of mice. The results thus suggested that

> the genetic and activity-dependent regulation of zif 268 may play a role in influencing learning performance and associated neurochemistry. (32, p. 566)

Although derived from an experimental paradigm with mice, the results may have potential implications for understanding the individuality of responses to learning perceptual-motor and perhaps motor tasks.

Strength training

Strength (resistance) training is associated with increments in strength in prepubertal children, adolescents, adults, and elderly individuals (30,31,38,74,77,98). As with the practice of motor skills, the question of interest is the role of genetic factors in the response to strength training. Unfortunately, data are not extensive at present. Among five pairs of MZ twins who were submitted to a 10-week isokinetic strength training program, there were as many interindividual differences in the response to training among members of any given pair of twins as between twin pairs (92). These limited results suggest that the response to strength training was independent of the genotype.

Aerobic training

Maximal aerobic power has limited trainability in children under 10 years of age; subsequently, in older children, adolescents, young adults, and older adults, $\dot{V}O_2$max and endurance performance are clearly trainable phenotypes in both sexes (8,58,85). The evidence indicates considerable individual differences in the response of these phenotypes to exercise training. Among young adults, some individuals exhibit a pattern of high response, while others present a pattern of no or minimal response, with a broad range of response phenotypes between the extremes. Hence, the key question relates to the individuality in responses to training.

Results of three studies of the trainability of $\dot{V}O_2$max in MZ twins are summarized in table 15.8. The rationale for the studies was that the pattern of response in MZ twins to the training program could be quantified in individuals having the same genotype (within pairs) and for subjects with different genotypes (between pairs). There is about six to nine times more variance between genotypes (pairs of twins) than within genotypes (within pairs of twins) in the response of $\dot{V}O_2$max to standardized training protocols. The similarity of the training response among members of the same MZ twin pairs is illustrated in figure 15.7. In the results of the experiment illustrated in the figure, 10 pairs of male MZ twins were submitted to a standardized, laboratory-controlled training program for 20 weeks. Gains in absolute $\dot{V}O_2$max showed almost eight times more variance between pairs of twins than within pairs of twins (5,76).

A similar experiment evaluated the response to training at a constant energy and nutrient intake in seven pairs of male MZ twins (12). The twins exercised twice daily following a standardized protocol and were kept on a constant daily nutrient intake for three months. Maximal oxygen uptake increased by about 13%, although the exercise program was of moderate intensity and not designed to improve $\dot{V}O_2$max. The intrapair resemblance in gains in $\dot{V}O_2$max were significant.

The preceding study focused on $\dot{V}O_2$max. A related measure of aerobic performance is total work output during prolonged exercise. In six pairs of MZ twins, total power output during a 90-min maximal cycle ergometer test was monitored before and after 15 weeks of training (35). Resemblance in total power output within twin pairs was significant (intraclass $r = 0.83$), and the ratio of between-pair to within-pair variances was about 11.

Table 15.8
Effects of Training on Maximal Oxygen Uptake in MZ Twins and Twin Resemblance in Training Response

Experiments	Effect of training (F-ratio)	F-ratio of between- to within-pair variances in response	Intrapair resemblance in response
Endurance training (16 pairs)[a]	45.6*	5.8*	0.71*
Endurance and intermittent training (26 pairs)[b]	105.9*	9.4*	0.81*

[a]Data in liters of O_2 per minute were adjusted for pretraining value and gender [Hamel et al. (35), Prud'homme et al. (76)].

[b]Data in liters of O_2 per minute were adjusted for pretraining value, gender, and training programs [Boulay et al. (14), Hamel et al. (35), Prud'homme et al. (76)].

*$p < .001$

Adapted from Bouchard et al. (5).

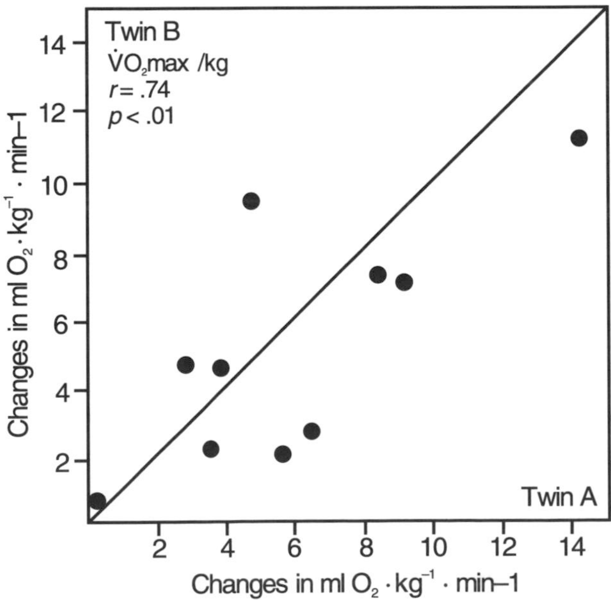

Figure 15.7 Intrapair resemblance in the magnitude of training changes in maximal aerobic power. From Prud'homme et al. (76).

Anaerobic training

Corresponding data dealing with the genetic determination of individual differences in response to anaerobic training are less extensive. In a study of the effects of 15 weeks of high-intensity, intermittent training with 14 pairs of MZ twins, the training response to short-term anaerobic performance (10-second power output) was minimally affected by genotype, while the training response to long-term anaerobic performance (90-second power output) was largely determined by genetic factors (82). The between-pair variance component of the trainability of the 90-second power output test per unit body weight amounted to about 70% of the training response, indicating that undetermined genetic characteristics underlie individuality of response to controlled exercise training programs.

Although these studies indicate large individual differences in response to exercise training programs of different intensities, subjects with the same genotype are more similar in responses than subjects with different genotypes. The evidence is consistent with the view that the genotype conditions, in part, the response to exercise training. However, the specific genetic characteristics that contribute to the individuality of responses to training are as yet undetermined.

Molecular Markers

A few reports on the topic of genetic markers and performance have appeared in the last 10 years or so. The study of elite performance by means of genetic markers was first conducted during the 1968 Olympic Games in Mexico City (22). The purpose of that study was to find whether there was any association between athletic participation in the Olympic Games and allelic variation in single-gene blood systems. Results indicated that participation in the 1968 Olympic Games was not associated with allelic variations in the red blood cell antigens or enzyme variants of red blood cells. A second effort to study elite performers was carried out during the 1976 Olympic Games in Montreal (18,20). In this study, a search for genetic markers of aerobic performance was done in a group of white athletes competing in endurance events. No significant differences were reported between elite endurance athletes and controls for genetic markers in red blood cell antigens and four red blood cell enzymes.

In sedentary human subjects, two reports indicated the lack of charge variants either in skeletal muscle enzymes of the tricarboxylic acid cycle (60) or in enzymes of the glycolytic pathway (7). This evidence supported the hypothesis that variations in the coding sequences of these genes were unlikely to account for the individuality of aerobic performance. Only one report has attempted to establish the relationship between inherited protein variants and the response of aerobic or anaerobic performance to exercise training (6). Enzyme variants were identified for creatine kinase (CK) and adenylate kinase (AK1) in skeletal muscle extracts of 295 male and female subjects by thin-layer isoelectrofocusing and specific staining with insoluble formazan. A muscle CK charge variant was detected in six individuals, for a gene frequency of 1%. Twenty-one individuals were heterozygotes for an inherited AK1 charge variant, which corresponds to an allele frequency of 3.5%. When paired on the basis of age, height, body weight, and subcutaneous fat with control subjects who presented the common muscle CK and AK1 phenotype, no significant differences were found in $\dot{V}O_2$max or endurance performance in the untrained state between carriers of the common and variant genes (6).

One study examined the presence of associations between mitochondrial DNA (mtDNA) sequence polymorphisms, $\dot{V}O_2$max in the sedentary state, and the response of $\dot{V}O_2$max to endurance training (25). Carriers of three mtDNA morphs—one due to a base change in the tRNA for threonine and two others caused by base substitutions in the subunit 5 of the NADH dehydrogenase (MTND5) gene—had a body mass–adjusted $\dot{V}O_2$max in the untrained state significantly higher than that found in the noncarriers. A low response in $\dot{V}O_2$max to the endurance training was also observed for three carriers of a variant in MTND5. Carriers of a KPNI morph due to a polymorphism in the mtDNA D-loop region presented a statistically significant higher training response than noncarrier cases. However, after adjustment for pretraining level and training programs, the difference in trainability was no longer significant.

Association and linkage studies were performed between a maximal 10-second ergocycle performance test phenotype and several polymorphic loci related to blood groups and red blood cell enzymes in a total of 250 subjects, including 55 pairs of brothers and sisters and 31 pairs of DZ twins (Ye et al., unpublished results). No evidence for association nor for linkage was found between the maximal 10-second ergocycle performance test and the polymorphism in ABO, Rh, MN, SS, KELL, and DUFFY erythrocyte antigenic systems; phosphoglucomutase-1; acid phosphatase; adenosine deaminase; adenylate kinase; and esterase D red blood cell enzymes.

The investigation of the molecular basis of human variation in performance and in trainability is still in its infancy. The databases necessary to undertake such studies are currently being established in a few centers, and a wealth of new data can be anticipated in the near future.

Summary

Estimates of the genetic contribution to individual differences in a variety of strength and motor tasks are derived largely from younger subjects and from twins. Many studies are limited to males, while others combine the sexes. The presently available data suggest a genetic effect, but of limited and unequal quantitative value. Hence, generalizations on the genetics of strength and motor performance are limited. On the other hand, flexibility measures show a somewhat greater familial similarity.

Gene sharing seems to be associated with covariation between biological relatives in submaximal and maximal aerobic performances. However, cohabitation is also related to these performances, as evident in data for foster parents and adopted children and spouses. The claims of earlier studies based on small samples of twins (e.g., 43,44) are unrealistically high compared to more comprehensively analyzed twin and familial data, which suggest heritabilities of PWC_{150}/kg of about 10% and of $\dot{V}O_2max$/kg in the range of 25% to 40%. In contrast, estimated heritability for total power output in 90 minutes per unit body weight is about 50% and perhaps more (8).

Results of path analyses suggest that variation observed in several performance phenotypes is mainly associated with nontransmissible environmental factors and that the contribution of the genotype is at best moderate. Several studies of motor learning, strength training, and aerobic training are also consistent with the view that the genotype modulates, in part, the response to specific practice and training programs.

Finally, the search for genetic markers of performance and of responsiveness to training has begun. Progress has been slow initially primarily because of the need to establish the complex database to address the issues. A variety of approaches and technologies are now being used to identify these polymorphic markers and genes.

References

1. Åstrand, P.O.; Rodahl, K. Textbook of work physiology: Physiological bases of exercise. 2nd ed. New York: McGraw-Hill; 1977.
2. Beunen, G.; Malina, R.M. Growth and physical performance relative to the timing of the adolescent sport. Exerc. Sport Sci. Rev. 16:503-40; 1988.
3. Birren, J.E.; Fisher, L.M. Aging and speed of behavior: Possible consequences for psychological functioning. Annu. Rev. Psychol. 46:329-53; 1995.

4. Bouchard, C. Genetic determinants of endurance performance. In: Shephard, R.J.; Åstrand, P.O., eds. Encyclopaedia of sports medicine. Vol. 2, Endurance in sport. Oxford: Blackwell Scientific; 1992:149-59.
5. Bouchard, C.; Boulay, M.R.; Dionne, F.T.; Pérusse, L.; Thibault, M.C.; Simoneau, J.A. Genotype, aerobic performance and response to training. In: Beunen, G.; Chesquiere, J.; Reybrouck, T.; Claessens, A.L., eds. Children and exercise. Enke Verlag: Band 4 Schriftenreihe der Hamburg-Mannheimer-Stiftung für Informationsmedizin; 1990:124-35.
6. Bouchard, C.; Chagnon, M.; Thibault, M.C.; Boulay, M.R.; Marcotte, M.; Côté, C.; Simoneau, J.A. Muscle genetic variants and relationship with performance and trainability. Med. Sci. Sports Exerc. 21:71-77; 1989.
7. Bouchard, C.; Chagnon, M.; Thibault, M.C.; Boulay, M.R.; Marcotte, M.; Simoneau, J.A. Absence of charge variants in human skeletal muscle enzymes of the glycolytic pathway. Hum. Genet. 78:100; 1988.
8. Bouchard, C.; Dionne, F.T.; Simoneau, J.A.; Boulay, M.R. Genetics of aerobic and anaerobic performances. Exerc. Sport Sci. Rev. 20:27-58; 1992.
9. Bouchard, C.; Lortie, G.; Simoneau, J.A.; Leblanc, C.; Thériault, G.; Tremblay, A. Submaximal power output in adopted and biological siblings. Ann. Hum. Biol. 11:303-9; 1984.
10. Bouchard, C.; Malina, R.M. Genetics of physiological fitness and motor performance. Exerc. Sport Sci. Rev. 11:306-39; 1983.
11. Bouchard, C.; Lesage, R.; Lortie, G.; Simoneau, J.A.; Hamel, P.; Boulay, M.R.; Pérusse, L.; Thériault, G.; Leblanc, C. Aerobic performance in brothers, dizygotic and monozygotic twins. Med. Sci. Sports Exerc. 18: 639-42; 1986.
12. Bouchard, C.; Tremblay, A.; Després, J.P.; Thériault, G.; Nadeau, A.; Lupien, P.J.; Moorjani, S.; Prud'homme, D.; Fournier, G. The response to exercise with constant energy intake in identical twins. Obes. Res. 2:400-410; 1994.
13. Boulay, M.R.; Hamel, P.; Simoneau, J.A.; Lortie, G.; Prud'homme, D.; Bouchard, C. A test of aerobic capacity: Description and reliability. Can. J. Appl. Sports Sci. 9:122-26; 1984.
14. Boulay, M.R.; Lortie, G.; Simoneau, J.A.; Bouchard, C. Sensitivity of maximal aerobic power and capacity to anaerobic training is partly genotype dependent. In: Malina, R.M.; Bouchard, C., eds. Sport and human genetics. Champaign, IL: Human Kinetics; 1986:173-81.
15. Brody, D. Twin resemblances in mechanical ability, with reference to the effects of practice on performance. Child Dev. 8:207-16; 1937.
16. Canada Fitness Survey. Fitness and lifestyle in Canada. Ottawa: Government of Canada, Fitness and Amateur Sport; 1983.
17. Carlier, M.; Beau, J.; Marchaland, C.; Michel, F. Sibling resemblance in two manual laterality tasks. Neuropsychologia. 32:741-46; 1994.
18. Chagnon, Y.C.; Allard, C.; Bouchard, C. Red blood cell genetic variation in Olympic endurance athletes. J. Sport Sci. 2:121-29; 1984.

19. Chatterjee, S.; Das, N. Physical and motor fitness in twins. Jpn. J. Physiol. 45:519-34; 1995.
20. Couture, L.; Chagnon, M.; Allard, C.; Bouchard, C. More on red blood cell genetic variation in Olympic athletes. Can. J. Appl. Sport Sci. 11:16-18; 1986.
21. Cratty, B.J. A comparison of fathers and sons in physical ability. Res. Q. 31:12-15; 1960.
22. deGaray, A.; Levine, L.; Carter, J.E.L. Genetic and anthropological studies of Olympic athletes. New York: Academic Press; 1974.
23. Devor, E.J.; Crawford, M.H. A commingling analysis of quantitative neuromuscular performance in Kansas Mennonite community. Am. J. Phys. Anthropol. 63:29-37; 1984.
24. Devor, E.J.; Crawford, M.H. Family resemblance for neuromuscular performance in a Kansas Mennonite community. Am. J. Phys. Anthropol. 64:289-96; 1984.
25. Dionne, F.T.; Turcotte, L.; Thibault, M.C.; Boulay, M.R.; Skinner, J.S.; Bouchard, C. Mitochondrial DNA sequence polymorphism, $\dot{V}O_2max$ and response to endurance training. Med. Sci. Sports Exerc. 23:177-85; 1991.
26. Doherty, T.J.; Vandervoort, A.A.; Taylor, A.W.; Brown, W.F. Effects of motor unit losses on strength in older men and women. J. Appl. Physiol. 74:868-74; 1993.
27. Engstrom, L.M.; Fischbein, S. Physical capacity in twins. Acta Genet. Med. Gemellol. 26:159-65; 1977.
28. Era, P.; Berg, S.; Schroll, M. Psychomotor speed and physical activity in 75-year-old residents in three Nordic localities. Aging Clin. Exp. Res. 7:195-204; 1995.
29. Fagard, R.; Bielen, E.; Amery, A. Heritability of aerobic power and anaerobic energy generation during exercise. J. Appl. Physiol. 70:352-62; 1991.
30. Fiatarone, M.A.; Marks, E.C.; Ryan, N.D.; Meredith, C.N.; Lipsitz, L.A.; Evans, W.J. High intensity strength training in nonagenarians: Effects on skeletal muscle. JAMA. 263:3029-34; 1990.
31. Fiatarone, M.A.; O'Neill, E.F.; Ryan, N.D.; Clements, K.M.; Solares, G.R.; Nelson, M.E.; Roberts, S.B.; Kehaylas, J.J.; Lipsitz, L.A.; Evans, W.J. Exercise training and nutritional supplementation for physical frailty in very elderly people. N. Engl. J. Med. 330:1769-75; 1994.
32. Fordyce, D.E.; Bjat, R.V.; Baraban, J.M.; Wehner, J.M. Genetic and activity-dependent regulation of zif 268 expression: Association with spatial learning. Hippocampus. 4:559-68; 1994.
33. Fordyce, D.E.; Farrar, R.P. Enhancement of spatial learning performance by physical activity in F344 rats and learning-associated alterations in hippocampal and cortical cholinergic function. Behav. Brain Res. 46:123-33; 1991.
34. Gesell, A.; Thompson, H. Learning and growth in identical infant twins: An experimental study by the method of co-twin controls. Genet. Psychol. Monogr. 6:1-120; 1929.

35. Hamel, P.; Simoneau, J.-A.; Lortie, G.; Boulay, M.R.; Bouchard, C. Heredity and muscle adaptation to endurance training. Med. Sci. Sports Exerc. 18:690-96; 1986.
36. Haubenstricker, J.; Seefeldt, V. Acquisition of motor skills during childhood. In: Seefeldt, V., ed. Physical activity and well-being. Reston, VA: American Alliance for Health, Physical Education, Recreation, and Dance; 1986:41-102.
37. Hilgard, J.R. The effect of early and delayed practice on memory and motor performance studied by the method of co-twin control. Genet. Psychol. Monogr. 14:493-567; 1933.
38. Ikai, M. The effects of training on muscular endurance. In: Kato, K., ed. Proceedings of the International Congress of Sport Sciences. Tokyo: University of Tokyo Press; 1966:145-58.
39. Ishidoya, Y. Sportfahigkeit der Zwillinge. Acta Genet. Med. Gemellol. 6:321-26; 1957.
40. Kasch, F.W.; Boyer, J.L.; Van Camp, S.P.; Verity, L.S.; Wallace, J.P. The effect of physical activity and inactivity on aerobic power in older men (a longitudinal study). Phys. Sports Med. 18:73-83; 1990.
41. Kimura, K. The study on physical ability of children and youths: On twins in Osaka City. Jinrui-gaku Zasshi (Anthropological Society of Nippon). 64:172-96; 1956.
42. Kimura, M.; Hirakawa, K.; Morimoto, T. Physical performance survey in 900 aged individuals. In: Kaneko, M., ed. Fitness for the aged, disabled, and industrial worker. Champaign, IL: Human Kinetics; 1990:55-60.
43. Klissouras, V. Heritability of adaptive variation. J. Appl. Physiol. 31:338-44; 1971.
44. Klissouras, V.; Pirnay, F.; Petit, J.M. Adaptation to maximal effort: Genetics and age. J. Appl. Physiol. 35:288-93; 1973.
45. Komi, P.V.; Klissouras, V.; Karvinen, E. Genetic variation in neuromuscular performance. Int. Z. Angew. Physiol. 31:289-304; 1973.
46. Kovar, R. Human variation in motor abilities and its genetic analysis. Prague: Charles University; 1981.
47. Kovar, R. Prispevek ke studiu geneticke podminenosti lidske motiriky. Doctoral dissertation, Charles University (Prague); 1974.
48. Kovar, R. Sledovani prdobnosti mezi rodici a jejich potomky v nekterych motorickych projevech. Teorie a Praxe Telesne Vychovy. 29:93-98; 1981.
49. Kuo, G.H. Physical fitness of the people in Taipei including the aged. In: Kaneko, M., ed. Fitness for the aged, disabled, and industrial worker. Champaign, IL: Human Kinetics; 1990:21-24.
50. Lesage, R.; Simoneau, J.A.; Jobin, J.; Leblanc, J.; Bouchard, C. Familial resemblance in maximal heart rate, blood lactate and aerobic power. Hum. Hered. 35:182-89; 1985.

51. Lortie, G.; Bouchard, C.; Leblanc, C.; Tremblay, A.; Simoneau, J.A.; Thériault, G.; Savoie, J.P. Familial similarity in aerobic power. Hum. Biol. 54:801-12; 1982.
52. Maes, H.H.M. Univariate and multivariate genetic analysis of physical characteristics of twins and parents. Unpublished doctoral dissertation, Catholic University of Leuven (Belgium); 1992.
53. Maes, H.; Beunen, G.; Vlietinck, R.; Lefevre, J.; Van den Bossche, C.; Claessens, A.; Derom, R.; Lysens, R.; Renson, R.; Simons, J.; Vanden Eynde, B. Heritability of health- and performance-related fitness: Data from the Leuven longitudinal twin study. In: Duquet, W.; Day, J.A.P., eds. Kinanthropometry IV. London: Spon; 1993:140-49.
54. Malina, R.M. Growth of muscle tissue and muscle mass. In: Falkner, F.; Tanner, J.M., eds. Human growth. Vol. 2, Postnatal growth, neurobiology. New York: Plenum Press; 1986:77-99.
55. Malina, R.M. Physical activity: Relationship to growth, maturation, and physical fitness. In: Bouchard, C.; Shephard, R.J.; Stephens, T., eds. Physical activity, fitness, and health: International proceedings and consensus statement. Champaign, IL: Human Kinetics; 1994:918-30.
56. Malina, R.M. Physical activity and fitness of children and youth: Questions and implications. Med. Exerc. Nutr. Health. 4:123-35; 1995.
57. Malina, R.M.; Bouchard, C. Genetic considerations in physical fitness. In: Assessing physical fitness and physical activity in population-based surveys. Hyattsville, MD: Department of Health and Human Services. National Center for Health Statistics; 1989:453-73.
58. Malina, R.M.; Bouchard, C. Growth, maturation and physical activity. Champaign, IL: Human Kinetics; 1991.
59. Malina, R.M.; Mueller, W.H. Genetic and environmental influences on the strength and motor performance of Philadelphia school children. Hum. Biol. 53:163-79; 1981.
60. Marcotte, M.; Chagnon, M.; Côté, C.; Thibault, M.C.; Boulay, M.R.; Bouchard, C. Lack of genetic polymorphism in human skeletal muscles enzymes of the tricarboxylic acid cycle. Hum. Genet. 77:200; 1987.
61. Marisi, D.Q. Genetic and extragenetic variance in motor performance. Acta Genet. Med. Gemellol. 26:197-204; 1977.
62. McGee, M.G. Human spatial abilities: Psychometric studies and environmental, genetic, hormonal, and neurological influences. Psychol. Bull. 86:889-918; 1979.
63. McGraw, M.B. Growth: A study of Johnny and Jimmy. New York: Appleton-Century; 1935.
64. McGraw, M.B. Later development of children specially trained during infancy: Johnny and Jimmy at school age. Child Dev. 10:1-19; 1939.
65. McNemar, Q. Twin resemblances in motor skills, and the effect of practice thereon. Ped. Sem. J. Genet. Psychol. 42:70-99; 1933.

66. Mirenva, A.N. Psychomotor education and the general development of preschool children: Experiments with twin controls. Ped. Sem. J. Genet. Psychol. 46:433-54; 1935.

67. Mizuno, T. Similarity of physique, muscular strength and motor ability in identical twins. Bulletin of the Faculty of Education, Tokyo University. 1:136-57; 1956.

68. Montoye, H.J.; Gayle, R. Familial relationships in maximal oxygen uptake. Hum. Biol. 50:241-49; 1978.

69. Montoye, H.J.; Metzner, H.L.; Keller, J.K. Familial aggregation of strength and heart rate response to exercise. Hum. Biol. 47:17-36; 1975.

70. Noble, C.E.; Baker, B.L.; Jones, T.A. Age and sex parameters in psychomotor learning. Percept. Mot. Skills. 19:935-45; 1964.

71. Pérusse, L.; Leblanc, C.; Bouchard, C. Inter-generation transmission of physical fitness in the Canadian population. Can. J. Sport Sci. 13:8-14; 1988.

72. Pérusse, L.; Leblanc, C.; Tremblay, A.; Allard, C.; Thériault, G.; Landry, F.; Talbot, J.; Bouchard, C. Familial aggregation in physical fitness, coronary heart disease risk factors, and pulmonary function measurements. Prev. Med. 16:607-15; 1987.

73. Pérusse, L.; Lortie, G.; Leblanc, C.; Tremblay, A.; Thériault, G.; Bouchard, C. Genetic and environmental sources of variation in physical fitness. Ann. Hum. Biol. 14:425-34; 1987.

74. Pfeiffer, R.D.; Francis, R.S. Effects of strength training on muscle development in prepubescent, pubescent, and postpubescent males. Phys. Sports Med. 14:134-43; 1986.

75. Potvin, A.R.; Syndulko, K.; Tourtellotte, W.W.; Lemmon, J.A.; Potvin, J.H. Human neurologic function and the aging process. J. Am. Geriatr. Soc. 28:1-9; 1980.

76. Prud'homme, D.; Bouchard, C.; Leblanc, C.; Landry, F.; Fontaine, E. Sensitivity of maximal aerobic power to training is genotype-dependent. Med. Sci. Sports Exerc. 16:489-93; 1984.

77. Pyka, G.; Lindenberger, E.; Charette, S.; Marcus, R. Muscle strength and fiber adaptations to a year-long resistance training program in elderly men and women. J. Gerontol. 49:M22-27; 1994.

78. Rijsdijk, F.V.; Boomsma, D.I.; Vernon, P.A. Genetic analysis of peripheral nerve conduction velocity in twins. Behav. Genet. 25:341-48; 1995.

79. Rogers, M.A.; Hagberg, J.M.; Martin, W.H.; Ehsani, A.A.; Holloszy, J.O. Decline in $\dot{V}O_2$max with aging in master athletes and sedentary men. J. Appl. Physiol. 68:2195-99; 1990.

80. Rose, R.J.; Miller, J.Z.; Dumon-Driscoll, M.; Evans, M.M. Twin-family studies of perceptual speed ability. Behav. Genet. 9:71-86; 1979.

81. Shay, K.A.; Roth, D.L. Association between aerobic fitness and visuospatial performance in healthy older adults. Psychol. Aging. 7:15-24; 1992.

82. Simoneau, J.A.; Lortie, G.; Boulay, M.R.; Marcotte, M.; Thibault, M.-C.; Bouchard, C. Inheritance of human skeletal muscle and anaerobic capac-

ity adaptation to high-intensity intermittent training. Int. J. Sports Med. 7:167-71; 1986.

83. Sklad, M. The genetic determination of the rate of learning of motor skills. Stud. Phys. Anthropol. 1:3-19; 1975.
84. Sklad, M. Rozwoj fizyczny i motorycznosc blizniat. Materialy i Prace Antropologiczne. 85:3-102; 1973.
85. Spina, R.J.; Ogawa, T.; Kohrt, W.M.; Martin, W.H.; Holloszy, J.O.; Ehsani, A.A. Differences in cardiovascular adaptations to endurance exercise training between older men and women. J. Appl. Physiol. 75:849-55; 1993.
86. Spirduso, W.W. Physical dimensions of aging. Champaign, IL: Human Kinetics; 1995.
87. Spirduso, W.W. Physical fitness, aging, and psychomotor speed: A review. J. Gerontol. 35:850-65; 1980.
88. Sundet, J.M.; Magnus, P.; Tambs, K. The heritability of maximal aerobic power: A study of Norwegian twins. Scand. J. Med. Sci. Sports. 4:181-85; 1994.
89. Szopa, J. Familial studies on genetic determination of some manifestations of muscular strength in man. Genet. Pol. 23:65-79; 1982.
90. Szopa, J. Genetic control of fundamental psychomotor properties in man. Genet. Pol. 27:137-50; 1986.
91. Szopa, J. Zmiennose oraz genetyczne uwarunkowania niektorych przejawow sily miesni u czlowieka wyniki badan rodzinnych. Materialy i Prace Antropologiczne. 103:131-54; 1983.
92. Thibault, M.C.; Simoneau, J.A.; Côté, C.; Boulay, M.R.; Lagassé, P.; Marcotte, M.; Bouchard, C. Inheritance of human muscle enzyme adaptation to isokinetic strength training. Hum. Hered. 36:341-47; 1986.
93. Vandenberg, S.G. Contributions of twin research to psychology. Psychol. Bull. 66:327-52; 1966.
94. Vandenberg, S.G. The hereditary abilities study: Hereditary components in a psychological test battery. Am. J. Hum. Genet. 14:220-37; 1962.
95. Venerando, A.; Milani-Comparetti, M. Twin studies in sport and physical performance. Acta Genet. Med. Gemellol. 19:80-82; 1972.
96. Welford, A.T. Aging and human skill. Oxford: Oxford University Press for the Nuffield Foundation (reprinted by Greenweed Press, Westport, CT, 1973); 1958.
97. Welford, A.T. Motor skill and aging. In: Nadeau, C.H.; Halliwell, W.R.; Newell, K.M.; Roberts, G.C., eds. Psychology of motor behavior and sport 1979. Champaign, IL: Human Kinetics; 1980:253-68.
98. Weltman, A.; Janney, C.; Rians, C.B.; Strand, K.; Berg, B.; Tippitt, S.; Wise, J.; Cahill, B.R.; Katch, F.I. The effects of hydraulic resistance strength training in pre-pubertal males. Med. Sci. Sports Exerc. 18:629-38; 1986.
99. Wilde, G.J.S. An experimental study of mutual behaviour imitation and person perception in MZ and DZ twins. Acta Genet. Med. Gemellol. 19:273-79; 1970.

100. Williams, H.G. Perceptual and motor development. Englewood Cliffs, NJ: Prentice Hall; 1983.

101. Williams, L.R.T.; Gross, J.B. Heritability of motor skill. Acta Genet. Med. Gemellol. 29:127-36; 1980.

102. Williams, L.R.T.; Hearfield, V. Heritability of a gross motor balance task. Res. Q. 44:109-12; 1973.

103. Wolanski, N. Assortative mating in the Polish rural populations. Stud. Hum. Ecol. 1:182-88; 1973.

104. Wolanski, N.; Kasprzak, E. Similarity in some physiological, biochemical and psychomotor traits between parents and 2-45 year old offspring. Stud. Hum. Ecol. 3:85-131; 1979.

105. Wolanski, N.; Tomonari, K.; Siniarska, A. Genetics and the motor development of man. Hum. Ecol. Race Hyg. 46:169-91; 1980.

CHAPTER 16

Molecular Medicine and Health-Related Fitness

A growing understanding of the structures and functions of the human genome will provide the tools to learn, perhaps everything, about individual genes. For example: How are genes regulated? How are they modulated during growth and aging? What is the effect of experience and nongenetic elements on the biology of genes? From the view of the exercise sciences, such advances in the biology of genes will have major consequences in two primary areas: first, for fitness and health and second, for performance, particularly high-performance sports. This chapter examines implications from the fitness and health perspectives. The domain of performance and athletic ability is considered in the next chapter.

It is not necessary to understand the rules of Mendelian inheritance or to know the basic notions of genetics to be aware that a family history for a specific disease is often associated with an increased probability for a relative to be affected by the same disease. Until recently, the prediction of the risk level for a biological relative of an affected person was based on computations that took into account a variety of indicators pertaining to the pattern of inheritance of the condition, that is, prevalence and level of penetrance; degree of immediate or remote consanguinity; the occurrence of the disease, especially among first-degree relatives; and other factors. Over the last few years, the prediction of level of risk in a biological relative of an affected person, or for a given person in a population, has increasingly taken advantage of information at the level of the gene responsible for the disease or at the level of markers for that gene. This progressive shift in the paradigm offers new opportunities that were seldom available in the past. However, the use of the resources of molecular medicine, which themselves are at an embryonic stage but are developing at a dazzling pace, has seemingly just begun. Progress in molecular medicine will eventually lead to a more sophisticated type of predictive medicine. The health-related fitness domain is one area that should benefit greatly from advances in understanding of the molecular and genetic basis of health and fitness and of the sensitivity to regular exercise or to chronic sedentarism.

Single-Gene Diseases

Although the clinical course is often very complex, a single-gene disease can be conceived as a condition in which a point mutation or some other molecular alteration is sufficient to cause a disease. While a change in a single base pair is often the molecular cause of the deficient phenotype, more complex molecular alterations in the coding or noncoding sequence of a gene can also be implicated. The latter are, in fact, more prevalent than originally thought.

Classic cases of disease caused by a simple but functionally significant mutation in a codon of the coding sequence of a gene include those in the beta chain gene of the hemoglobin molecule resulting in sickle cell anemia and those in the phenylalanine hydroxylase gene resulting in phenylketonuria. Hundreds of such point mutations with important implications for health and disease have been described over the last 50 years or so and are discussed in considerable detail in the major source book on this topic (21).

More recently, it has become evident that many single-gene diseases are caused by more complex genomic alterations. For instance, some diseases result from two independent mutational events in the promoter region of a gene (e.g., retinoblastoma) (20), and this pattern may be quite common in the apparent loss of control over cell replication, which is a central feature in growth of a tumor. Moreover, deletions or duplications of a gene segment have been found. For example, large deletions are observed in Duchenne muscular dystrophy (16). Insertion/deletion polymorphisms occur in the angiotensin I–converting enzyme gene (19) and in the apolipoprotein B gene (10). Such genomic features influence the biological activity of gene products.

Another type of genomic alteration is proving to be quite prevalent in several disease states. Unstable or heterogeneous triplet amplification has been identified in noncoding regions of the genes associated with myotonic dystrophy (variable number of CTG repeats in the myotonin kinase gene) (5,7,15). Unstable numbers of triplet repeats have also been observed in Huntington's chorea (CAG repeats), the fragile X syndrome, and other diseases, while polymorphism for a compounded dinucleotide repeat has been identified in the human glucokinase gene

(22), a gene that appears to play a role in the susceptibility to noninsulin dependent diabetes mellitus.

Chromosomal translocations have also been described. They are involved in some forms of leukemia (23). Mutations and deletions in the mitochondrial DNA genome have been implicated in a variety of disease states, including mitochondrial myopathies, myotonic dystrophy, optic atrophy, diabetes, and deafness (1,17).

Common Multifactorial Diseases

Although single-gene diseases account for a substantial fraction of human diseases, they are dwarfed, by comparison, with common multifactorial diseases. In complex multifactorial diseases, genetic susceptibility is generally conferred by several genes and by relevant environmental conditions and lifestyle features, which probably vary from disease to disease. These several factors must come together to cause the disease.

Most health problems that are of interest to those involved in the fitness domain fall within this category of disease phenotypes, which includes hypertension, NIDDM, coronary heart disease, stroke, peripheral claudication, obesity, some types of dyslipoproteinemia, some forms of cancer, and others as well. These common diseases have proven very difficult to study from a genetic perspective in the past. In a typical case, after having established that the condition exhibited familial aggregation and after having perhaps found statistical evidence for the contribution of a single major gene in addition to a multifactorial transmissible component, no further progress could be made, as the trait could not be investigated at the gene level. Progress in genetics and molecular biology over the last decade has, however, altered the picture. There are now good reasons to believe that the dissection of genetic and environment effects on complex diseases can be achieved (12,18). This work has already begun as discussed in several earlier chapters.

Complex multifactorial diseases are complex in terms of pathophysiology and causation, and they are generally quite common in the adult population. These diseases affect most individuals who live beyond retirement age, are responsible for most deaths, and require the largest fraction of health care expenditures. Because of these characteristics, it is obviously important to understand the complex network of genetic and nongenetic factors associated with an increased susceptibility to the premature advent of these diseases or with a resistance to them. The genetic susceptibility is likely to result from polymorphisms of several genes, with the most-susceptible individuals carrying the deficient alleles at all loci, while the least-susceptible individuals may have only a few or none of the susceptibility alleles. However, an important feature of these diseases is that the genes of the susceptibility genotypes cannot be considered in isolation. They must be seen in the context of the environmental and lifestyle conditions specific to the individual, family, or population studied. For example, failure to consider indicators of environment and lifestyle may in some cases limit understanding of why a highly genetically susceptible individual does not develop overt manifestations of a disease. Similarly, the converse situation also occurs, that is, the presence of clear symptoms for a disease in a person who is genetically among the least susceptible.

Common multifactorial diseases are, therefore, entities of utmost interest, but they are also the type that cannot be easily reduced to simple classification schemes. They are also not resolved with simple models and research strategies. Nonetheless, advances over the last few years in genetic modeling and molecular biology have given new impetus to the study of the genetic basis of complex multifactorial diseases of interest to those in the health-related fitness area. The genes responsible for susceptibility to hypertension, insulin resistance and NIDDM, atherosclerotic diseases, obesity, and other such diseases can and will be identified with time. The reasons for such optimism have been elaborated in several recent publications (3,9,11,12,14,18).

The Era of Molecular Medicine

Advances in genetics and molecular biology, which climaxed in the launching of the Human Genome Project in the early 1990s, have already had and will continue to have an even more profound influence on understanding human biology, health, and disease in the future. The progress in DNA technology—restriction endonuclease, Southern blotting, isolation of a gene or DNA fragment and molecular cloning, construction of a genomic library, techniques to produce cDNA from mRNA, DNA sequencing, polymerase chain reaction, ligase chain reaction, mutation detection methods such as mismatch cleavage, single-strand conformation polymorphism, denaturing gradient gel electrophoresis, allele-specific oligonucleotide hybridization, fluorescence *in situ* hybridization, and other technologies—has been instrumental in the emergence of the field of "molecular medicine" (6).

These and other technologies have been employed in familial and large pedigree studies, association and linkage analyses, quantitative trait locus mapping studies based mainly on informative mouse intercrosses, positional cloning, transgenic animal experiments, and a variety of other designs. The current yield is already impressive. A good number of susceptibility genes have been identified for NIDDM, hypertension, dyslipoproteinemia, atherosclerosis, obesity, breast cancer, and several other diseases. Although much remains to be done, there is a confluence of factors that allows one to be optimistic about the increase in knowledge to be expected from molecular medicine.

This optimism is related in part to the enormous progress made in the identification of genes and other molecular markers. Figure 16.1 depicts changes in human gene mapping data since 1973, when 64 genes were available on the human gene map. In 1994 alone, over 40,000 genes and markers were submitted for inclusion in the Genome Data Base. In the not-too-distant future, a very detailed genetic map of the human genome with a resolution level of the order of 0.1 to 0.2 cM can be expected. Likewise, a physical map with molecular markers at about every 1,000 base pairs is likely to be produced. DNA sequencing is bound to proceed at a faster and more reliable pace in the near future, such that several

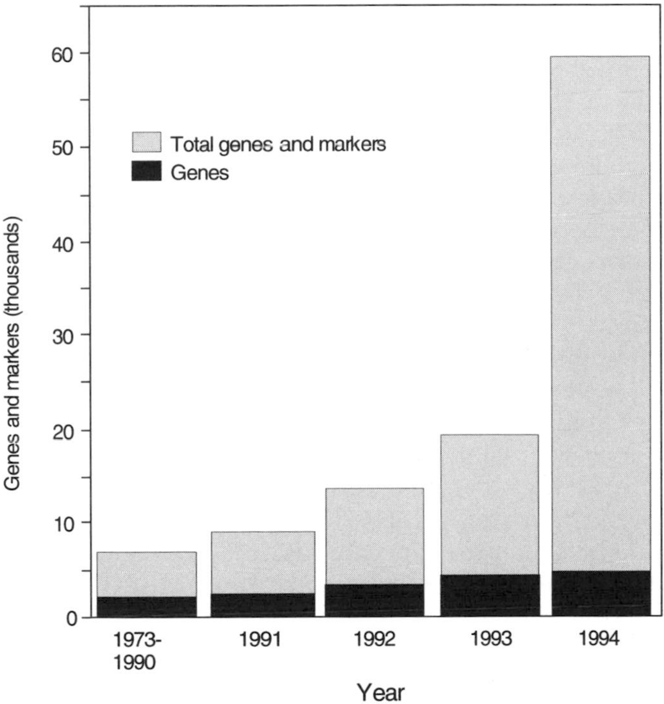

Figure 16.1 Growth in data about genomic markers from 1973 to 1994. In 1994, nearly 60,000 genes and markers were registered in the Genome Data Base. Redrawn from *Human Genome News* (8).

millions of base pairs of the human genome will have been sequenced in three to five years. These advances will provide the impetus for a growing influence of genetics and molecular biology in the diagnostic and eventually in the therapeutic aspects of medicine.

Definition of Susceptibility and Risk at the Molecular Level

The field of medicine will make increasing use of concepts and technologies of genetics and molecular biology. The applications will include contributions pertaining to the assessment of the susceptibility and risk of exposure to a variety of environmental, occupational, and lifestyle factors; identification of addictive genotypes with respect to tobacco and alcohol products; detection of those individuals with potentially adverse response to therapeutic drugs; and production of resources for various diagnostic and therapeutic needs, to name but a few. A detailed description of the anticipated developments in these areas is beyond the scope of this volume, but the issue of susceptibility and risk at the molecular level is considered because of implications for preventive medicine and health-related fitness.

An important aim of epidemiology and prospective medicine is to quantify levels of risk to which populations and subgroups of a population are exposed, based on relevant personal, demographic, environmental, lifestyle, and other information. Risk assessment is valid for the population or subpopulation but is less discriminant for individuals. In the emerging field of molecular epidemiology, which incorporates concepts from epidemiology, genetics, and molecular biology, the degree of susceptibility or level of risk is by definition estimated for each individual having the same genetic characteristics. However, even though the identification of high-risk cases and of level of susceptibility for each person is possible, it is still necessary to interpret the results in terms of probability rather than certitude. Indeed, constant molding of the phenotype by environment and by lifestyle and gene-gene interactions are likely to alter the expectations and course of events.

From a practical point of view, it will be possible in the near future to genotype individuals for their susceptibility to common multifactorial diseases. The approach will be of considerable interest for diseases such as hypertension, NIDDM, cardiovascular disease, obesity, some cancers, and others. Establishing the level of genetic risk for a given individual will likely be a cost-effective measure at some point in the future, but probably more for those with a positive family history for one of these diseases. The application of molecular biology technologies to genotype individuals will be particularly advantageous if genes with large effects on host susceptibility are present. The cost effectiveness of the approach will diminish if a large number of genes, each with only minor effects, are involved.

It will eventually be possible to genotype individuals for genetically determined host susceptibility to a large number of factors whose life exposure could be partly or totally voluntarily controlled. For instance, host susceptibility to each macronutrient and micronutrient, air pollutants, alcohol, tobacco products, cold weather, hot and humid weather, drugs, various allergic reactions, and so on could be determined with appropriate molecular tests. There is no reason why these tests could not be done soon after birth, as they will require only a few drops of blood or some other somatic cells. As a matter of fact, there may be compelling reasons from a preventive medicine point of view to know very early in life as much as possible about the inherited susceptibilities of the newborn.

Implications for Health-Related Fitness

The bulk of the evidence suggests that regular physical activity has, by and large, favorable consequences on fitness and health. However, it must be appreciated that such a conclusion is based on average effects observed in groups of men and women or in elderly subjects of both sexes. Data for children are equivocal. The point is that these influences documented at the level of a group may not fully apply to each member of the group. Indeed, there are considerable individual differences in the response to regular physical activity, even when all members of the exercising group are exposed to the same volume of physical activity at the same relative intensity (2,4,13).

Health is the culmination of many interacting factors, including genetic constitution. Given genetic individuality, an equal state of health and of physical and mental well-being is unlikely to be achieved for all individuals even under similar environmental and lifestyle conditions. Some will thrive better than others and will remain free from disabilities for a longer period of time. Allowing for such individuality, it should come as no surprise that a minority of adults remain relatively fit by most common definitions in spite of a sedentary lifestyle.

Genetic differences do not operate in a vacuum. They constantly interact with existing cellular and tissue conditions to provide a biological response commensurate with environmental demands. Genes constantly interact with everything in the physical environment and with lifestyle characteristics of the individual that translate into signals capable of affecting cells of the body. For instance, overfeeding, a high fat diet, smoking, and regular endurance exercise are powerful stimuli that may elicit strong biological responses. However, because of inherited differences at specific genes, the amplitude of adaptive responses varies from one individual to another. Inheritance is one of the important reasons why individuals are not equally prone to become diabetic or hypertensive or to die from a heart attack. It is also one major explanation for individual differences in the response to dietary intervention or regular physical activity.

Genetic individuality is important in the present context because it has an impact on the physical activity, fitness, and health paradigm. Thus, there are inherited differences in level of habitual physical activity and in most components of health-related fitness. There is also highly suggestive evidence that genetic variation accounts for most of the individual differences in the response of health-related fitness components and of various risk factors for cardiovascular disease and NIDDM to regular exercise. It is important not only to recognize that there are individual differences in response to regular physical activity, but also to recognize that there are some members of the population who do not respond. Typically, there is a 3- to 10-fold difference between low responders and high responders, depending upon the phenotype considered, as a result of exposure to the same standardized physical activity regimen.

An appreciation of the critical role of DNA sequence variation in human responses to a variety of challenges and environmental conditions is essential to those interested in the physical activity, fitness, and health paradigm. This appreciation can only augment understanding of human individuality. It should also demand more caution when defining fitness and health benefits that may be anticipated from a physically active lifestyle. Incorporating biological individuality into the various paradigms will increase the relevance of the available data to the true human situation. This will become even more important as knowledge of the genetic and molecular basis of health-related fitness are more refined.

An understanding of the true value of a physically active lifestyle requires the recognition that physical activity cannot be viewed in isolation. It is only one element in the lifestyle of an individual. Nutritional habits, smoking, alcohol consumption, and leisure activities deserve particular attention in this regard. Individual differences in level of habitual physical activity are superimposed on heterogeneity stemming from other lifestyle components and from the physical and social environments. Such individual differences are variously modulated from person to person when they interface at the organ, tissue, and cellular levels because of biological individuality. In other words, caution and restraint are paramount in pronouncements about the values and virtues of a physically active lifestyle for all. Some are likely to benefit much, others somewhat less, and perhaps a few not at all from a physically active lifestyle in terms of risk reduction and time free from disability.

A Vision of the Year 2500

Centuries from now, the controversy regarding the appropriateness of sequencing all the DNA base pairs of the human genome and of defining the extent of within- and between-population and within– and between–ethnic group DNA heterogeneity will have subsided. Society will have long recognized that there is inborn inequality with respect to disease susceptibility, life duration free from disability, resistance to infection, personality characteristics, personal achievement potential in arts, science, and all areas of human endeavor, and so on. Since consensus had been reached a long time ago on the multifaceted dangers brought about by tobacco, the substance had been banned and eradicated from the surface of the earth. Those with susceptible genotypes to direct or indirect exposures to tobacco products were, therefore, free from any of the detrimental effects that had affected earlier generations of human beings. This had become a classic example of a public health measure applied at the level of the whole world that had totally alleviated the health risk caused by genetic differences in response to exposure.

Progress in molecular epidemiology initially had been slow. But with advances in understanding of the biology of the gene and of the human genome, it became progressively clear that the benefits associated with the determination of genetic susceptibilities at the molecular level in all newborns were enormous for the individual and for society as a whole. Determination of genetic susceptibilities was now done routinely for hundreds of genes. Parents, family physicians and medical specialists, nutritionists, and kinesiologists were using that information all the time. Kinesiologists were particularly enthusiastic about these advances, as they were now able to prescribe exercise programs specifically designed to take into account individual differences in the needs of people of all ages, while targeting the biological systems, metabolic pathways, and psychosocial characteristics most likely to benefit from a physically active lifestyle. Kinesiologists have thus become key members of the preventive medicine network.

Summary

The era of molecular medicine is here. Rapid changes will occur in the field of predicting levels of risk for a variety of common diseases. These advances will have an impact on opportunities available to exercise scientists and kinesiologists involved in the area of health-related fitness and preventive medicine. The most sensible way to prepare for the challenges and opportunities arising from progress in the identification of the genetic and molecular basis of health and disease is to become familiar with this field and to learn its tools.

References

1. Ballinger, S.W.; Shoffner, J.M.; Hedaya, E.V.; Trounce, I.; Polak, M.A.; Koontz, D.A.; Wallace, D.C. Maternally transmitted diabetes and deafness associated with a 10.4 kb mitochondrial DNA deletion. Nat. Genet. 1:11-15; 1992.
2. Bouchard, C. Genetics of aerobic power and capacity. In: Malina, R.M.; Bouchard, C., eds. Sport and human genetics. Champaign, IL: Human Kinetics; 1986:59-89.
3. Bouchard, C. Genetics of obesity: Overview and research directions. In: Bouchard, C., ed. The genetics of obesity. Boca Raton: CRC Press; 1994:223-33.
4. Bouchard, C. Human adaptability may have a genetic basis. In: Landry, F., ed. Health risk estimation, risk reduction and health promotion. Proceedings of the 18th Annual Meeting of the Society of Prospective Medicine. Ottawa: Canadian Public Health Association; 1983:463-76.
5. Brook, J.D.; McCurrach, M.E.; Harley, H.G.; et al. Molecular basis of myotonic dystrophy: Expansion of a trinucleotide (CTG) repeat at the 3' end of a transcript encoding a protein kinase family member. Cell. 68:799-808; 1992.
6. Caskey, C.T. Molecular medicine: A spin-off from the helix. JAMA. 269:1986-92; 1993.
7. Fu, Y.H.; Pizzuti, A.; Fenwick, R.G., Jr.; et al. An unstable triplet repeat in a gene related to myotonic muscular dystrophy. Science. 255:1256-58; 1992.
8. Genetic map goal met ahead of schedule. Hum. Gen. News. 6:1-14; 1994.
9. Goldbourt, U.; de Faire, U.; Berg, K., eds. Genetic factors in coronary heart disease. Dordrecht, Netherlands: Kluwer Academic; 1994.
10. Hixson, J.E.; McMahan, C.A.; McGill, H.C.J.; Strong, J.P.; The Pathobiological Determinants of Atherosclerosis in Youth (PDAY) Research Group. Apo B insertion/deletion polymorphisms are associated

with atherosclerosis in young black but not young white males. Arterioscler. Thromb. 12:1023-29; 1992.

11. King, R.A.; Rotter, J.I.; Motulsky, A.G. The approach to genetic bases of common diseases. In: King, R.A.; Rotter, J.I.; Motulsky, A.G., eds. The genetic basis of common diseases. New York: Oxford University Press; 1992:3-70.
12. Lander, E.S.; Schork, N.J. Genetic dissection of complex traits. Science. 265:2037-48; 1994.
13. Lortie, G.; Simoneau, J.A.; Hamel, P.; Boulay, M.R.; Landry, F.; Bouchard, C. Responses of maximal aerobic power and capacity to aerobic training. Int. J. Sports Med. 5:232-36; 1984.
14. Lusis, A.J.; Rotter, J.I.; Sparkes, R.S. Molecular genetics of coronary artery disease. Monographs in human genetics, vol. 14. Basel: Karger; 1992.
15. Mahadevan, M.; Tsilfidis, C.; Sabourin, L.; et al. Myotonic dystrophy mutation: An unstable CTG repeat in the 3' untranslated region of the gene. Science. 255:1253-55; 1992.
16. Multicenter Study Group. Diagnosis of Duchenne and Becker muscular dystrophies by polymerase chain reaction. JAMA. 267:2609-15; 1992.
17. Poulton, J. Mitochondrial DNA and genetic disease. Arch. Dis. Child. 63:883-85; 1988.
18. Report of the Expert Panel on Genetic Strategies for Heart, Lung, and Blood Diseases. Bethesda, MD: National Institutes of Health; 1993.
19. Rigat, B.; Hubert, C.; Alhenc-Gelas, F.; Cambien, F.; Corvol, P.; Soubrier, F. An insertion/deletion polymorphism in the angiotensin I–converting enzyme gene accounting for half the variance of serum enzyme levels. J. Clin. Invest. 86:1343-46; 1990.
20. Sakai, T.; Ohtani, N.; McGee, T.L.; Robbins, P.D.; Dryja, T.P. Oncogenic germ-line mutations in Sp1 and ATF sites in the human retinoblastoma gene. Nature. 353:83-86; 1991.
21. Scriver, C.R.; Beaudet, A.L.; Sly, W.S.; Valle, D. The metabolic basis of inherited disease, vol. 1 and 2. 6th ed. New York: McGraw-Hill; 1989.
22. Tanizawa, Y.; Matsutani, A.; Chiu, K.C.; Permutt, M.A. Human glucokinase gene: Isolation structural characterization, and identification of a microsatellite repeat polymorphism. Mol. Endocrinol. 6:1070-81; 1992.
23. Yunis, J.J. The chromosomal basis of human neoplasia. Science. 221:227-36; 1983.

CHAPTER 17

Genes and High-Performance Sports

World-class athletes in any sport are a special breed of individuals in the sense that not only they are talented for a particular sport activity, but they also have been willing to train and compete until world-class status was attained. The price for such prominence is very high in terms of human dedication and financial involvement.

The field of contemporary high-performance sport deals with the most-talented individuals from a biological, physical, and psychological point of view. Elite athletes do not simply happen to execute a national or world-class performance by chance. In previous chapters of this book, evidence for the role of inherited differences for a good number of the physiological and metabolic determinants of sports performance has been reviewed. In this chapter, data are summarized and gaps in the knowledge base are identified. Issues related to procedures to identify talented individuals, sometimes at a young age, are discussed. Finally, several dramatic changes to be anticipated from advances in genetics and molecular biology applied to performance, and important related ethical issues arising from these developments are presented.

Science and High-Performance Sports

It was possible up until about the middle of this century to become a national or a world-class athlete in a given athletic event without being among the most-talented individuals of a nation or the world at the time. The selection process was less stringent and level of competition was not as demanding as it is today. With the continuous expansion of the pool of young participants and competitors and with the growing sophistication of training, psychological preparation, equipment, and facilities, the level of competition has increased to the point that only those individuals who are highly gifted can expect to reach elite status. Moreover, the process is ongoing, particularly with the growth in the number of participants in developing countries. This will result in an expanded pool of individuals, and therefore a larger array of genotypes, who potentially have the chance to test themselves in a given sport or event within a sport and perhaps to experience success so that they may nurture the interest and motivation to train seriously for the activity.

Although much remains to be investigated, a reasonable body of knowledge has accumulated about the physical, biomechanical, physiological, metabolic, psychological, and social determinants of high-performance sport in several individual and team activities. Among these, it is fair to say that more is probably known about the determinants of endurance performance than of any other type of performance. Unfortunately, the exercise science and sports medicine research domains are not particularly well funded. As a result, progress is slow. As a matter of fact, most countries of the world do not even have exercise science research initiatives. It therefore should not be surprising that the knowledge base about the genetic and molecular foundations of human sports performance is embryonic.

On the other hand, science has much to contribute to sport, especially high-performance sport. The public, media, business and industry, coaches, and performers are generally aware that scientific advances can result in better training conditions and improved performances. On the other hand, fraud and cheating often result as a consequence of deviant use of existing or new knowledge. The field of sport is not unique in this sense, however.

Current Understanding of the Role of Genes

Two main conclusions can be reached from previous chapters. First, the elite athlete is probably an individual with a favorable profile in terms of the morphological, physiological, metabolic, motor, perceptual, biomechanical, and personality determinants of the relevant sport. Second, the elite athlete is a highly responsive individual to regular training and practice. The evidence for a role of genes in modulating the status of performance determinants in the sedentary population is slowly accumulating. This is generally estimated using the methods of genetic epidemiology, considering one phenotype at a time. Data are available for a variety of aerobic and anaerobic performance phenotypes as well as for selected determinants, such as body size, body composition, heart size, muscle fiber–type distribution, glycolytic and aerobic-oxidative markers of skeletal muscle metabolism, lipid mobilization from adipose cells, indicators of substrates oxidized, and others. Heritability of these phenotypes is generally low (about 25% or less) and rarely exceeds 50%.

The genetic effect seems to be polygenic, with no evidence reported to date for single-gene effects, with the notable exception of body composition and fat distribution phenotypes.

The issue of the heterogeneity of response to regular exercise has also been examined. Evidence clearly indicates considerable individual differences in the capacity to adapt and benefit from an exercise training program. For instance, the response of $\dot{V}O_2$max to a standardized program ranges from no progress at all (0%) to a doubling of initial $\dot{V}O_2$max (100% increase). Similar individual differences have been observed for other phenotypes of relevance to sports performance. The topic of the role of genes in determining the response to training was also considered. Several training studies were conducted with pairs of identical twins (training both members of each pair) in an attempt to understand the importance of genetic similarity and genetic heterogeneity in trainability when exposed to endurance training, high-intensity intermittent training, or strength training. The message from this series of studies is reasonably coherent. Individuals with the same genotype respond more similarly to training than those with different genotypes. The response of $\dot{V}O_2$max to training was monitored in four different experiments and the F-ratios of the between-genotype variances to the within-genotype variances ranged from 6 to 9. Thus, the genotype appears to be a very important determinant of trainability of $\dot{V}O_2$max. Though less extensive, results for high-intensity intermittent training point in the same direction.

Finally, studies dealing with the identification of genes responsible for these apparent genetic effects have begun to appear. The task is enormous, and the first reports focus on $\dot{V}O_2$max, body composition, and metabolic phenotypes in the untrained state and their responses to training. The search for genetic markers of trainability status will likely be more productive than the investigation of molecular markers of the performance phenotype in the untrained state for two reasons. First, the genetic effect appears to be more important when a response phenotype is considered, and second, the limits of adaptation become more clearly delineated when various biological systems are challenged by the demands of regular exercise. One multicenter study, the Heritage Family Study, is specifically designed to address these issues for a variety of health-related fitness and performance phenotypes (2).

The search for molecular markers is currently carried out using a battery of candidate genes, DNA microsatellites, and other genomic markers. Inherited differences may arise from sequence variation in nuclear and mitochondrial DNA, and both are being investigated. It will eventually be possible, using a series of probes, to identify the pool of gifted and highly trainable individuals, especially if genes with large effects are implicated. This goal will probably be attained initially for endurance activities. While much remains to be done, the genetic and molecular tools are available and the necessary database is being established so that sport scientists can reach this goal within a reasonable period of time.

Talent Detection

Two views are commonly encountered in the world of sport concerning the identification of the gifted performer (1,4). The first holds that the talented athletes emerge from the sport pyramid. A large number of participants (the base of the pyramid) have the opportunity to take part in the activity, engage in competition, and—if they perform well and are interested—get the chance to reach higher levels, depending on talent, interest, motivation, and economic circumstances. In this approach, a large base of recreational participants, particularly among children and adolescents, is a desirable characteristic. This has been the dominant approach up until the late 1960s and is still the basic philosophy of sport governing bodies in many countries.

The second view is based on the notion that there are individuals who are more endowed than others in terms of the basic skills and physique characteristics commonly associated with success in a given competitive sport. The approach is then to try to identify the children who exhibit these favorable traits. In practice, large numbers of children and adolescents are measured and tested with a battery of tests that tend to be sport specific. Stature, body mass, maturity level, body composition, muscular strength and endurance, speed, agility, and aerobic and anaerobic skill tests are commonly included in these talent-detection test batteries. The most talented subjects, as defined by these tests, are then typically offered the opportunity to be part of a special school curriculum in which time as well as human and physical resources are available to learn the skills of the particular sport and to train at a high level. The youths who succeed are then nurtured in the developmental system of the particular sport (1).

The two approaches are obviously not mutually exclusive, particularly in a democratic society in which anyone who can meet the standards or attain the performance levels of the elite can expect to be invited to join and be offered training conditions conducive to the development of his or her talent. A limiting factor, however, is opportunity during childhood and youth, which may prevent potentially talented individuals from being identified (4).

An issue related to talent detection is the manipulation of talented youth for high-performance sport. Such manipulation takes several forms. Social manipulation includes, for example, holding a child back a grade in school so that he or she can experience an additional year of growth and in turn have a larger body size compared with grade peers. Dietary manipulation includes maintenance of elite young gymnasts on marginal energy intake to maintain the body weight believed optimal for performance. Chemical manipulation includes the use of anabolic steroids and perhaps synthetic growth hormone to enhance performance and body size (3). The involvement of major investment and management firms in the careers of talented teenage athletes is another form of manipulation: commercial.

Dramatic Changes to Be Expected

With progress in the biological sciences, particularly in molecular biology, and with advances in the exercise sciences, the present talent-detection practices will eventually be altered, perhaps in a dramatic manner. While specific "athletic genotypes" will never be sufficient by themselves to predict who will become an elite athlete, since there are many other variables that must be considered, it is probably safe to assume that there will be few elite performers in the future who will not enjoy favorable genetic characteristics. It is therefore likely that probes will be eventually used to identify the carriers of DNA sequence variations desirable for sports performance, particularly if a small number of genes are found to have a substantial influence.

In a forum held under the patronage of the International Olympic Committee, it has been suggested that nothing will prevent parents, sport leaders, coaches, or entrepreneurs from using genetic probes in children and then in infants for the purpose of identifying potentially talented individuals (1). There is also the possibility that athletic-minded and overly ambitious parents, with the help of forceful entrepreneurs, may later take advantage of the progress made in the biology of reproduction to advocate embryo selection based on specific athletic probes to allow parents or surrogate parents to fulfill their dreams. On the surface, this scenario may appear like science fiction; however, it is not. All of the technologies necessary to make this possible are presently available and are commonly used in a number of specialized centers around the world in a variety of circumstances. The only missing element for the application of these advanced technologies in the context of sport is knowledge about the important genes upon which athletic genotypes can be defined and embryo selection be designed.

Cryopreservation of gametes and fertilized eggs is common. Elite athletes are likely to be proven carriers of desirable genetic polymorphisms relative to high-performance sport. It may be predicted that some of these elite male and female athletes will be asked and will accept to serve as gamete donors for the purpose of subsequent embryo selection. Specific drug treatments may also be recommended to the female donor so that multiple eggs can be recovered for the purpose of fertilizing several embryos. Single-cell sampling from fertilized embryos at an early stage of development would allow DNA extraction and subsequent amplification of critical sequences. This would make it possible to screen for each embryo's relevant DNA sequences in order to select one or several that carry all or the highest number of desired polymorphisms, including sex of the embryo. Implantation of the selected embryo in the uterus of the female donor, or of a surrogate mother, could then be attempted. The rationale underlying this approach would be that a child carrying the allelic versions of the genes or of DNA sequences that are associated with high-performance sport would have a greater likelihood of becoming an elite performer if exposed to the proper environmental conditions (1).

It is likely that this phenomenon will occur within the next 10 to 15 years. The world of high-performance sport is so competitive, and for some nations the stakes are so high, that the use of such technologies will simply constitute a means to attain a goal. The determination of some ambitious parents who would like to see one of their offspring on the Olympic podium should not be underestimated. Some would probably be prepared to rely on egg and sperm donors if they knew that it would significantly increase the chances of having an athletically gifted child that they could nurture as their own for this dream to be fulfilled (1).

This would certainly be an expensive undertaking from an economic point of view. Because the technologies are common and will likely be even more so in the future, private laboratories and clinics will eventually move aggressively in this direction as people, sport organizations, or nations are prepared to pay the price for such selective services. It may be anticipated that such practices will be outlawed in most countries of the world. Nevertheless, there will always be safe havens somewhere in the world where determined entrepreneurs could establish their operations.

Needless to say, systems for identifying talented individuals for high-performance sport will experience a major revolution in decades to come. This will in all likelihood begin when the first key genes (i.e., genes with large effects) associated with level of performance, determinants of performance, or trainability are uncovered. When this happens, the world of sport will be faced with the daunting task of addressing the ethical issues associated with the new opportunities and the ensuing changing environment of sport.

Summary

The field of high-performance sport is likely to be affected soon by the advances in the understanding of the genetic and molecular basis of performance, its determinants, and trainability of the human organism. Although specific genes have yet to be identified, it is anticipated that some will be defined in the near future, and eventually all of the relevant genetic polymorphisms will be known. These advances are likely to affect the way talented individuals are identified and nurtured in sports developmental programs. Some may be tempted to use the information to conceive embryos carrying the maximum number of favorable polymorphisms in order to increase the probability of nurturing a child gifted toward success in high-performance sport. Unprecedented ethical issues are likely to ensue as a result of anticipated advances in the genetic and molecular basis of performance.

References

1. Bouchard, C. Quelques réflexions sur l'avèvement des biotechnologies dans le sport. In: Landry, F.; Landry, M.; Yerlès, M., eds. Proceedings of the International Symposium on Sport: The third millennium. Quebec: Les Presses de l'Université Laval; 1991:455-64.
2. Bouchard, C.; Leon, A.S.; Rao, D.C.; Skinner, J.S.; Wilmore, J.H.; Gagnon, J. The Heritage Family Study: Aims, design, and measurement protocol. Med. Sci. Sports Exerc. 27:721-29; 1995.
3. Malina, R.M. The young athlete: Biological growth and maturation in a biocultural context. In: Smoll, F.L.; Smith, R.E., eds. Children and youth in sport. Dubuque, IA: Brown and Benchmark; 1996:161-86.
4. Malina, R.M. Youth sports: Readiness, selection and trainability. In: Duquet, W.; Day, J.A.P., eds. Kinathropometry IV. London: E & FN Spon; 1993:285-301.

CHAPTER 18

Ethical Issues

The authors hope that this book has conveyed the message that exciting developments are currently occurring in the biological sciences, specifically genetics and molecular biology. These advances are beginning to affect the understanding of complex and common multifactorial phenotypes, including those of interest for health-related fitness and performance. The contribution of genetics and molecular biology is no longer limited to single-gene phenotypes. The impetus provided by the ambitious and far-reaching Human Genome Project initiative has created a new environment in which advances in technology, study designs, and informatics are constantly improving capabilities for investigating multifactorial phenotypes to an extent that could not have been contemplated only a few years ago.

While these exciting advances are responsible for new research opportunities, they are also a source of uneasiness. The sheer power of the technological and scientific advances are a matter of concern. In addition, the kinds of questions that are being addressed with the help of the new technologies have become a matter of serious and, at times, emotional debate. Defining a genetic deficiency or genetic susceptibility for a given individual obviously affects the individual and his or her family. The study of behavioral phenotypes—such as personality characteristics, psychiatric disorders, or sexual preference—and of racial or ethnic variation has always been and will likely continue to be a controversial research activity that generates passionate debate. With time, particularly when the whole genome is fully sequenced and the biology of the genes is well understood, the debate and controversies may become less acute. Perhaps in another century or two, some of the discussion taking place today will appear strange and perhaps nonproductive. Recognizing the importance of the unprecedented ethical issues arising from the revolution in genetic and molecular biology, the human genome initiatives of all countries involved include funding to address the ethical, social, and legal implications.

In spite of current controversies and various levels of resistance encountered from some individuals and special interest groups, progress in the identification of the deficient and susceptibility genes continues unabated. The potential of this type of research for the health and well-being of humanity is enormous, and it is now impossible to try to stop these scientific initiatives. There are already literally hundreds of laboratories around the world that are involved. In this context, several pressing ethical questions are discussed. These issues will confront society in general, and exercise scientists and sport governing bodies in particular, as the expected developments unfold.

Implications for Society

Over the last few years, questions have been raised primarily in relation to genetic and DNA technologies, confidentiality of the information generated by the genetic tests, protection from abuse and discrimination that may arise with the availability of the genetic information, and inequality of access to the benefits brought about by these advances. Discussion on these issues has been fostered by human genome initiatives in various countries (4,8,11) and in international commissions (9). Legislation pertaining to one aspect or another of the gene technologies and their use has been enacted in several countries, including Austria, Belgium, France, Germany, Norway, Sweden and Switzerland (9).

DNA technologies and manipulations

Discussion about the risks of DNA recombinant techniques and other gene technologies has been going on for a long time. It began in the 1960s, well before the advent of the human genome initiative. The scientific community and regulatory agencies of the various countries involved in these efforts recognized the need for some form of regulation over these types of research early in the process. Oversight systems were put into place to ensure that research on genetic materials of all kinds of organisms posed no risk to the laboratory personnel or to the public. Scientists were able to continue their work, provided that they used the proper level of confinement and adhered to the safety guidelines imposed by funding and regulatory agencies. It turned out that the risks associated with the DNA recombinant technologies were generally less than anticipated. In a substantial segment of society, one major concern remains to this day: the risk posed by newly engineered animals (e.g., transgenic animals) or plants (e.g., genetically altered tomatoes) released by accident or by design into the environment. Elaborate oversight mechanisms and a case-by-case, in-depth discussion of the merits and risks continue to be essential in such instances to protect the public and the environment.

Confidentiality of the information and privacy

As revealed in a recent comment in *Science* (12), genetic testing is set for take-off. Screening for deficient or susceptibility genes is already offered in clinical centers and also by private companies but will likely become routine over the next decade. With the availability of so much genetic information relevant to health and disease risk for an individual and also, by inference, for his or her relatives, confidentiality becomes a crucial issue (9). There is consensus among all parties involved, except perhaps the health insurance industry, that genetic information about an identifiable person should not be disclosed without the consent of that individual. Elaborate and safe systems are needed to store the genetic information and for its secure dissemination. At the present time, however, there is no agreed-upon set of standards governing the management of DNA and other tissue banks or of genetic databases (8). However, it is clear that internal review boards of academic and hospital institutions are imposing strict confidentiality rules on the use of the information generated from these resources.

Protection from discrimination

Genetic information about risk of disease or behavioral phenotypes can potentially be used against an individual in a variety of circumstances. It could lead to employment discrimination (8) or to denial of or reduced access to health care or life insurance plans. The U.S. National Center for Human Genome Research has addressed the latter issue in a 1993 report prepared by its Task Force on Genetic Information and Insurance (13). Obviously, children and adults who are physically handicapped or mentally retarded are the most vulnerable from this point of view, but the availability of genetic information on apparently healthy individuals will result in new and subtle forms of potential discrimination that must be addressed by society.

Equality of access to benefits

Knowing more about oneself should eventually translate into better life choices, new opportunities, reduced disease load, and access to resources. As for many other scientific advances, the less-developed nations and the poor or minorities in developed countries are not likely to have easy access or to benefit from the opportunities arising from gains in knowledge derived from genetic and molecular biology. The fear that genetic testing could increase social inequality has already been expressed (9).

Gene therapy and gene transfer

A special case of genetic engineering technologies that are generating some controversies and posing ethical problems is that of human gene therapy and human gene transfer. The topic is not yet of concern to those interested in health-related fitness or high-performance sport, but it will undoubtedly appear on the agenda at some point in the future. Gene alteration can be divided into four broad categories (1,14):

- Somatic cell alteration to treat a disease (i.e., gene therapy) (2)
- Germ cell alteration to correct a genetic deficiency and prevent a disease (2,14)
- Somatic cell alteration for the purpose of enhancement, improvement, or augmentation in healthy individuals (1,14)
- Germ cell alteration for the purpose of enhancement of a trait in healthy individuals (1,14)

There is a general consensus among scientists and clinicians that somatic cell alteration to treat a severe genetic or acquired disease is acceptable. There is also broad public support for gene therapy in this specific context (2,7,10). More than 40 clinical trials of gene therapy were already approved by 1993 by the Recombinant DNA Advisory Committee of the National Institutes of Health (5).

The second category, altering a gene in the gametes to prevent a disease, raises more controversies. Altering the reproductive cells would ensure that future generations would inherit the functional gene. It is therefore a preventive strategy both for the fertilized embryo to be produced as a result of the gene alteration and for the future offspring of that particular individual (2). Considerable technical obstacles must be overcome before carefully targeted germ cell gene modification can be used with human cells (14). However, with the progress being progressively acquired in mouse gamete cells for transgenic experiments, it can be assumed that the appropriate technologies will be available in the future. There are several ethical arguments in favor of and against the use of germ cell–line gene alteration in humans (14). They are summarized in table 18.1. Consensus has not emerged yet, but the basic research continues.

The third and fourth categories may more readily get the attention of the exercise scientist or the coach, but they are the most controversial forms of genetic intervention. Somatic or germ cell gene alteration would be done solely for the purpose of improvement of an already healthy and competent genome. Since most of the phenotypes that could be targeted for such gene modifications are complex multifactorial phenotypes, such gene transfer interventions are well beyond present technological capabilities. However, progress is being made at an astonishing pace, so biologists will eventually be in a position to make such gene alteration interventions. Augmentation intervention designed to enhance muscularity, beauty, intelligence, height, skeletal muscle oxidative potential, exercise power and capacity, longevity, and other similar traits can be speculated and would undoubtedly receive much

Table 18.1
Some Ethical Arguments For and Against Germ-Line Gene Modification

Arguments in favor

- Moral obligation of health professions to use best available treatment methods
- Parental autonomy and access to available technologies for purposes of having a healthy child
- Germ-line gene modification more efficient and cost-effective than somatic cell gene therapy
- Freedom of scientific inquiry and intrinsic value of knowledge

Arguments against

- Expensive intervention with limited applicability
- Availability of alternative strategies for preventing genetic diseases
- Unavoidable risks, irreversible mistakes
- Inevitable pressures to use germ-line gene modification for enhancement

Reprinted, by permission, from N.A. Wivel and L.R. Walters, 1993, "Germ-line gene modification and disease prevention: Some medical and ethical perspectives," *Science* 262:533-38. Copyright 1993 American Association for the Advancement of Science.

attention in a democratic society and a free market economy where the consumer is the driving force. As argued by Wivel and Walters (14), there are precedents for interventions in conditions that do not meet the definition of a disease. Thus, recombinant human growth hormone was first used to treat dwarfism secondary to human growth hormone deficiency. However, recombinant human growth hormone recently has been administered to children of short stature but with no evidence of growth hormone deficiency. Here, the progressive shift has been from treatment to enhancement. One of the concerns is that the same pattern may occur for gene alteration. The ethical implications of gene alteration operations will be thoroughly discussed in the coming decades. They cannot be undertaken at the present time and will not be feasible in the immediate future.

Considerations for the Exercise Scientist

Early in the third millennium, exercise scientists will be confronted by the advances in genetics, molecular biology, and the biology of reproduction as well as by the progress made in understanding the genetic basis of health-related fitness and performance phenotypes. Chances are that the ethical questions will be associated with the use of molecular biology and genetic technologies for two purposes: (1) to screen populations in order to identify those who are more likely to benefit from regular exercise or those who are carriers of an athletic genotype in one form or another and (2) to enhance or improve a trait pertaining to fitness or high performance.

These critical issues will not suddenly arise. Exercise scientists have the time to prepare for this new environment. It will be necessary for them to become familiar with the basic concepts and the techniques of genetics, molecular biology, and the biology of reproduction. It will also be essential that they be exposed to the ethical questions posed by these technologies and the specific implications for their domain. Courses on these topics will have to become an integral part of the exercise science curriculum. Professional and scientific societies, such as the American College of Sports Medicine, will have to provide a forum for debating these issues and for informing members and the public about the dangers and benefits of these new opportunities.

It is difficult to predict exactly what the most important issues will be at a particular moment in the future. It is, however, quite certain that all of the problems that have been identified in this and in previous chapters, as well as other problems, will appear on the agenda of the exercise science community in the coming decades. They can be met adequately only by increased education in these areas so that exercise scientists' level of expertise and awareness becomes commensurate with the complexity of the issues.

Considerations for Sport Governing Bodies

The *Citius, Altius, Fortius* spirit prevailing at the origin of the modern Olympic Games is still there, but it has already assumed a new meaning and will be dramatically influenced by the anticipated developments described in the preceding chapters. The world of sport is not immune to and cannot be shielded from the developments in genetics and molecular biology. Many of these advances will seem aberrant to those who still have faith in the Olympic ideal and in the de Coubertin credo. They will occur anyway, and the ethical issues associated with their advent will be numerous and complex (3). Some of these issues are briefly described. Note, however, that the enhancement of performance with a variety of unethical means has been a regular practice for some time (6).

First, a new ethic of high-performance sport will be needed. Delineation of acceptable and unacceptable practices, standards of conduct, banned genetic interventions, illegal performance enhancement measures, and other topics will have to be addressed. Athletes, parents, geneticists, coaches, sport scientists and physicians, sport leaders, representatives from sport governing bodies, the Olympic movement, jurists, philosophers, ethicists, and the public at large should be involved in the discussion about a sport ethic adapted to the changing times and technologies.

Second, the impact of knowing as a child that one is a carrier of a series of genetic variants associated with a higher probability of experiencing success in high-performance sport should be considered on the agenda of ethical issues (3). This knowledge is potentially useful for the youth, but it may also be a source of problems. It may exert a lasting influence on career choice and be anxiogenic in many ways. The right to know may have to be balanced with the right not to know, which in itself is a challenging issue!

Third, considerable pressure could be exerted by those who have the genetic information on youths who are carriers or noncarriers of desirable genetic characteristics for high-performance sport (3). They may intervene unduly to try to convince the individual to enroll or to drop out of a sport development program. Those with access to the genetic information will, therefore, have critical responsibilities in the sport system. Current legislation on protection of human subjects, civil liberties, individual rights, and other related topics promulgated in several countries will be insufficient to properly regulate these activities.

Fourth, society will have to consider whether this new and augmented level of knowledge will decrease sports participation in general, especially among those who find that they have a less favorable genetic profile for high performance (3). Will the impact of these advances be to increase performance but to decrease sports participation? On a larger scale, will it engender greater sedentarism in the population?

These problems are intended as examples of the types of issues that will confront the world of sport with the advent of the new genetics of sports performance. A proscience or an antiscience attitude is not a useful position in this debate. It will be important to let reason prevail, because many of the scientific and technological developments described in this volume have the potential to dramatically alter the foundation upon which the entire world of high-performance sport is based and perhaps ultimately to destroy international sports competition, especially world championships and the Olympic Games as presently known.

Summary

Scientists recognized early on that advances in human genetics and molecular medicine would have considerable ethical and social implications. This chapter addressed some of the issues brought about by this increase in knowledge. Some of the concerns about DNA manipulation, confidentiality of information, protection from discrimination, and persistent social inequalities in the presence of these technological developments were highlighted. The topic of gene therapy and gene transfer was introduced and discussed in the context of fitness- and performance-related situations. Finally, considerations for the exercise scientist and sport governing bodies were introduced.

References

1. Anderson, W.F. Uses and abuses of human gene transfer. Hum. Gene Ther. 3:1-2; 1992.
2. Baird, P.A. Altering human genes: Social, ethical, and legal implications. Perspect. Biol. Med. 37:566-75; 1994.
3. Bouchard, C. Quelques réflexions sur l'avèvement des biotechnologies dans le sport. In: Landry, F.; Landry, M.; Yerlès, M., eds. Proceedings of the International Symposium on Sport: The third millennium. Quebec: Les Presses de l'Université Laval; 1991:455-64.
4. Durfy, S.J. Ethics and the human genome project. Arch. Pathol. Lab. Med. 117:466-69; 1993.
5. Goldspiel, B.R.; Green, L.; Calis, K.A. Clinical frontiers. Clin. Phar. 12:488-505; 1993.
6. Hoberman, J. Mortal engines: The science of performance and the dehumanization of sport. New York: Free Press; 1992.
7. Hoeben, R.C.; Valerio, D.; Van der Eb, A.J.; Ormondt, H.V. Gene therapy for human inherited disorders: Techniques and status. Crit. Rev. Oncol. Hematol. 13:33-54; 1992.
8. Juengst, E.T. Human genome research and the public interest: Progress notes from an American science policy experiment. Am. J. Hum. Genet. 54:121-28; 1994.
9. Knoppers, B.M.; Chadwick, R. The Human Genome Project: Under an international ethical microscope. Science. 265:2035-36; 1994.
10. Macer, D.R.J. The "far east" of biological ethics. Nature. 359:770; 1992.
11. Macer, D.R.J. Public acceptance of human gene therapy and perceptions of human genetic manipulation. Hum. Gene Ther. 3:511-18; 1992.
12. News and comment: Genetic testing set for takeoff. Science. 265:464-67; 1994.
13. NIH-DOE Task Force on Genetic Information and Insurance. Genetic information and health insurance (NIH-DOE Working Group on Ethical, Legal, and Social Implications of Human Genome Research). Bethesda, Maryland: National Institutes of Health; 1993.
14. Wivel, N.A.; Walters, L.R. Germ-line gene modification and disease prevention: Some medical and ethical perspectives. Science. 262:533-38; 1993.

Glossary

Adenine (A)—A nucleotide base found in RNA and DNA. Adenine is complimentary (pairs) with thymine.
Adoption studies—Studies comparing the resemblance between biological and adoptive relatives living in the same family environment.
Allele—Genes that occupy the same position (locus) on corresponding chromosomes. Each gene is represented twice in normal cells (a pair of alleles).
Anaphase—The stage during cell division in which chromosomes move to opposite poles of the cell.
Animal model—An animal model serves as an analog for humans in studying a particular disease.
Anticodon—The three-nucleotide sequence in a tRNA molecule complimentary to a three-nucleotide sequence (codon) in mRNA.
Association analysis—A method used to test for the co-occurence of a specific allele at a marker locus and a trait in a population.
Autosome—Any chromosome other than a sex chromosome (X or Y chromosome).
Basal metabolic rate—Oxygen used by an individual during minimal physiologic activity while awake.
Biometrics—The statisical analysis of biological data.
Centromere—The area of a chromosome where its two arms (chromotids) join. The centromere is the point where the spindle fibers attach during mitosis and meiosis.
Chromatin—The basic material of chromosomes consisting of DNA and histones.
Chromatid—One of the two DNA strands of a duplicated chromosome.
Chromosome—Long strand of genes in the form of coiled filaments compacted to a tight cylinder, segmented into two arms (chromatids) by the centromere. Chromosomes (normally 46 in humans) are found in the cell nucleus.
Chromosome mapping—A technique to diagram the relative position of genes along a chromosome.
Chromosome pairing—The process whereby homologous chromosomes (members of a chromosome pair) align side by side during meiosis.
Codon—A triplet of bases in mRNA that specifies an amino acid or a signal for the initiation or termination of the polypeptide synthesis.
Commingling analysis—A method used to test whether the distribution of a quantitative phenotype is better characterized by a mixture of distributions (commingling) rather than by a single distribution.
Cosegregation—A condition that occurs when two loci are relatively close to each other on the same chromosome and are passed together from parents to offspring.
Co-twin control—An extension of the twin method that compares twins discordant for exposure to a factor or factors in the environment.

Crossing over—The exchange of genetic material between two paired chromosomes.
Cultural heritability—The proportion of phenotypic variance attributable to shared familial environmental effects. It measures the importance of nongenetic transmission across generations.
Cytokinesis—The cytoplasmic division of one cell into two daughter cells.
Cytosine (C)—A nucleotide base found in RNA and DNA. Cytosine is complimentary (pairs) with guanine.
Daughter cells—Cells resulting from the division of a parent cell.
Deoxyribose—A sugar molecule that is found in DNA.
Deoxyribonucleic acid (DNA)—Nucleic acids that contain deoxyribose as the sugar component. Specific genes are composed of a particular sequence of DNA molecules.
Dizygotic twins—Twins derived from two separate zygotes. Commonly referred to as fraternal twins.
Dominant allele—An allele which is expressed to the exclusion of the contrasting allele (recessive).
DNA polymerase—An enzyme that catalyzes the joining of nucleotides, thereby forming DNA strands.
DNA sequencing—A technique in which the sequence of nucleotides (A, T, G, C) is determined. With this sequrence information, the complimentary amino acid sequence and protein product of the gene can be predicted.
Diploid—Referring to a normal or full complement number (2N) of chromosomes (46 chromosomes in humans, 2N). One member of each chromosome pair is derived from the father and one from the mother.
Dyad—From a genetic perspective, the pair of sister chromatids resulting from the separation of tetrads during the first meiotic division.
Electrophoresis—A technique used to detect protein polymorphisms in which protein charge or mass variants can be distinguished by migration in an electric field.
Exon—DNA that codes for mRNA and is therefore expressed (translated into protein).
Family studies—Studies of biological relatives to investigate the genetic basis of quantitative phenotypes.
Frameshift mutation—A mutation resulting from the insertion or deletion of one or more bases in the DNA causing a shift in the sequence. After the mutation site, the sequence of triplets is modified and the gene product is changed, often with dramatic consequences.
Gamete—A sex cell, either an ovum (egg) or sperm. In humans, gametes contain one-half of the full complement of chromosomes (22 autosomes and one sex chromosome).
Gene—A sequence of DNA which encodes a polypeptide or an RNA product.
Gene–environment interaction—A condition arising when the response of a phenotype to environmental changes is dependent on the individual's genotype.

Gene expression—Refers to the translation (expression) of a particular gene into its protein product and eventually into a functional phenotype.
Gene knockout—A procedure used to inactivate a particular gene and examine its effect in a cellular or an animal model.
Genetic heritability—The fraction of phenotypic variance that is genetic; it represents a population estimate of the degree of genetic determination.
Genetic map—A schematic array of gene loci on a chromosome.
Genome—The total set of genes in the nucleus of a cell.
Genotype—The genetic makeup of an individual. It may also apply to a specific locus or to a combination of genes.
Guanine (G)—A nucleotide base found in RNA and DNA. Guanine is complimentary (pairs) with cytosine.
Haploid—The number (1N) of chromosomes in sperm or ovum, one-half of the total number of chromosomes in somatic (diploid or 2N) cells.
Heterozygote—An individual that has two different versions of a gene (different alleles) at a given locus.
Heterozygous—Having different allelles at a given locus on homologous chromosomes.
Homologs—Denotes chromosomes of the same pair, one being inherited from the father and the other from the mother.
Homologous region—Identical regions along a chromosome.
Homozygous—Having identical alleles at one or more loci.
Interphase—The stage between cell division.
Intron—Intervening sequence of DNA that lies between two exons. Such sequences are not translated into proteins.
Karyotype—A photomicrograph of metaphase chromosomes.
Kilobases (kb)—A unit used to measure a DNA sequence. One kb equals 1000 nucleotide bases.
Linkage analysis—A mathematical procedure used to map loci along a chromosome. Two genetic loci that are inherited together as a consequence of tehir close proximity on a chromsome are said to be linked.
Locus (loci)—The chromsomal location of a gene or a specific DNA sequence.
Major locus—A locus having a major effect on a quantitative phenotype and segregating according to Mendelian expectations.
Marker—A polymorphic DNA sequence or protein variant used in genetic mapping and other genetic studies.
Measured genotype approach—An approach used to study the genetic basis of quantitative phenotypes that uses genetic variation in random genetic markers or in candidate genes and attempts to evaluate the impact of variation at the DNA level on the quantitative phenotype under study.
Meiosis—The formation of gametes (sperm or ova) by two cell divisions. The result is four gametes, each with one-half the number of chromosomes found in somatic cells.
Mendelian genetics—Relating to the basic principles of heredity developed by Gregor Mendel, usually referring to genetic transmission of a single-locus (gene) trait.

Messenger RNA (mRNA)—Template RNA that reflects a specific nucleotide sequence in DNA.
Metaphase—The phase of mitosis or meiosis when chromosomes align in the middle of the cell (along the equatorial plate). Metaphase occurs immediately before anaphase.
Missense mutation—A base change or substitution in the DNA sequence that gives rise to an amino acid change.
Mitosis—The division of a cell that results in the formation of two daughter cells, having the same number of chromosomes as in the parent cell.
Monozygotic twins—A single fertilized ovum divides and gives rise to two individuals with the identical genetic constitution. Commonly known as identical twins.
Mutation—A change in DNA that is passed along in subsequent divisions of the cell.
Nature—In genetic terms, nature relates to inherited genetic traits.
Nonsense mutation—A mutation resulting in a stop codon.
Nurture—Nurture refers to environmental influences upon a trait.
Nuclear membrane—The membrane that surrounds the nucleus.
Nucleotide—A base molecule of DNA composed of a purine (adenine or guanine) or pyrimidine (cytosine or thymine) and a phosphorylated ribose sugar.
Nucleus—An organelle within the cytoplasm of a cell that contains DNA.
Oocyte—Immature ovum.
Path analysis—A statistical approach commonly used to estimate the components of phenotypic variance.
Pedigree—A family tree. Used in genetics to determine trait inheritance.
Phenotype—Manisfestion of a trait resulting from the expression of relevant genes and nongenetic influences.
Point mutation—A substitution, deletion, or insertion of a single nucleotide.
Polymerase chain reaction (PCR) amplification—An in vitro method to amplify selectively a specific DNA sequence.
Polymorphism information content (PIC)—A measure of the degree of polymorphism of a marker used to evaluate how informative a marker is for mapping purposes.
Polypeptides—A chain of amino acids held together by peptide bands.
Prophase—The first stage of mitosis or meiosis.
Quantitative trait locus (QTL) mapping—An approach that identifies loci influencing quantitative phenotypes.
Recombinant DNA technology—Techniques to join DNA from different organisms permitting detailed analysis of genetic variation at the DNA level.
Recombination—A process in which corresponding parts of homologous chromosomes are exchanged.
Resting metabolic rate (RMR)—Energy expended in the resting state.
Restriction enzyme—An enzyme that cuts DNA at a specific place, called a restriction site.
Restriction fragment—A piece of DNA generated by a restriction enzyme.
Ribonucleic acid (RNA)—Ribonucleosides connected by phosphate molecules.
Ribose—The sugar molecule found in ribonucleic acid.

Segregation analysis—Method used to test a hypothesis about the mode of inheritance of a trait or a disease within families.
Sex chromosomes—The chromosomes (X and Y) responsible for sex determination. XX designates a female and XY a male.
Silent mutation—A mutation that does not change the final gene product and makes no distinguishable change in phenotype.
Somatic cells—The cells of an organism with the exception of the sex cells.
Southern blotting—A technique to detect DNA sequence variation.
Spermatogenesis—The process by which spermatogonial cells divide and differentiate into sperm.
Spindle fibers—Fibers which attach at the centomere of a chromosome and "pull" the chromosome to either side of the cell before cytokinesis.
Susceptibility gene—A gene that increases susceptibility or predisposition or risk level for a disease, but one that is not necessary for the expression of the disease phenotype.
Telophase—The final stage of mitosis or meiosis that begins when migration of chromosomes to the poles of the cell is completed.
Thymine (T)—A nucleotide base found in RNA and DNA. Thymine is complimentary (pairs) with adenine.
Transcription—A process by which a working copy of the gene (messenger RNA or mRNA) is made. Other types of RNA are also transcribed (ribosomal RNA and transfer RNA).
Translation—A process in which the mRNA is translated into a particular sequence of amino acids. This chain of amino acids is processed and folded to become a protein.
Transfer RNA (tRNA)—A type of RNA molecule to which an amino acid is bound and which recognizes a specific base sequence in the mRNA. The tRNA is thus responsible for the incorporation of the proper amino acid to the growing polypeptide chain.
Transgenic animals—Animals whose genetic information is permanently modified by the transfer of a gene or genes into their germ line.
Twin-family method—An extension of twin studies that includes data on the spouses and offspring of adult twins.
Twin studies—Studies used to assess the importance of genetic factors in traits or diseases by comparison of identical (monozygotic) and fraternal (dizygotic) twins.
Two-dimensional electrophoresis—Technique to separate proteins according to charge and mass.
Unmeasured genotype approach—An approach used to study the genetic basis of quantitative phenotypes based on statistical analysis of the distribution of phenotypes in individuals and families.
Uracil (U)—A nucleotide base found in ribonucleic acid in place of thymine.

Index

D

G

H

I

J

K

L

S

T

U

V

W

Y

Z

About the Authors

Claude Bouchard, one of the world's leaders in the study of genetics and physical activity, is professor of exercise physiology and director of the Biology of Physical Activity Research Group at Université Laval in Sainte Foy, Quebec. He has an MSc in exercise physiology and a PhD in anthropological genetics.

For more than 20 years, Dr. Bouchard's research has focused on the genetic and molecular basis of human variation in body composition, physical fitness, and performance. This research has been supported by the Medical Research Council of Canada and other Canadian granting agencies as well as the U.S. National Institutes of Health.

Dr. Bouchard has authored or coauthored several books, including *Growth, Maturation, and Physical Activity,* and has published more than 500 professional and scientific papers. He received the Canadian Atherosclerosis Society Sandoz Award in 1996 at the Royal College of Physicians and Surgeons of Canada meeting. In 1997 Dr. Bouchard gave the J.B. Wolffe Lecture at American College of Sports Medicine Clinic meeting in Denver.

Dr. Bouchard is a Fellow of the American College of Sports Medicine and the American Society of Human Genetics. He recently has been elected as a foreign member of the Royal Academy of Medicine of Belgium.

Robert M. Malina is professor of physical education and exercise science at Michigan State University and director of the Institute for the Study of Youth Sports. Before assuming his position at MSU, Dr. Malina was professor of kinesiology and anthropology at the University of Texas at Austin for 28 years. He has a PhD in both physical education and physical anthropology and has done extensive research on the growth, maturation, and performance of children and youth in the United States and abroad.

Dr. Malina has served as president of the Human Biology Council and the American Academy of Physical Education, editor of the *Yearbook of Physical Anthropology*, and section editor for growth and development for the *Research Quarterly for Exercise and Sport* and the *Exercise and Sport Sciences Review*. He is editor-in-chief of the *American Journal of Human Biology*. Dr. Malina is coauthor of the book, *Growth, Maturation, and Physical Activity*, and has published more than 300 professional and scientific papers.

In 1989, Dr. Malina was awarded an honorary doctorate by the Institute of Physical Education, Faculty of Biomedical Sciences, Katholieke Universiteit Leuven, Belgium. He is a fellow of the American College of Sports Medicine, the American Association for the Advancement of Science, and the Human Biology Council, and an elected foreign member of the Polish Academy of Sciences.

Dr. Louis Pérusse is research scholar and associate professor in the Department of Physical Education at Université Laval in Sainte Foy, Quebec. He has an MSc and PhD in physical activity sciences from Université Laval. He has also completed postdoctoral work in human genetics at the University of Michigan at Ann Arbor.

Dr. Pérusse's research has focused on the genetic epidemiology of various determinants of physical fitness. He has published his findings in more than 75 scholarly journals and books, including several papers on the role of genes on body composition and blood pressure.

Dr. Pérusse is a member of the American College of Sports Medicine and the American Society of Human Genetics.